Contributions to Environmental Sciences & Innovative Business Technology

Editorial Board

Allam Hamdan, Ahlia University, Manama, Bahrain

Wesam Al Madhoun, Air Resources Research Laboratory, MJIIT, UTM, Kuala Lumpur, Malaysia

Mohammed Baalousha, Department of EHS, Arnold School of Public Health, University of South Carolina, Columbia, SC, USA

Islam Elgedawy, AlAlamein International University, Alexandria, Egypt

Khaled Hussainey, Faculty of Business and Law, University of Portsmouth, Portsmouth, UK

Derar Eleyan, Palestine Technical University—Kadoori, Tulkarm, Palestine, State of

Reem Hamdan, University College of Bahrain, Manama, Bahrain

Mohammed Salem, University College of Applied Sciences, Gaza, Palestine, State of

Rim Jallouli, University of Manouba, Manouba, Tunisia

Abdelouahid Assaidi, Laurentian University, Sudbury, ON, Canada

Noorshella Binti Che Nawi, Universiti Malaysia Kelantan, Kota Bharu, Kelantan, Malaysia

Kholoud AL-Kayid, University of Wollongong, Leppington, NSW, Australia

Martin Wolf, Center for Environmental Law and Policy, Yale University, New Haven, CT, USA

Rim El Khoury, Accounting and Finance, Notre Dame University, Loauize, Lebanon

Editor-in-Chief

Bahaaeddin Alareeni, Middle East Technical University, Northern Cyprus Campus, Kalkanlı, KKTC, Turkey

Contributions to Environmental Sciences & Innovative Business Technology (CESIBT) is an interdisciplinary series of peer-reviewed books dedicated to addressing emerging research trends relevant to the interplay between Environmental Sciences, Innovation, and Business Technology in their broadest sense. This series constitutes a comprehensive up-to-date interdisciplinary reference that develops integrated concepts for sustainability and discusses the emerging trends and practices that will define the future of these disciplines.

This series publishes the latest developments and research in the various areas of Environmental Sciences, Innovation, and Business Technology, combined with scientific quality and timeliness. It encompasses the theoretical, practical, and methodological aspects of all branches of these scientific disciplines embedded in the fields of Environmental Sciences, Innovation, and Business Technology.

The series also draws on the best research papers from EuroMid Academy of Business and Technology (EMABT) and other international conferences to foster the creation and development of sustainable solutions for local and international organizations worldwide. By including interdisciplinary contributions, this series introduces innovative tools that can best support and shape both the economical and sustainability agenda for the welfare of all countries, through better use of data, a more effective organization, and global, local, and individual work. The series can also present new case studies in real-world settings offering solid examples of recent innovations and business technology with special consideration for resolving environmental issues in different regions of the world.

The series can be beneficial to researchers, instructors, practitioners, consultants, and industrial experts, in addition to governments from around the world. Published in collaboration with EMABT, the Springer CESIBT series will bring together the latest research that addresses key challenges and issues in the domain of Environmental Sciences & Innovative Business Technology for sustainable development.

Anand Nayyar · Mohd Naved ·
Rudra Rameshwar
Editors

New Horizons for Industry 4.0 in Modern Business

Editors
Anand Nayyar
School of Computer Science
Duy Tan University
Da Nang, Vietnam

Mohd Naved
Amity International Business School
Amity University
Noida, India

Rudra Rameshwar
Thapar University
Patiala, India

ISSN 2731-8303 ISSN 2731-8311 (electronic)
Contributions to Environmental Sciences & Innovative Business Technology
ISBN 978-3-031-20442-5 ISBN 978-3-031-20443-2 (eBook)
https://doi.org/10.1007/978-3-031-20443-2

© The Editor(s) (if applicable) and The Author(s), under exclusive license to Springer Nature Switzerland AG 2023

This work is subject to copyright. All rights are solely and exclusively licensed by the Publisher, whether the whole or part of the material is concerned, specifically the rights of translation, reprinting, reuse of illustrations, recitation, broadcasting, reproduction on microfilms or in any other physical way, and transmission or information storage and retrieval, electronic adaptation, computer software, or by similar or dissimilar methodology now known or hereafter developed.

The use of general descriptive names, registered names, trademarks, service marks, etc. in this publication does not imply, even in the absence of a specific statement, that such names are exempt from the relevant protective laws and regulations and therefore free for general use.

The publisher, the authors, and the editors are safe to assume that the advice and information in this book are believed to be true and accurate at the date of publication. Neither the publisher nor the authors or the editors give a warranty, expressed or implied, with respect to the material contained herein or for any errors or omissions that may have been made. The publisher remains neutral with regard to jurisdictional claims in published maps and institutional affiliations.

This Springer imprint is published by the registered company Springer Nature Switzerland AG
The registered company address is: Gewerbestrasse 11, 6330 Cham, Switzerland

Preface

Industry 4.0 means the fourth industrial revolution and commotion. This period of the industrialization cycle is, as well as the three past stages, overpowered by specific headways. While mechanization and zap of gathering processes have incited the underlying two current unrests, the third stage, which is depicted by an augmentation of information and automatization, is transforming which is impeccably changing into the accompanying present-day revolution. In the modern world and advances, Industry 4.0 is separated by a particular compromise of cyber-physical systems in gathering and arranged tasks processes as well as the usage of the Internet of Things and Services in current cycles. New advances will arbitrarily influence regard creation, work affiliation, downstream organizations, and game plans of associations. At the forefront of all Industry 4.0 developments, the possibility of a Smart Factory expects a gigantic part in trimming the vision of another cutting-edge age. In current composition and legitimate journals, experts are referring to a complete have an impact on in context in collecting with respect to Industry 4.0. It is said that a decentralized, self-figuring out, and versatile creation environment will replace the old style, halfway controlled creation request.

This book presents the most recent evaluation viewpoints on how the new horizons for Industry 4.0 in Modern Business point of view are attempting the course of mechanical and fundamental change and what the redesign of the economy means for key change. It correspondingly investigates the effect of quickly making degrees of progress on the distinction in financial and standard frameworks and sees whether covered-up and creative change can convey sensible cash-related headway and work. Further, the book presents the focal new developments (new turns of events, materials, energy, and so on) and current philosophies that can incite such a fundamental change. It also dissects Industry 4.0 from a worldwide point of view, zeroing in on the most progressive economies. Nevertheless, it also brings together the choices for future modern improvement that is mechanically conceivable. Gives a subjectively new catalyst to financial development. It likewise gives various situations of execution of trend-setting innovations, materials, and energies on unambiguous economies, addressing various landmasses and realities.

The book comprises of 14 chapters. Chapter titled "Artificial Intelligence Powered Automation for Industry 4.0" introduces deep concepts of Artificial Intelligence (AI)-Powered Automation for Industry 4.0 and its scope of applications. The chapter enlightens the basic cum essential concepts and ideas of Automation for Industry 4.0 and stresses the importance of automation in modern industry followed by applications of AI in Smart Industry and also explores different concepts of AI algorithms used in Smart Industry.

Chapter titled "Business Sustainability and Growth in Journey of Industry 4.0-A Case Study" discusses orientation and implementation of Industry 4.0, analyzes firms' operational performance using computer vision in smart Manufacturing process, examines implementation of IIoT and Robotics in Manufacturing and also reviews the challenges and opportunities of the firms applying tools of Industry 4.0 with management perspective.

Chapter titled "Foundation Concepts for Industry 4.0" elaborates fundamental aspects of Industry 4.0 with its history and significance and explores how India is ready to adopt Industry 4.0 and proposes a conceptual model of Industry 4.0 for sustainable development.

Chapter titled "Challenges and Opportunities for Mutual Fund Investment and the Role of Industry 4.0 to Recommend the Individual for Speculation" explores the best tools and technology to identify the right mutual fund for investors and proposes safety and stability parameters for investment opportunities via right selection of mutual funds with optimal returns. In addition, the chapter also explores three case studies with regard to Three Mutual Funds-Tata Infrastructure Fund, Aditya Birla Sun Life Insurance Fund, and LIC MF Infrastructure fund.

Chapter titled "Implementing Digital Age Experience Marketing to Make Customer Relations More Sustainable" analyzes the sustainable resources in digital market in terms of experimental marketing and evaluates the online characteristics that influence the customer experience. In addition, this chapter evaluates the application of Experimental Marketing in Industry 4.0.

Chapter titled "Prospective Application of Blockchain in Mutual Fund Industry" stresses the importance of Blockchain technology implementation in Mutual Fund Industry and Industry 4.0 and also explores blockchain exchange-traded fund.

Chapter titled "Industry 4.0 Internet of Medical Things Enabled Cost Effective Secure Smart Patient Care Medicine Pouch" explores the concept of the Internet of Medical Things and Industry 4.0 and also proposes a novel architecture for a low-cost IoMT-based medical pouch system. The proposed system is also compared with existing techniques using the Proteus software tool and experimental results state that proposed methodology is better as compared to existing techniques.

Chapter titled "Design and Automation of Hybrid Quadruped Mobile Robot for Industry 4.0 Implementation" proposes a novel Hybrid Quadruped Mobile Robot with minimal complexity and high modularity and experimental results state that robot has strong capacity to traverse over long distance in Industry 4.0 environment.

Chapter titled "To Trust or Not to Trust Cybots: Ethical Dilemmas in the Posthuman Organization" highlights the theoretical elements of digital trust in the posthuman condition of Industry 4.0, enlightens Michel Foucault's Ethics of Curiosity and also elaborates two Foucauldian Scenarios: The Centrality of Care:

The "Bad" Scenario in the Posthuman Organization and Not-Yet-post humanized trust: The "Good" Scenario in Industry 4.0.

Chapter titled "Hydrogel Based on Alginate as an Ink in Additive Manufacturing Technology—Processing Methods and Printability Enhancement" explores the possibilities of using hydrogel inks based on sodium alginate in additive manufacturing technologies for biomedical technologies and also focuses on the factors that influence the change in characteristics during fabrication processes and also elaborates approaches to increase the durability of alginate-based hydrogel structures.

Chapter titled "Coal Fly Ash Utilization in India" presents a review of the current production and utilization of coal fly ash in India for Years 2018 and 2019 and also elaborates additional information regarding its usage in various industries, regulations, and initiation of fly ash utilization in India.

Chapter titled "3D Printing Pathways for Sustainable Manufacturing" explores the current status of 3D printing in various industrial applications leading to sustainable manufacturing technology and also reveals some shortcomings to adoption of 3D printing technology.

Chapter titled "Role of 3D Printing in Pharmaceutical Industry" elaborates the role of 3D printing in pharmaceuticals and explores complications, limitations, and importance of its Implementation.

Chapter titled "3D Printing: A Game Changer for Indian MSME Sector in Industry 4.0" provides basic understanding of 3D printing including background, types of technologies, materials, and process, presents applications of 3D printing during COVID-19 and also explores operational constraints and future challenges of AM in India.

The purpose of this book is to thus confer the preliminaries of the cutting-edge smart technology-driven production maneuver, the Industry 4.0, mainly to determine and verify its potential as a practice which endorses sustainability to ultimately revolutionize the competitiveness of businesses, and regions. The highlighting feature of the proposed book is that it will proffer basics to the beginners and at the same time serve as a reference study to advance learners. Also, the book is not country-specific or audience-specific. It will be of interest to academia, industry practitioners and researchers globally.

Da Nang, Vietnam	Anand Nayyar
Patiala, India	Rudra Rameshwar
Noida, India	Mohd Naved

Contents

Artificial Intelligence Powered Automation for Industry 4.0 1
Dennise Mathew, N. C. Brintha, and J. T. Winowlin Jappes

Business Sustainability and Growth in Journey of Industry 4.0-A Case Study 29
Gouranga Patra and Raj Kumar Roy

Foundation Concepts for Industry 4.0 51
Bhakti Parashar, Ravindra Sharma, Geeta Rana, and R. D. Balaji

Challenges and Opportunities for Mutual Fund Investment and the Role of Industry 4.0 to Recommend the Individual for Speculation 69
Sanjay Kumar, Meenakshi Srivastava, and Vijay Prakash

Implementing Digital Age Experience Marketing to Make Customer Relations More Sustainable 99
Amrita Baid More

Prospective Application of Blockchain in Mutual Fund Industry 121
Sonali Srivastava

Industry 4.0 Internet of Medical Things Enabled Cost Effective Secure Smart Patient Care Medicine Pouch 149
Sourav Singh, Sachin Sharma, Shuchi Bhadula, and Seshadri Mohan

Design and Automation of Hybrid Quadruped Mobile Robot for Industry 4.0 Implementation 171
Sivathanu Anitha Kumari, Abdul Basit Dost, and Saksham Bhadani

To Trust or Not to Trust Cybots: Ethical Dilemmas in the Posthuman Organization 189
Arindam Das and Debojoy Chanda

Hydrogel Based on Alginate as an Ink in Additive Manufacturing Technology—Processing Methods and Printability Enhancement 209
Magdalena B. Łabowska, Ewa I. Borowska, Patrycja Szymczyk-Ziółkowska, Izabela Michalak, and Jerzy Detyna

Coal Fly Ash Utilization in India 233
Dipankar Das and Prasanta Kumar Rout

3D Printing Pathways for Sustainable Manufacturing 253
Granville Embia, Bikash Ranjan Moharana, Aezeden Mohamed,
Kamalakanta Muduli, and Noorhafiza Binti Muhammad

Role of 3D Printing in Pharmaceutical Industry 273
Rajeshwar Kamal Kant Arya, Dheeraj Bisht, Karuna Dhondiyal,
Meena Kausar, Hauzel Lalhlenmawia, Pem Lhamu Bhutia,
and Deepak Kumar

**3D Printing: A Game Changer for Indian MSME Sector in
Industry 4.0** 295
Nidhi U. Argade and Hirak Mazumdar

Artificial Intelligence Powered Automation for Industry 4.0

Dennise Mathew, N. C. Brintha, and J. T. Winowlin Jappes

1 Introduction

In the last decades, there has been an immeasurable technological evolution and the procreation of automation in human society. Today we experience many services provided by industries and institutions as digital services. Industry 4.0 is the present proclivity of automation to create "Smart Industry" with different technologies such as Internet of Things, Big Data, Cloud Computing, Cyber Security etc. Industries are in urge to increase their revenue by understanding the advantages of smart factories and smart manufacturing. Driven by emerging technologies, the routine of our lifestyle as different roles is being remodeled across the world, so does Industry 4.0, which is based on three driving forces: Intelligence, Connectivity and Automation. These sights are guiding Industry 4.0 for a digital transformation with a myriad of profit for the stakeholders. It is considered as the advanced generation of automation which allows industrialists to optimize the industries and improve productivity by analyzing the data pattern they use. Artificial Intelligence, AI can be defined as a study and design of intelligent systems which act on the environment to do some actions and maximize its probability of success rate and so this technology is considered as the brain of Industry 4.0. Though the application areas of AI are enormous, some of the prominent

---

D. Mathew · N. C. Brintha (✉) · J. T. W. Jappes
Department of Computer Science and Engineering, Kalasalingam Academy of Research and Education, Krishnankoil, India
e-mail: n.c.brintha@klu.ac.in

D. Mathew
e-mail: dennisemathew@gmail.com

J. T. W. Jappes
e-mail: winowlin@klu.ac.in

© The Author(s), under exclusive license to Springer Nature Switzerland AG 2023
A. Nayyar et al. (eds.), *New Horizons for Industry 4.0 in Modern Business*, Contributions to Environmental Sciences & Innovative Business Technology,
https://doi.org/10.1007/978-3-031-20443-2_1

areas where AI is excelling are Healthcare, Logistics and Transportation, Food Technology, Manufacturing, Banking and Financial Services, Travel, Retail and E-commerce, Real Estate, Entertainment etc. Artificial Intelligence algorithms with industry analysis and advanced modeling techniques will improve efficiency and optimize the process to improvise profitability. Neural Networks, Deep-Learning and Reinforcement Learning along with Cognitive technologies can be used to design intelligent machines with natural language processing, thus empowering the efficiency of the system. Artificial Intelligence when used in combination with Industry 4.0 makes the dream of converging man and machine together for digital transformation of the industry.

AI powered Automation in the industry floor will change the way by which employees work, empower employees, transform and grow businesses with building resilience with AI. AI algorithms optimize the supply chain process of manufacturing and provide better response to the changing market. AI in Industry 4.0 uses real time data analysis for Intelligent Manufacturing. Robotics and Automation plays a major role in Industry 4.0. But the intelligent brain behind the Industrial Revolution is AI. SMEs use AI to automate their manufacturing, increase the quality of production and prediction can be done to foresee the future requirements of their consumers.

Economists are of the view that Industry 4.0 empowered with AI will bring out the true digital transformation across the world and yield profit. Technological innovations are done to automate many manual processes and intelligent software solutions are designed with Artificial Intelligence to handle large volumes of data to make the manufacturing process more efficient. Digital representation of a tool in industry will ensure to meet all its quality requirements at the manufacturing phase. AI powered Industry 4.0 is evolving rapidly across the globe embarking on the next generation industrial revolution. It is estimated that, the global smart manufacturing market which was valued at USD 195 Billion in 2020, will reach USD 314 Billion in 2026 at a CAGR growth rate of 8.4% and further taking it to USD 576 Billion by 2028 with the increased numbers of robots in the process of industrial automation and digital transformation. The goal is to empower the existing legacy system, instead of replacement, aiding the transition to an ecosystem of connected plants and enterprises. Industry 4.0 with AI has got off the ground with lots of innovation, creating opportunities and challenges in industry to redesign the business process.

The Objectives of the chapter are:

- To provide the basic idea of Industry 4.0 with a real time application and its challenges identified while implementation,
- To elaborate new trends on Industry 4.0 with focus on Artificial Intelligence, and
- And, to highlight a few AI algorithms implemented in the flourishing smart industries.

Organization of Chapter

The chapter is organized as: In Sect. 2, general concepts of Automation for Industry 4.0 with its roles, Challenges, Technologies used with examples, their advantages and future in making smart industries are presented. In Sect. 3, the role of Artificial Intelligence Algorithms in Industry 4.0 Automation is illustrated along with its implementation results on the required dataset. It also spotlights the Comparison of Performance metrics of different AI Algorithms with its discussion. Section 4 stresses on results and discussion. Section 5 concludes the chapter with future scope.

2 Automation for Industry 4.0

With smart Production, we are in the midst of a new industrial revolution. Smart Production is not a procurement of mechanization, but an information system which automates the task more efficiently at an affordable budget than humans. Automation for Industry 4.0 is done with the Industrial Internet of Things, to connect, control and monitor any devices, machines or robots. This automation will optimize production with minimal human intervention.

2.1 What is Industry 4.0?

Three industrial revolutions laid the path for smart production and automation using Artificial Intelligence. In the 1970s, Industry 1.0 was used to increase automatic production and aimed at the agricultural sector. Industry 2.0 mainly focused on improving the bulk production in industries. Industry 3.0, along with the Digital Revolution, improved human life in the 20th century. Industry 4.0 is concerned with smart industries or factories, smart cities, smart vehicles etc. [1]. The illustration of the industrial revolution is depicted in Fig. 1.

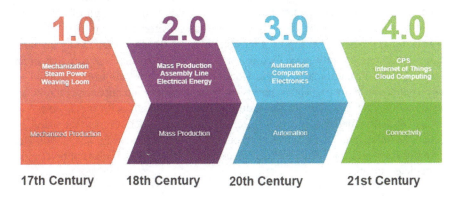

Fig. 1 Industrial automation

2.2 Merging the Virtual and Real World

The environment that makes the user get immersed, which has features that are similar to those found in the actual world is called Virtual Reality, where it exceeds the boundaries of the real world. Virtual Reality merged with Industry 4.0 resulted in the increase in quality of product and decrease in the cost invested in design and production. Virtual Reality in Industry 4.0 plays a key role in training employees, revamping the current plant, Safety Inspections etc.

The term "Digital Twin" can be defined as a detailed virtual image developed based on a plan to develop a new product or new plant with the help of digital tools. It acts as a combination of motion-based simulations with Data Analytics in a virtual world. The concept of Digital Twin consists of three forms: 1. Digital Twin of the product, 2. Digital Twin of the production 3. Digital Twin of the product and the production. Digital Twins require fewer prototypes when compared with traditional methods of planning in the industry. The innovations based on Digital Twin are reliable [1]. The data required for the Data Analytics process is created more when the product is produced based on the planning in a virtual environment. The performance in each stage of production is collected and analyzed for future developments of the industry. Digital Twin is considered as the accurate virtual model of a production or a product. It throws insights on the machine operators to predict the behavior and performance compared with their previous working model.

2.3 The Role of Automation in Industry 4.0

Industrial Automation is considered as a pillar in all manufacturing sectors. The role of automation in Industry 4.0 is to boost quality, reliability and production by reducing the overall cost of production. Industry 4.0 along with Automation is no longer focused on its efficiency and profitability, but on increasing flexibility and improvement of process quality, reducing the margin of errors [2]. Digital Twins play a major role by making the organization exercise virtual models to make efficient decisions. The human carried out processes with a margin of error 10% could be reduced to 0.00001% by the process automation platform.

The modern industry setup uses the following levels for implementing automation in Industry 4.0. I/O level focuses on input and output devices installed in an industry. The Programmable Logic Controller [PLC] level controls the logic of all machines in industry. The Supervisory level monitors and supervises the information from the machines in the industry. The MES deals with management of information flow in the production system [3]. The level ERP refers to the planning related to sales, purchasing, human resource, quality, engineering, production etc. The illustration of the various levels of automation in modern industrial revolution is depicted in Fig. 2.

Fig. 2 Levels of automation in modern industry

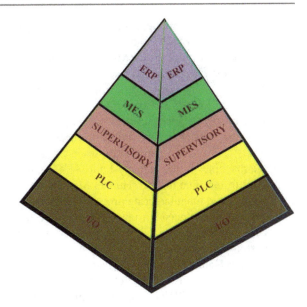

2.4 Automation Challenges in Industry 4.0

The key automation challenges in Industry 4.0 include getting businesses to think about the future of their businesses. Employees and business operations will face new problems as a result of the digital transition in Industry 4.0. The Challenges faced by Industry 4.0 in automating their business models are considered a huge investment, change in existing infrastructure, strategic planning in investing of machinery, so as to be safe, secure and efficient, Intelligent Workforce for Automation [3].

The main Automation Challenges in Industry 4.0 are listed below:

i. **Huge Investment**: There is a huge demand of investing for a smooth transition of traditional methods to Automation in Industry. Also, a huge investment for training the existing workforce. The top areas of industry are investing a lot for the requirement of data in automation. They are ready to invest in digital transformation with an expectation as automation will understand their process, improvise performance, decrease manufacturing cost and increase productivity. Huge investments made in industries face unprecedented challenges by COVID 19 impacted business growths, profitability and their rate of production. There is no denying that investment in automation will reap incredible rewards, but the manufacturers must invest in the right technology at the right time.

ii. **Considering new Business Models**: All industries face challenges in their real-time process, product and client engagement when new Business Models are integrated to their working environment. According to Salesforce, for new Industrial Paradigm, there is a need in their interaction style with customers to know their business needs, and how customers interact with the end products. AI powered CRMs are used in Zoho to learn from past decisions and historical patterns to increase their sales. IoT assisted devices are used by Siemens and CISCO to learn how customers use products or looking for the launch of new devices.

iii. **Flexibility**: Detailed analysis is done to examine the working culture of the company and ensure flexibility to meet the automation requirements. Flexibility is ensured in industries when the software from computers controls the machines in the industry floor [4]. Commands are created and entered as code to machine through Human Machine Interfaces. The industries adopted with a rigid manufacturing process to develop a single product and if the process defined is needed to be changed, it faces flexibility problems. In certain cases, to ensure flexibility in the industry floor, tools to be changed, more customization and more money is invested.

iv. **Process Reorganization**: For automation, integrate the vertical and horizontal plans of the industry. The potential adoption rate is stunning by any measure: the McKinsey Global Institute estimates that, using process reorganization, more than 81% of predictable physical work, 69% of data processing, and 64% of data-collection activities could feasibly be automated. In process reorganization, it should be carefully analyzed by the types and the input and output required which seems to be a tedious process.

v. **Workforce**: According to the World Economic Forum's Global Trends study from 2018, recruiters and hiring managers believe Industry 4.0 will have a significant impact on the recruiting industry, with 76% belief it will. With the shortage of skilled workers, industries are shifting faster to technologies to maintain their operations. With the automation growth, over the next 10 to 15 years, the workforce must be skilled to interact with smarter machines.

vi. **Standardization**: Industries must require norms and standards to ensure the digital transition of all individual components and interoperability. The world is run by Industrial Automation. The Industrial Automation Standards are necessary for the proper functioning of industrial automation. The International Organization for Standardization (ISO) has published many standards detailing industrial automation systems and integration, testing, management, and interfaces. The importance of implementing standardization is to increase phenomenal efficiency, decreased prices, and significantly improved reliability with respect to both output quantity and quality, industrial automation depends on proper design and implementation in order to continue delivering on its track record.

vii. **Data Management**: Need methodologies to deal with the quality of large datasets generated on real time production. Managing data in industries in the 21st century is a key challenge across the globe to improve efficiency and boost productivity. Managing a higher amount of live data generated in the industry floor makes data management a tough process. In certain cases, errors in data will make proactive decision making impossible.

viii. **Competition**: Increasing Competition to integrate customers and manufactures. The industries are dedicated to remain as market leaders with their innovation in product manufacturing and identifying market strategy. In the competitive strategy, industries compete to meet servicing their old and future customers.

2.4.1 Industry 4.0 Technologies

The technologies incorporated with Industry 4.0 for its giant growth towards automation and data exchange are:

- The Internet of Things (IoT)
- The industrial Internet of Things (IIoT)
- Cyber-Physical Systems (CPS)
- Smart Manufacturing
- Smart Factories
- Cloud Computing
- Cognitive Computing
- Artificial Intelligence

2.4.2 Internet of Things (IoT)

Keyur and Sunil [5] defined IoT as the Internet of three things which are categorized as humans to humans, humans to machines, machines to machines. Luigi et al. [6] suggests that people will trust IoT only when IoT is robust, resilient, secure and easy to implement. Cisco defined the network of different physical objects that are connected with different sensors, software and different technologies as The Internet of Things. When these things are connected over the internet, they are capable of interchanging real-time data over all devices. These IoT devices interconnected with automated systems will make the devices learn the process and act intelligently based on the changing requirements of the industries. ITU-T Y.2060 provides the definition of IoT as IoT can be thought of as a global information community environment that provides advanced services by linking (physical and virtual) things using existing and evolving interoperable information and communication technologies (ICT). The illustration of the Internet of Things in Smart Industry and Automation is depicted in Fig. 3.

2.4.3 The Industrial Internet of Things (IIoT)

Sam [7] recommended that implementing IIoT plays a major role in the future insights of an enterprise based on their turnover, location, performance etc. In the aspects of Khana et al. [8], IIoT brings cost reduction in Capital and Operating Expenses in Smart manufacturing sectors. The network of interconnected sensors, devices and equipment interconnected with industrial applications such as manufacturing and energy units is known as The Industrial Internet of Things. The difference between IoT and IIoT is IoT makes us live smart and easier life while IIoT increases safety and efficiency in industry floor. In the midst of 2021, it is estimated that over 30 billion of the world's devices will be connected in using IIoT. This networking had a great influence on data collection and analysis in real

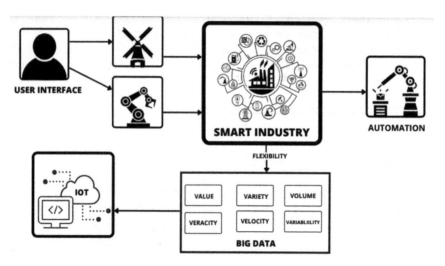

Fig. 3 Internet of things (IoT)

time facilitated merits such as increased productivity, efficiency with notable economic benefits. Industry 4.0 can also be referred to as IIoT because it transforms industry into a smart industry with Artificial Intelligence, Machine Learning and Big Data to access the real-time insights about processes, customers and products in the industry floor. The main challenge in implementing IIoT is security and privacy. The Components of Industrial IoT is illustrated in Fig. 4.

2.4.4 Cyber-Physical Systems (CPS)

An Intelligent System monitored or controlled by Computer based algorithms is called Cyber-Physical Systems (CPS). The difference between IoT and CPS is, in IoT machines and systems are connected with the Internet. In CPS, the communication is done by integration of computation, networks and systems. CPS mainly focuses on enabling smart and connected worlds. Khana et al. [8] represented CPS as all the digital computers that exist in the real world. In such systems, the hardware and software components are deeply linked with each other to operate effectively in a dynamically changing world in different operating modes. The key technologies required for CPS are Robotics, Cybernetics, Mechatronics, Embedded systems programming, Cyber Security, Machine Learning, Statistics, Cryptography etc. Hu et al. [9] realized that an intelligent factory is based on Cyber-Physical Systems which is the essence of Industry 4.0. Ribeiro et al. [10] suggested that Cyber-Physical Systems (CPS) are used in the industrial sector to merge the physical and virtual worlds. CPS plays a major role in the domains such as Transportation, Manufacturing, Health, Education, Sustainability etc. The illustration of Cyber Physical Systems (CPS) is illustrated in Fig. 5.

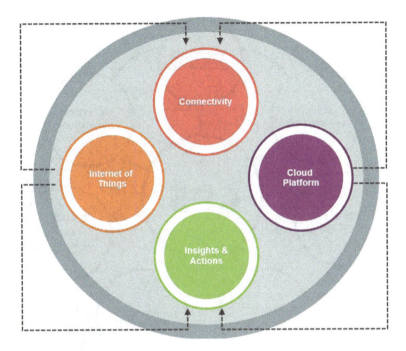

Fig. 4 Components of industrial IoT

2.4.5 Smart Manufacturing

Sameer et al. [11] defined Smart Manufacturing as a set of manufacturing technologies that work based on data from sensors over communication technologies for coordinating manufacturing processes. Ray et al. [12] proposed a systematic framework for Smart Manufacturing based on smart design, smart machining, smart monitoring, smart control and smart scheduling to rethink Industry 4.0 in various aspects to upgrade the smart manufacturing process. Computer-integrated manufacturing with good adaptability to rapid changes in industries manufacturing and design processes with a flexible workforce is called Smart Manufacturing. Smart Manufacturing with IoT speeds up the response time to make best business decisions [32–34]. When it is interconnected with Machine Learning and Artificial Intelligence, it will lead to better decisions while minimizing implementation cost and downtime. The illustration of Smart manufacturing is illustrated in Fig. 6.

2.4.6 Smart Factories

Sam et al. [7] viewed smart factories as dependent on robots and autonomous vehicles for transporting goods, to detect malfunctioning of systems to incorporate quality, and to optimize the work on the industry floor. Ribeiro et al. [10] come up with the basic principles of smart factories as completing complex tasks in less time, reduced budget, and increased production with high quality. Matthew Potts, Project Sales Engineer at HMK described the foundations of a smart factory as a

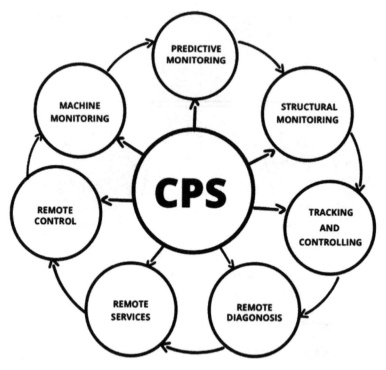

Fig. 5 Cyber physical systems (CPS)

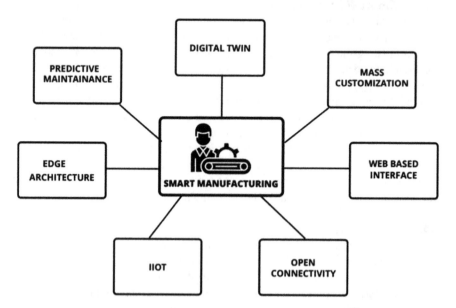

Fig. 6 Smart manufacturing

Fig. 7 Smart factories

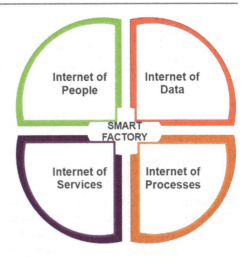

communication among all the industry floor's devices to ensure the functioning of equipment as well as for monitoring the productivity. The Internet of things along with Artificial Intelligence is the efficient way to prototype Smart Factories. Smart manufacturing allows factory managers to automatically collect and analyze data to make better-informed decisions and optimize production. The data from sensors and machines are communicated to the Cloud by IoT connectivity solutions deployed at the factory level. The illustration of Smart Factories is illustrated in Fig. 7.

2.4.7 Cloud Computing

Sam et al. [7] addressed Cloud Computing as a means for deployment flexibility in industrial AI. Cloud Computing plays a major role in meeting all demanding conditions of the industries. Lee et al. [13] in his work defined Cloud Computing as a catalyst in order to create and deploy Industry 4.0. Tay et al. [1] in his work concluded that Cloud Computing is used by most of the companies to manage large volumes of data in a secured manner. Cloud Computing with Industry 4.0 had a great impact on performance, storage optimization, efficiency, security, and privacy. While establishing smart factories, Cloud Computing with Artificial Intelligence will make the networks smarter, dynamic and affordable. The customizable enterprise applications with cloud optimization and migration will help in the development of mobile applications to make the industry smarter. The illustration of Cloud Computing is illustrated in Fig. 8.

2.4.8 Cognitive Computing

In the views of Monika et al. [14], when a machine possesses human characteristics, it will incorporate cognitive capabilities. Cognitive Computing when built with Industry 4.0 will support human decision making for abnormal and complex business activities. Industry 4.0 with Cognitive Computing has evolved cognitive

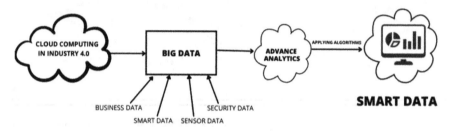

Fig. 8 Cloud computing

manufacturing to throw various insights across the entire change of design to support smarter factories and industries. Ribeiro et al. [10] suggested that for cognitive automation and analytical data analysis, RPA is implemented. In the field of industrial automation, Cognitive Computing has created a great revolution to analyze and process a large volume of data. Cognitive Computing overlaps with AI and uses the same technologies to develop cognitive applications like neural networks, expert systems, robotics etc. It also includes data mining, pattern recognition and Natural language Processing. Cognitive Computing is mainly used to assist people in decision making by identifying the patterns hidden in the data. The illustration of Cognitive Computing is illustrated in Fig. 9.

2.4.9 Artificial Intelligence

Lee et al. [13] defined Artificial Intelligence as a cognitive science that enables machines to explore like humans with logical reasoning and thinking in an intelligent manner. Artificial Intelligence with Industry 4.0 is not bound to the production floor. AI algorithms are optimized to make the industries to provide optimized response to the changes in the market [15]. According to a survey by Capgemini, Industry 4.0 with robotics, computers connected to the Internet of Things, and between 2018 and 2022, AI algorithms are predicted to add $500 billion–$1.5 trillion to the global economy. Hu et al. [9] believed that the new era of internet and Artificial Intelligence plays a vital role in intelligent manufacturing. Artificial Intelligence with big data addresses the problems faced by the industry when it shifts to the smart industry [5]. The illustration of Core Areas of AI is illustrated in Fig. 10.

2.5 Example of the Industry 4.0 Revolution

Let's explore some real-time examples of the industry 4.0 Revolution. First, we can consider the revolution of Industry 4.0 with big data Analytics. Consider a factory floor which has small sensors in machines of that factory. The machine related information collected in small sensors are implemented with machine learning algorithms. This Big data can be used for predictive maintenance i.e., the

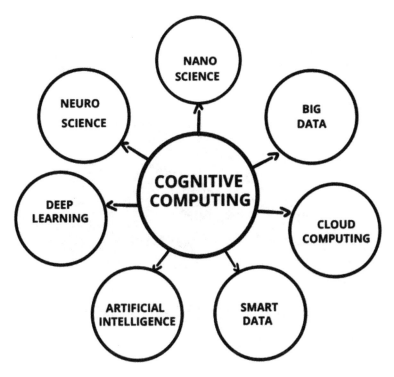

Fig. 9 Cognitive computing

results of this algorithm can be used to prepare schedules for periodic equipment maintenance [6, 15].

Let's dive into another example on Autonomous Robots. It is used in the manufacturing sector for shifting heavy machines and producing items [16]. Autonomous Robots play a major role in providing security to human power. It is built with the ability to transport products by analyzing the shortest path within the factory floor even with multiple orders of pick up at the same time. The main advantage of Autonomous Robots is its continuous production without break downs. BMW, a well-known vehicle manufacturer, announced plans for a manufacturing sector based on NVIDIA's Isaac robotics platform in 2020. BMW is creating AI robots as part of its Industry 4.0 strategy to handle and deliver their products in less time. The new technology, according to Jurgen Maidl, BMW's SVP of Logistics, is the future of factory automation. He also mentioned that 3D modeling, simulation, visualization, and deep learning systems are important components of the company's Industry 4.0 strategy and a method [17] to improve manufacturing and logistics processes.

Another example of Industry 4.0 is as Simulation tool in a manufacturing company. Using Simulation, we can view the entire industry floor in virtual space [18]. This Simulation tool can also be used to make employees explore and train on new

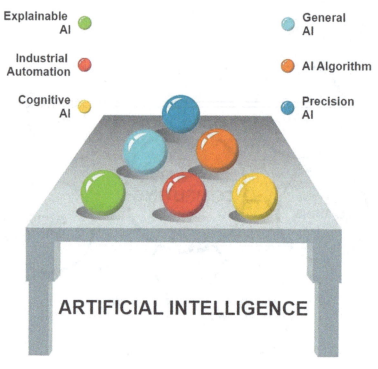

Fig. 10 Core areas of AI

machines when installed with less training cost and high level of security. Also, it allows employees to explore deeper into machine parts to predict their life span and maintenance cost. The other important revolution is in the field of industry R&D, manufacturing and sales to make everyone operate dynamically. The usage of 3D graphical simulation tools will help in analyzing real time problems and optimize the system operation.

In industry 4.0, Cybersecurity is another real time application in the manufacturing sector [7]. It provides security to their intellectual products and patents, factory data and equipment, their products including their IoT systems. Augmented Reality with Industry 4.0 can be used for employee training with a high security level. AR provides a virtualized environment to make technicians explore dangerous equipment in a safe manner [17].

Optimizing the factory floor and implementing industry 4.0 technologies can be done sometimes without retraining the employees. Bossard Smart Factory Logistics optimize the factory floor in which no learning curve is required by employees, but the factory becomes smarter. In the vision of Hyundai and Kai, smart factory deals with automation of the production process [19]. The core objectives of Hyundai in innovation of manufacturing are flexible manufacture, high degree of automation, Human-robot collaboration, Built in order, intelligent factory, immaculate quality control, sales innovation etc.

The plants of Morris Garages are fully automated with advanced robot spot welding, robotic roller hemming, robotic brazing to achieve good weld quality and dimensional consistency [20]. Robotic Automation is also used in coating to gain good finishing color quality. In assembly lines, the automated guided vehicles are used in various assembly processes. The core objectives of Morris Garages are autonomous technology, fully internet car, smart co-pilot, voice enabled and interactive systems [21].

The Jaguar Land Rover uses intelligent agents to sustain its success for a long decade. It uses technologies such as Artificial Intelligence, Cognitive Computing to capture a large volume of data, and applies machine learning automation to extend their business solutions across the globe [22]. The data analytics team play a role in providing insights to increase their profits and reveal new opportunities.

2.6 Challenges and Issues of Implementing Industry 4.0

The main challenges and issues of implementing automation in Industry 4.0 are listed as follows in Table 1.

2.7 Advantages of Automation in Industry 4.0

The main advantages of automation in Industry 4.0 are listed as follows:

- *Cost efficiency*: It reduces employee cost as automating processes will reduce the compilation in recruiting highly skilled employees [23]. Virtual and Augmented reality technologies will improve the learning process of employees and improve productivity with higher revenues.
- *Competitive advantages*: Automation in Industry 4.0 makes industries operate 24/7. Automation procedures will increase productivity, reduce maintenance cost with less operating time.
- *Scalability and Flexibility*: Installing new equipment on the industry floor will require training for employees with huge investments. In the automated industry, Robots and devices can be programmed accurately by reduced execution time and cost.
- *Time reduction*: Automation will reduce processing time. The automated platform with automated processes has good storage capacity with high processing speed.
- *Utmost Safety*: The manufacturing line with many autonomous machines installed will reduce the risk factor of employees. Cyber security measures are implemented with equip components and people working in industry with high levels of security [24, 25, 27].

- **Improved Control**: The automated process will be monitored and recorded. This big data will be analyzed to predict future requirements.
- **Increased knowledge sharing and Collaborative working**: In traditional manufacturing slots, there is minimal knowledge sharing or collaborative working among different employees. Industry 4.0 with automation will allow manufacturing sectors and different departments to communicate irrespective of time zone, location and other factors. This will allow knowledge learned by the sensor in one machine to be disseminated throughout all the departments in the organization with less human interventions.
- **Better Customer Experience**: Many services provided by Industry 4.0 will improve customer experience by automated tracking and resolving problems. Automation with Industry 4.0 increases product availability, product quality and provides customers with more choice to improve the profitability of the sector.
- **Creates Innovation Opportunities**: Industry 4.0 with automation will provide us a deeper knowledge on their design process, design patterns, business performance etc. It provides a lot of innovative opportunities to develop new products based on new innovative design patterns, improving profit and more [27].

Table 1 Challenges and issues of implementing industry 4.0

Authors	Industry 4.0: challenges and issues				
	Organizational complexity	Big data and analytics	Cybersecurity	Additive manufacturing	Investment issues
Kushik et al. [23]	No strategy to coordinate actions	Lot of missing talents	Depends on third-party providers	No technologies included	Lot of uncertainties in financial benefits
Nilufer et al. [24]	Lot of human engineering	Require more added and ethical engineers	No security	Life cycle management implemented	Financial investments is more
Stock and Seliger [2]	The increasing organizational complexity in the manufacturing system cannot be managed by a central instance from a certain point	The workers increasingly have to monitor the automated equipment, are being integrated in decentralized decision-making, and are participating in engineering activities as part of the end-to-end engineering	Highly complicated	This allows designing more complex, stronger, and more lightweight geometries as well as the application of additive manufacturing to higher quantities and larger scales of the product	The equipment will be able to flexibly adapt to changes in the other value creation factors, e.g., the robots will be working together collaboratively with the workers on joint tasks

(continued)

Table 1 (continued)

Authors	Industry 4.0: challenges and issues				
	Organizational complexity	Big data and analytics	Cybersecurity	Additive manufacturing	Investment issues
Shiyong Wang et al. [25]	No centralized coordination	There is a need to create a modularized and smart conveying unit that can dynamically reconfigure the production routes	The IWN network used today can't provide enough bandwidth for heavy communication and transfer of high volume of data but it is superior to the weird network in the manufacturing environment	The research is still going on for complex systems	Lot of financial investments
Nayyar et al. [26]	Could be adopted easily for larger firms but needs expertise to implement in smaller firms	It is a challenge to ensure high quality and integrity of the data recorded from the manufacturing system	With the increased connectivity and use of standard communications protocols that come with Industry 4.0, the need to protect critical industrial systems and manufacturing lines and system data from cyber security threats increases dramatically	The research is still going on for complex systems	Dynamically increasing due to their requirements

- ***Recovering the return on Investment***: Industry 4.0 with automation is transforming the industry across the world. The investment on Industry 4.0 is truly a return on Investment. To stay competitive, Automation is the next stage of the journey in Industry 4.0 to equip industry to meet all the future requirements.

2.8 The Future of Industry 4.0

The pillars of Industry 4.0 are Internet of Things (IoT), Big Data, Data Analytics, Augmented Reality, Cyber security, Robotics, Cloud Computing, Artificial Intelligence and 5G networks. Many companies across the globe have not yet developed

Fig. 11 Thinking ahead to the future industry

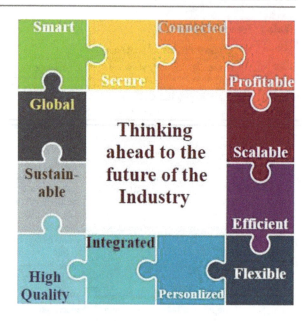

to implement Industry 4.0. There are many booming trends in the future of Industry 4.0, i.e., Industry 5.0.

The future of Industry 4.0 is more focused on personalization, integrating customers and meeting their requirements immediately. The illustration of thinking ahead to the Future Industry is depicted in the above Fig. 11. The main intention of Industry 4.0 is bridging the gap between people and machines instead of replacing humans. The main objective in the future is to develop an intelligent factory by educating human society. Humans must be trained and qualified to meet the future enhancement of Industry 4.0.

3 AI in Automation for Industry 4.0

In earlier decades, AI was considered as a concept divided into many application fields such as natural language processing, automatic programming, robotics, computer vision, intelligent agent, etc. Nowadays these application areas are considered as new fields, due to its massive growth in implementing real time AI applications. AI is today viewed as a collection of concepts with numerous applications. AI in Automation for Industry 4.0 will make industries smart, to complete complex tasks with reduced cost, with improved quality of goods and services. AI technologies are spreading throughout industries to merge the physical and virtual world. The use of AI to automate industries will address major issues of industry such as customizable requirements, reduction in delivery time of products to market, implementing automation in industries with the help of various sensors. Robotics with AI also plays a major role in automation for Industry 4.0.

3.1 Role of AI in Industry 4.0

Artificial Intelligence in Industry 4.0 leads to advanced growth in the global economy. An IBM study "The Global Race for AI" illustrates 82% of Spanish companies are already working on their floor with AI Automation. An IDC survey on global industries illustrates that around 25% of industries are already using AI solutions. AI in industries will improvise production with increasing process quality with the growing competitiveness in the market. Machine Learning and Deep Learning with AI techniques had a major impact on industries' Return on Investment (ROI). AI with ML algorithms on Smart Manufacturing will allow employees to gather data during production and the changes are automatically adapted. The role of AI techniques in industry 4.0 can be illustrated in Fig. 12.

In the manufacturing sector, artificial intelligence (AI) is required to manage and organize the vast amount of live streaming data acquired from factory floors via sensors. For admin-based jobs and responsibilities, with the goal of innovating and deploying AI, manufacturing companies must develop integrated automated platforms that use fractal science technologies. Fractal technology reduces the amount of computational power required for infrastructure, making it considerably easier and faster to install. Unstructured datasets, such as handwritten notes and inventory checklists, are better served by Intelligent Automation Systems with Fractal Technology. The ultimate goal is to improve the overall operational efficiency of the company.

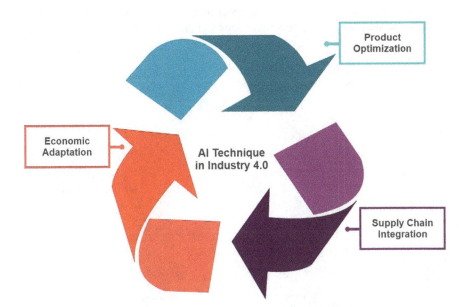

Fig. 12 The role of AI techniques in industry 4.0

3.2 Impact of AI in Industry 4.0

Artificial intelligence's impact in Industry 4.0 is organized into 5 broad areas as depicted in Fig. 13.

3.2.1 Predictive Quality and Yield

In manufacturing sectors, production losses and process inefficiencies must be reduced. In the market, as the demand of the product increases, competition also increases. On another end, customer expectation also increases due to the global knowledge of emerging technologies in Industry 4.0. Various surveys come with the information that the global population will increase by 25% by 2050 which in turn increases 200,000 additional mouths to be fed every day.

Based on the recent survey, it is also noticed that customers are not moving to alternatives if their product is not available in the market. The manufacturer believes that process inefficiencies and losses can no longer be tolerated. Loss that occurs in terms of waste, yield or quality must be measured properly to make production processes more efficient.

Another challenge on the industry floor is how complex processes are optimized. Industrial Artificial Intelligence used with Predictive Quality and Yield can be used to sort the hidden causes for production losses. This is accomplished by using machine learning algorithms that have been trained to analyze each specific manufacturing process. The "Supervised Learning" machine learning technique is used to educate the computer to recognize data patterns.

A recommendation system can be installed to generate automated recommendations and alerts to warn production teams and process engineers about problems

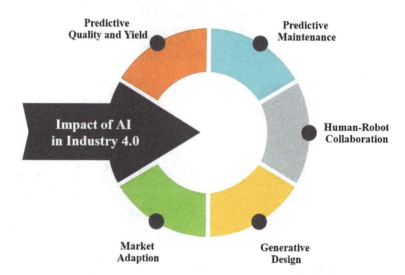

Fig. 13 Impact of AI in industry 4.0

on the floor and act as intelligent agents to exchange knowledge on how to avoid losses before they occur.

3.2.2 Predictive Maintenance

Another important application of Industrial AI is Predictive maintenance. In traditional systems, maintenance is performed according to a predetermined schedule. When predictive maintenance is used on the industry floor, the algorithms can predict component or machine failure, alerting engineers to perform maintenance to prevent failures without wasting production time.

Predictive Maintenance System inbuilt on Machine Learning techniques with different supervised and unsupervised algorithms can work on more predictions by reducing maintenance cost and eliminating the need for planned shut down in many instances.

Predicting failure of machines with various machine learning algorithms, will make industries function without unnecessary shutdowns. Maintenance will be informed to technicians prior to the need for inspecting particular components in the machine, their replacement and methods to be followed to rectify the faults in a cost-effective manner. It also increases the Remaining Useful Life (RUL) of machines and its components by diagnosing its secondary failure earlier to perform proactive maintenance.

3.2.3 Human-Robot Collaboration

In 2020, the International Federation of Robotics (IFR) estimated that roughly 1.64 million industrial robots will be in use in industry. Robots are not a replacement for human power in industries. It can make industries more efficient by human-robot collaboration, offering training to humans on higher level programming, design and maintenance with high safety for man power.

The implementation of robotics in industry 4.0 will improve efficiency of the industry by improvising the manufacturing quantity. AI will help safeguard human safety by assigning robots to vital activities and making process optimization decisions based on simulated data collected on the factory floor [35].

3.2.4 Generative Design

In the design phase of industries, Artificial Intelligence plays a major role. The input data must be clearly defined. Designers and engineers can utilize generative design software, an AI system, to create possible solutions using live data obtained from the factory floor.

In the design phase, the material types and its restrictions, production methods, time limits and budget constraints are defined. The solutions explored by generative design software can be tested using different machine learning algorithms. This will provide information about which design ideas can help to make better decisions. The usage of machine learning algorithms will improve solutions by making optimal solutions.

Autodesk uses Generative Design with machine learning algorithms which will guide to take optimal decisions on feasible designs for their products like

Fusion360. To deploy machine learning models, Autodesk uses Amazon SageMaker which handles workflow, labelling the data, training the model, tuning and predicting results. Autodesk built this entire technology in Amazon Web Service to meet customer demands, security and cost effectiveness.

3.2.5 Market Adoption

Artificial Intelligence (AI) isn't just for the factory floor. It is used throughout Industry 4.0 ecosystem. It is used to optimize the supply chains and to adjust with the changing market conditions, i.e., market adoption. For market demand estimation, algorithms consider datasets based on different features such as date, location, attributes related to social and economic factors, behavioral data, political status, weather factors and more. This estimation can be used by industries to optimize their inventory control, recruiting employees, and purchasing raw materials based on financial decisions.

3.3 Industrial Artificial Intelligence

Industries are in the requirement of a systematic structure for implementing AI in their industrial floor. Industrial Artificial Intelligence is a systematic strategy to creating, validating, and deploying various machine learning algorithms for high-performing industries. It can make factories smarter with high fault tolerance. Based on the definition of industrial AI, the applications are more related to manufacture of products, supply chain, warehouse which deals with storing and transporting products, etc. The four Industrial Artificial Intelligence technologies that is used for smart manufacturing are as follows [13]:

- Platform Technology
- Data Technology
- Analytic Technology
- Operation Technology

Globally large companies are using Industrial Artificial Intelligence to create a competitive environment with greater efficiency and high quality. Let us explore some examples:

- In the last five years, Amazon had invested $775 to form Amazon Robotics. This is the heavy investment made by Amazon in AI techniques. Amazon works on increasing the ties between industries and academic robotic communities to provide solutions for big problems with automation.
- J.B. Hunt, Supply Chain providers invested $500 to develop technologies to connect transporters by using real-time data and artificial intelligence.
- Boeing acquires Liquid Robotics and partnered with Microsoft to use Azure Cloud Platform to improve its efficiency in designing aircrafts.

- Bosch invested €300 million for Bosch Center for Artificial Intelligence. Dr. Volkmar Denner, chairman of Bosch said that in the next ten years, Bosch products with Artificial Intelligence will be scarce, Intelligence with automation will play a major role in the development of Bosch products.

3.4 Applications of AI in Smart Industry

With the tremendous growth of the Internet, the universal availability of sensors, big data, information communication, interconnection of data and knowledge derived leads to the emergence of new technologies to the new phase of AI: AI 2.0. The main applications of AI are the evolution of smart cities, medical care, transportation, intelligent robots, autonomous vehicles, smart phones, smart industries to meet growing market demand. AI is used in industries about 29% for manufacturing and 27% to increase the quality of the product.

Industrial statistics divulge the increase of interest among industries to AI applications during this COVID 19 pandemic lock down. The different applications of AI in Smart Industry are listed below:

- Digital Twins
- Product Development
- Design Customization
- Shop floor Performance Improvement
- Logistics optimization
- Predictive Analytics
- Predictive Maintenance
- Generative Design
- Price Forecasting of raw material
- Robotics
- Computer Vision
- Edge Analytics
- Quality assurance
- Inventory Management
- Process Optimization
- 24 * 7 Production
- Safety
- Cost Reduction
- Faster Decision Making
- Production and Logistics Automation

Digital Twins

Digital Twin Technology is the core technology for CPS. The digital twin technology played a key role for NASA in manufacturing the NASA's Apollo Program [28, 31]. Twins are representing the prototype or model of the real situation using simulation. In manufacturing, digital twin technology refers to a virtual world created using augmented reality to process real-time data based on realistic configurations of actual products, industrial processes, or equipment. Digital Twins are used to improve manufacturing operations such as product design, process optimization, quality management, supply chain management, predictive maintenance, cross-discipline collaboration and analyze the future of manufacturing. In 2025, it is estimated that 89% of IoT platforms include digital twins.

Predictive Maintenance

Predictive Maintenance avoids machine failures on the floor with increased productivity. It decreases 50% downtime with a 20% increase in production. Predictive maintenance, according to a McKinsey analysis, will save companies $630 billion by 2025. The leading AI transformation projects in manufacturing initiatives today are machinery maintenance and quality. Caterpillar's Marine Division is saving $400 K per ship per year after analyzing data on how often hulls should be cleaned for maximum efficiency. The BMW Group uses artificial intelligence to assess component photos in live production lines in order to detect deviations from the norm in real time.

Computer Vision

In the smart world, Computer Vision plays a major role in understanding images and videos more accurately than humans [29]. The various application areas of Computer Vision are Optical Character Recognition, Face Detection, Smile Detection, 3D of various images, Object Recognitions, Vision based biometrics, Special effects on images, Smart cars, Sports, Smart Factories, VR Games, Industrial Robots etc....Computer Vision is being used by Amazon in their retail industry. When things are taken from shelves and placed in carts, facial recognition cameras powered by CV keep track. CV systems track customers as they walk around the store, recording every item they choose. The customer then departs the store once they've finished their shopping. The store's technology can link the customer to their Amazon account, allowing them to be automatically billed for their purchases.

3.5 AI Algorithms Used in Smart Industry

Let us explore some of the commonly used AI Algorithms in Smart Industry for designing various process:

A. *Support Vector Machine (SVM)*

The supervised learning algorithm, Support Vector Machine (SVM) can be used to solve any problems that arise in the industry floor. The main goal of SVM is for predicting problems. SVM will design a model based on the training dataset and predict the test data. SVM requires a large dataset to predict future insights with maximum accuracy. The main advantage of SVM is it will work in an incomplete dataset and speed the classification. The disadvantage of SVM is the speed of learning is slow and it lacks the explanation ability of humans [30].

B. *Decision Tree*

This machine learning algorithm is easily followed by humans by its graphical representation. The main challenge with decision trees is to find the optimal decision tree for different training datasets. The advantage of using it is it can explore all the critical features of prediction problems with high speed of learning. The main disadvantage is it cannot be used to solve nonlinear problems [30].

C. **Rule-Based Learners**
It is also called expert systems. It is used in the form of machine learning combining data mining features. It extracts information based on statistical techniques. The main disadvantage is when implemented in unsupervised learning environments, it lacks classification accuracy. But the classification accuracy can be improved by including all the characteristics and knowledge of the background. The main advantage of rule-based learners is that the system can give clear explanations about the process involved in creating results and periodically check the entire datasets [30].

D. **K-Nearest Neighbors**
K-NN is the machine learning algorithm mainly for classification and pattern recognition. It acts as a prediction algorithm by forming clusters and comparing similarities among the data. This algorithm is very dependent on data and leads to wrong predictions due to missing data, noisy, fuzzy, irrelevant and redundant data which also slows the processing. This also gives low accuracy. Data interpretation is difficult as mostly it uses an unstructured dataset [30].

E. **Naïve Bayesian**
It is an ML algorithm where the model is represented as an acyclic graph. It has one parent with many children. The parent represents the unobserved node and children represent the observed node. It is used to solve problems related to classification, clustering, regression etc. The major advantage is it requires only less memory space for training and classification [30].

F. **Artificial Neural Network**
It is a ML algorithm for classification and regression problems in various fields of manufacturing. It allows the system to perform supervised, unsupervised and reinforcement learnings. It deals with high dimensional data and works with hidden patterns and throws insights just as the human brain does. The main usage of this model is to optimize the model with less human intervention [30].

4 Results and Discussion

The major finding of this study is Industry 4.0 uses Internet of Things and sensors to extract live streaming data which is an important component for industrialists [24]. The main features of industry 4.0 are smart manufacturing, Cyber Systems, implementing Sensors, Interconnection of systems, Supply Chain etc. Industry 4.0 is mainly focused on digital enhancements and reengineering of products [33, 38]. The challenges while implementing industry 4.0 can be rectified by continuous innovation and learning by employees on the industry floor.

Artificial Intelligence is defined as cognitive science that enables humans to explore processes with high end sensing and reasoning. Industrial AI is defined as a systematic approach that enables engineers to develop and deploy AI algorithms to automate industries [34]. Nowadays Artificial Intelligence is attracting industrialists, researchers and government authorities for investing huge money for developing innovative machine learning technologies and applications [35]. AI algorithms are used for predictions and making faster decisions for all business activities. In the manufacturing sectors, robots can work along with humans, and in home environments, refrigerators can order veggies and provisions for their master based on their preference etc.

The innovative manufacturers are using artificial intelligence to tackle real time challenges. Industrial AI can be applied in industries to implement artificial intelligence in various application areas [36]. In generalized artificial intelligence, it can be used to computerize systems to perform tasks based on human intelligence. Industrial AI is more concerned in addressing design processes, pattern designing, customer value creation, increase in productivity with high quality, future insights etc. [23]. AI applications can be viewed in a negative aspect as it decreases the jobs of humans and causes social and ethical impact. But in industry, it has a more positive impact as unstoppable production which increases their economy and drastic improvement in the process design [37]. As a result, leading industries are moving to artificial intelligence. More experiments are carried out with different technologies such as digital connectivity of machines, artificial intelligence, machine learning, automation, robotics in manufacturing sectors [38]. This impact of transformation is referred to as industry 4.0 or fourth industrial revolution. This revolution in innovation is not concentrated more on productivity and efficiency, but more on business value strategies to meet the competitive environment by establishing an efficient supply chain [25].

5 Conclusion and Future Scope

The use of artificial intelligence to solve real-world problems has ushered in the industrial systems of the future, known as Industry 4.0. This article aims at defining Industrial AI, its benefits and challenges in Industry 4.0, as well as the future of the industry 4.0 paradigm. It also provides an overview of Industrial AI in today's manufacturing sector and the realization of the Industrial AI system. The book chapter concludes with the main objectives of Artificial Intelligence in Industry 4.0 for optimization of the industry processes. AI techniques play a major role in the revolution of industry 4.0 making humans experience a fusion of different technologies such as Internet of Things, intelligent automation, cyber security, intelligent process design etc. The emergence of all these technologies affects the normal workflow of the industry by improvising digitization. Nowadays along with artificial intelligence, industries are incorporating RPA to make industries reach the next level of intelligence in the automation process.

Industrial AI is a set of applications that create insights, provide guidance, and automate and democratize knowledge to create more value for the process industries. In future, to increase the efficiency of automation, Hybrid Cloud can be embedded with technologies like AI, edge computing, Cyber Security in which speed and latency of data is considered as major factor to work. The book chapter concludes that when automation is implemented in industry 4.0 in the manufacturing process, it will meet the immediate requirements of investors across the globe.

References

1. S.I. Tay, T.C. Lee, N.A. A. Hamid, A.N.A. Ahmad, "An Overview of Industry 4.0: Definition, Components, and Government Initiatives", Journal of Adv Research in Dynamical & Control Systems, Vol. 10, 14-Special Issue, 2018
2. T. Stock, G. Seliger, "Opportunities of Sustainable Manufacturing in Industry 4.0.", 13th Global Conference on Sustainable Manufacturing - Decoupling Growth from Resource Use, Procedia CIRP, No: 40, pp: 536–541, 2016
3. Cesar Da Costa, Cleiton Mendes, Raphael Osaki, "Industry 4.0 In Automated Production", IEEE, 978-1-5386-2278-0/17, 2017
4. Stecke K.E, Parker R.P. (2000), "FLEXIBLE AUTOMATION", In: Swamidass P.M. (eds) Encyclopedia of Production and Manufacturing Management. Springer, Boston, MA. https://doi.org/10.1007/1-4020-0612-8_343
5. Keyur K Patel, Sunil M Patel, "Internet of Things-IOT: Definition, Characteristics, Architecture, Enabling Technologies, Application & Future Challenges", Volume 6, Issue 5, International Journal of Engineering Science and Computing, May 2016
6. Luigi Atzori, Antonio Iera, Giacomo Morabito, "Understanding the Internet of Things: definition, potentials, and societal role of a fast-evolving paradigm", Ad Hoc Networks · December 2016
7. Charrington, S. (2017), Artificial Intelligence for Industrial Applications. Cloudpulse Strategias
8. W. Z. Khana, M. H. Rehmanb, H. M. Zangotic, M. K. Afzald, N. Armia and K. Salahe, "Industrial Internet of Things: Recent Advances, Enabling Technologies and Open Challenges", Computers & Electrical Engineering November 2019
9. Li, B. H., Hou, B. C., Yu, W. T., Lu, X. B., & Yang, C. W. (2017), Applications of artificial intelligence in intelligent manufacturing: a review. Frontiers of Information Technology & Electronic Engineering, 18(1), 86-96
10. Ribeiro, J., Lima, R., Eckhardt, T., & Paiva, S. (2021), Robotic process automation and artificial intelligence in industry 4.0–a literature review. Procedia Computer Science, 181, 51-58
11. Sameer Mittal, Muztoba Ahmad Khan, David Romero, Thorsten Wuest, "Smart manufacturing: Characteristics, technologies and enabling factors", Proceedings of the Institution of Mechanical Engineers Part B Journal of Engineering Manufacture April 2019
12. Ray Y. Zhong, Xun Xu, Shohin Aheleroff, "Smart Manufacturing Systems for Industry 4.0: A Conceptual Framework", CIE47 Proceedings, 11–13 October 2017, Lisbon/Portugal
13. Jay Lee, Jaskaran Singh, Moslem Azamfar, "Industrial Artificial Intelligence", 2018 Society of Manufacturing Engineers (SME). Published by Elsevier Ltd.
14. Monika Gadre, Aruna Deoskar, "Industry 4.0-Digital Transformation, Challenges and Benefits", International Journal of Future Generation Communication and Networking, Vol. 13, No. 2, (2020), pp. 139–149
15. Nayyar, A., Rameshwar, R. U. D. R. A., & Solanki, A. (2020). Internet of Things (IoT) and the Digital Business Environment: A Standpoint Inclusive Cyber Space, Cyber Crimes, and Cybersecurity. The Evolution of Business in the Cyber Age, 10, 9780429276484-6
16. Solanki, A., & Nayyar, A. (2019). Green internet of things (G-IoT): ICT technologies, principles, applications, projects, and challenges. In Handbook of Research on Big Data and the IoT (pp. 379–405). IGI Global

17. Singh, K. K., Nayyar, A., Tanwar, S., & Abouhawwash, M. (2021). Emergence of Cyber Physical System and IoT in Smart Automation and Robotics: Computer Engineering in Automation. Springer Nature
18. Krishnamurthi, R., Kumar, A., Gopinathan, D., Nayyar, A., & Qureshi, B. (2020). An overview of IoT sensor data processing, fusion, and analysis techniques. Sensors, 20(21), 6076
19. Rathee, D., Ahuja, K., & Nayyar, A. (2019). Sustainable future IoT services with touch-enabled handheld devices. Security and Privacy of Electronic Healthcare Records: Concepts, Paradigms and Solutions, 131, 131–152
20. Nayyar, A. (2019). Handbook of Cloud Computing: Basic to Advance research on the concepts and design of Cloud Computing. BPB Publications
21. Mamad Mohamed, "Challenges and Benefits of Industry 4.0: An overview", International Journal of Supply and Operations Management, August 2018, Volume 5, Issue 3, pp. 256–265
22. Kaur, A., Singh, P., & Nayyar, A. (2020). Fog Computing: Building a Road to IoT with Fog Analytics. In Fog Data Analytics for IoT Applications (pp. 59–78). Springer, Singapore
23. A. Kushik, "Smart Manufacturing", Int J Prod Res, vol: 56, pp: 205–217, 2017
24. Nilufer Tuptuk, Stepen Hailes, "Security of smart manufacturing systems", Journal of Manufacturing Systems 47(2018) 93–106
25. Shiyong Wang, Jiafu Wan jiafuwan, Di Li, and Chunhua Zhang, "Implementing Smart Factory of Industrie 4.0: An Outlook", International Journal of Distributed Sensor Networks, Sage Journals, January 2016, https://doi.org/10.1155/2016/3159805
26. Nayyar, A., Puri, V., & Le, D. N. (2017). Internet of nano things (IoNT): Next evolutionary step in nanotechnology. Nanoscience and Nanotechnology, 7(1), 4–8
27. K.R. Sowmia, S. Poonkuzhali, "Artificial Intelligence in the field of Education: A Systematic Study of Artificial Intelligence Impact on Safe Teaching Learning Process with Digital Technology", Journal of Green Engineering, Vol 10, Issue 4, April 2020, pp. 1566–1583
28. Nayyar, A., & Kumar, A. (Eds.). (2020). A roadmap to industry 4.0: Smart production, sharp business and sustainable development. Springer
29. Kumar, A., & Nayyar, A. (2020). si 3-Industry: A Sustainable, Intelligent, Innovative, Internet-of-Things Industry. In A Roadmap to Industry 4.0: Smart Production, Sharp Business and Sustainable Development (pp. 1–21). Springer, Cham
30. Chih-Wen Chang, Hau-Wei Lee, Chein-Hung Liu, "A Review of Artificial Intelligence Algorithms Used for Smart Machine Tools", Inventions
31. Fei Tao, "Digital Twins and Cyber-Physical Systems towards Smart Manufacturing and Industry 4.0, Engineering 5(2019), 653–661
32. A. Kushik, "Smart Manufacturing must embrace big Data", Nature 2017:544(7648):23–5
33. Fei Tao, Qinglin Qi, "Data-driven Smart Manufacturing", Journal of Manufacturing Systems, 2018
34. Yeou Ren Shiue, Ken Chuan Lee, Chao Ton Su, "A Reinforcement Learning Approach to Dynamic Scheduling in a Product-Mix Flexibility Environment", IEEE Access 2020
35. L. Jiang, H. Huang, Z. Ding, "Path Planning for Intelligent robots based on deep Q-Learning with experience replay and heuristic knowledge", IEEE/CAA 2019
36. Hui Yang, Soundar Kumara, Sathish T.S, Fungee Tsung, "The internet of things for smart manufacturing: A Review", IIE Transactions, 2019, https://doi.org/10.1080/24725854.2018.1555383
37. M. Feurer, J. Springenberg, and F. Hutter, "Initializing Bayesian hyperparameter optimization via meta-learning", In Proc of AAAI's, No 15, pp: 1128–1135, 2015
38. L.D. Xu, W. He and S. Li, "Internet of Things in Industries: A survey" IEEE Transaction on Industrial Informatics, vol 10, no 4, pp 2233–2243, 2014

Business Sustainability and Growth in Journey of Industry 4.0-A Case Study

Gouranga Patra and Raj Kumar Roy

1 Introduction

Revolution of market is going on; we are significantly observing the last two decades, it is changing very fast. With the changing of time and requirement, Industries are also changing their operational pattern. Mass market converted into customized market and product systems are now more customized than mass. Therefore, there is a need for changes in manufacturing and market process within and outside of industry. Industrial revolution is an old concept but with changing of time it changes the shape and now it is the time for revolution of industry 4.0 in term of production, process, outcome and system at large. The transformation of electrical to electronics to digital automated operation process are making industry process more advance and providing high sustainability where customers can easily access with effective and efficient way [1]. The primary objective of Industry 4.0 is to customized the need and requirement in term of ordering, enquiry, delivery to the utilization and recycling of the product for environmental safety [2]. To make the production and manufacturing system more customized, the Process has been divided into various units and each unit is connected with their next respective unit for managing efficient production system [3]. The industry 4.0 is a conversion of regular machine to self-operated machine to improvise the efficiency level of production system with the help of new techniques and technology like IoT, Artificial intelligence, Big data analytics and other forms which develop surrounding interaction [4]. It also helps to monitors production and manufacturing system with the help of market requirements by using real time data monitoring, tracking as well as give instruction for controlling production process [5].

G. Patra (✉) · R. K. Roy
Department of Management, Adamas University, Kolkata, West Bengal 700126, India
e-mail: gourangapatra13@gmail.com

© The Author(s), under exclusive license to Springer Nature Switzerland AG 2023
A. Nayyar et al. (eds.), *New Horizons for Industry 4.0 in Modern Business*, Contributions to Environmental Sciences & Innovative Business Technology,
https://doi.org/10.1007/978-3-031-20443-2_2

Digitalization is the transformation and perfect, effective and efficient innovation of all the process for Industry 4.0 by collecting structured data, make it meaningful through process and use it for operational decision process in a particular time. Maximizing revenue and minimization of risks and costs are required of accurate and consistent decision making by the firms [6]. The turning of operation decision helps industry to identify gap and do better business forecasting, which turn into sustained business outcome and make competitive advantage in the market. The modern process (Industry 4.0) using AI, Machine learning, IOT vision make the industry smarter in area of operation, market understanding, market prediction, resource mobilization and cost control.

Through the application of industry 4.0 and digital transformation in production system, firms allow customers to actively participate in their innovation process like product design, development and process, there are few companies in US follow user generated content where user are being allowed to take part in the product design and development and make more customized user preference product [7]. Not only in manufacturing, firms also redesign their business model transforming value generated process through smart manufacturing, smart product through Industrial internet of things [8, 9]. With the help of industry 4.0, the firms are reshaping their global value chain by making close proximity between manufacturing and market with environmentally sustainable results [10, 11]. Application of industry 4.0 not only develops market, manufacturing process and supply chain in the industry, it is also making a strong competitive advantage in the market within the competitors in term of efficiency, differentiation and innovation. Firms that adapt the new process in their internal operations in manufacturing and markets and exploit the advantage of new industrial revolution can achieve positive result in front of competitors and make sustainable market [12], It is needless to say that application of industry 4.0 makes significant changes of the firm in their financial performance. In our present study we will try to investigate the firm performance with the help of new techniques of Industry 4.0 and how it effectively uses to run business and reduce human effort and cost.

Objectives of the Chapter

The objectives of the chapter are:

1. To discuss the orientation and implementation of Industry 4.0 in present context,
2. To analyze the firm's operational performance after using computer vision in the smart manufacturing process of Siemens,
3. To examine the implementation of IIOT and Robotics in manufacturing process with reference to Cognizant and Tesla, and
4. To review the challenges and opportunities of the firms' applying tools of industry 4.0 in respect of management perspective.

Organization of Chapter

The chapter is organized as: Sect. 2 elaborates literature review. Section 3 enlightens Industry 4.0 with regard its characteristics, importance and implementation in Indian Industry. Section 4 highlights proposed work. Section 5 concludes the chapter.

2 Literature Review

In first and second industrial revolution, we do not understand much in the outside change but third industrial revolution mainly focuses on electronics and IT revolution and faster production process.

Goldhar and Jelinek [13] discussed that changes in the application of computers have altered the manufacturing process. The use of computer added design and computer added manufacturing bring greater flexibility, shorter production cycles, more customized products, faster responses to changing market demands, better control and accuracy of processes.

Kagermann et al. [14] discussed strategic implementation of industry 4.0. It applies both corporate and government organization and make the market more dynamic, flexible and develop competitive advantage with help of artificial intelligence, machine learning to make their operational process smoother and more effective.

Chesbrough [15] described that how open innovation can unlock the latent economic value in a company's ideas and technologies. Author mentioned that with the passing of time, most of the organizations are adopting the process automatically. It combines operational, design, development and related services in a parallel platform and value chin process become change.

Charles [16] discussed the scenario of application of industry 4.0 and industrialization in digital age. It can be stated that Industry 4.0 is an entity of data exchange and automation in manufacturing technology which develop smart factory, process, product and reduce human intervention.

Bhat [17] discussed how smart devises make impact on manufacturing and control system. Author evaluated the Indian scenario. The paper illustrated the negative side of industry 4.0 in job market and it was concluded that India needs a new policy to incentivize adoption of digital technologies, develop requisite digital ecosystem, augment competitiveness, and allow leapfrogging into the digital 21st century, while meeting the requisite skill gaps and ensuring jobs for millions entering the job market.

Michael and Wirtz [18] explained Digitally connected industrial production promises faster and more efficient processes—in development and production, services, marketing & sales and for adapting entire business models. The study highlighted that fulfilling the customer expectation is big hurdle for marketer but due to I4.0 technology, company provide best value chain support and develop of integrated automated production process which link customers, suppliers and manufacturer in single phase through cloud computing.

Stock and Seliger [11] reported that how big data helps marketer more accurate. As the markets now based on data driven, therefore, big data analytics makes marketer more predictive rather than descriptive in their market understanding in terms of customers need and requirement.

Hermann et al. [19] and Burmeister et al. [20] discussed the integration of the internet of everything into the industrial value chain. Industries through their revolution of digitalization are adopting new technology which is related to the organization value chain through connecting between physical and virtual world and explain the industry smooth functioning using this technology. I4.0 also helps the faster production process, increase capacity utilization and adopt faster market.

Nagy et al. [21] conducted research to know how companies operating in Hungary interpret the phenomenon of Industry 4.0, what Internet of Things (IoT) tools they use to support their processes, and what critical issues they face during adaptation. The study concluded that most of companies reported financial gain after adopting Industry 4.0 in their business, on other side it reduces the cost of lead times, improve assets utilization and improve production quality [22]. Therefore, I4.0 have huge impact on business performance and it penetrates the value chain system which are based on production and logistics operation of the firms and enhances firms for market capturing and predict more customized based market environment. In India, it is also observed that there are few sectors like automobile, healthcare; FMCG etc. are trying to adopt this concept in partly way.

Ghadge et al. [23] explained the progress of supply chain network using industry 4.0 technology and concluded the it directs a new way in the supply chain and make the business process more transparent which increase survivable of industry.

In a study Andreas et al. [24] investigated the impact of digital transformation in manufacturing industry. The study highlighted dynamic capabilities of economic and social impact on industry. Authors proposed interrelation model which helps practitioners for developing sustainable strategy for manufacturing firms.

Ramos and Brito [25] investigated application of industry 4.0 in tourism industry. The study explained that technological drivers change the tourism market. The study also revealed that with the help of innovative technology like tourism management information systems make better customer satisfaction. The new technology integrates design, production and marketing to meet the demands inherent in the new model.

Paul et al. [26] examined the concept of healthcare 4.0 and the way how to use of IoT, AI, Big data analytics and blockchain technology in healthcare firms. The study approved that using this technology, helps firms to design innovative approach which is effective to dragonize and treated patients in better way.

Singhal [27] examined the status quo of Industry 4.0 in Indian context in manufacturing industries. The study concluded that there is limited application of industry 4.0 in India but the study explained digitalized manufacturing increase higher productivity, low cost of production, lower labor cost and higher customer satisfaction.

Barzotto and Propis [28] explained the company's performance on emerging manufacturing using Industry 4.0 with the help of technology disrupting manufacturing model. The outcome of the study state that collaboration of new technology (I4.0) with local suppliers increase firms' productivity and efficiency. It maintains a balance of firm, place, global drive improves performance of manufacturing sector.

Chauhan et al. [29] explained the challenges aspect of use of modern techniques in industry. It is said industry 4.0 techniques improve operational efficiency of the firm but contingency effects act as an extrinsic barrier on firm performance towards digitalization.

Sima et al. [30] explained the influences of industry 4.0 on human capital development and consumer behavior and who are the drivers for this. The result of the study highlighted that education 4.0 is the way to develop human capital through vocational education, entrepreneurial education, financial education, and digital education and IOT, AI, ML are the key drivers to develop better understanding of consumer.

In a study Rüßmann [31] stated the impact of nine pillar of industry 4.0 like Big data analytics, Cloud, IIOT, Cybersecurity and others, which make a significant impact on increasing revenue of manufacturing sector, employment and investment. The study also clarified that it is a process of development of more customized market.

In another study, Tippayawong et al. [32] discussed the implementation of industry 4.0 in Indian SME sectors. The study revealed that there is need to generate more awareness with mind of labor forces and develop management strategy for supporting new technology and the benefit of using technology.

Cioffi et al. [33] explained application of AI and ML in manufacturing industry. It is observed from research that these tools increase transparency in process, make accurate demand forecasting, improve response time and reduce production failure which make market prompter and give better customer delivery. Artificial intelligence change manufacturing process with the help of core technology like intelligence manufacturing.

Verma et al. [34] investigated the application of AI technology in marketing practices and how it optimizes marketing process. The study revealed AI helps to increase customer satisfaction and experience and increase firms marketing outcomes. Chatterjee et al. [35] and Nayyar et al. [36] worked on similar lines and explained the importance of AI to help understand customer habits and purchase intention. In another study, Sha and Rajeshwari [37] claimed the impact of AI on e-commerce business. Findings of the study stated that Advanced AI-enabled machine could be able to track five human senses and in turn leads to improved e-commerce business.

3 Revolution of Industry 4.0

During the old times, the products were basically implemented and designed by any group or organization and were using tools which is basic required for production by the micro workshop [38]. The first evolvement was done during the First Industry Revolution (Industry1.0) and this was started at United Kingdom in the 17th Century and later United stated adopted the innovation at 18th century. This revolution showed a great impact in many industries Like Agriculture, Coal and Diamond Mines [39].

The Cost of materials and the time of production were decreased. This brought a huge change with disadvantages like Machines were heavy and the workers used to work for long hours with risks on another side.

Industry 2.0: This manufacturing shift was brought to enhance the quality of Railways, Roadways to help people travel places in short time. Evolvement of 2.0 gave huge Business growth with great productivity and minimized a decent percentage of unemployment because machines were started by providing good quality products by manufacturing [40].

Industry 3.0: Third industrial revolution mainly an era of IT and automate manufacturing. This era introduced functions of computers, chips and internet and emphasis more on R&D for commercial reason [41]. In this era computer integrated manufacturing (CIM), computer aided design (CAD), computer aided manufacturing (CAM), and flexible manufacturing system make production and operation system faster and more effective and this era introduced advanced manufacturing technologies for flexible operation, customized product, faster response on market requirement and demand.

Industry 4.0: It can be said as advance and traditional manufacturing process taken place by industries using smart and advance technology [42]. The principle of industry 4.0 are Interconnection, Information Transparency, Technical Assistance and Decision which is decentralized so that the machines and devices can help people for communicating and travelling from one place to another, this is possible with Internet of things and artificial intelligence. Due to high accuracy the interconnectivity it should be more advance and which can allow operators to collect large amount of data and excessive information from all points so that the research can be more useful for manufacturing industry [43].

Hence this kind of facility will always require technical assistance to assist in decision making and problem solving [15]. This will allow all the system to make decision on their own and to perform autonomously so that the performance of the systems can also work at higher level. The graphical representation of Industry 4.0 is given in Fig. 1.

3.1 Characteristics of Industry 4.0

Vertical Networking: The cyber-physical production system which is merging the virtual and real world with a environment of Internet of Things, Internet of services, Internet of people and Internet of data, concludes a smart industry and brings

Fig. 1 Industry 4.0 diagrammatic representation

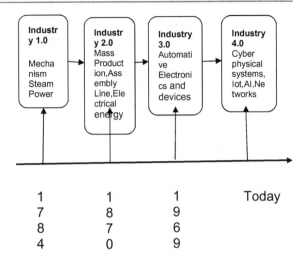

a smart infrastructure like smart mobility, smart technologies and smart homes building and exponential technologies like, Biotech, neurotech, nanotech, mobile technology, sensing, robotics and artificial intelligence [44]. The use of Cyberphysical Production system in vertical networking is to enable the manufacturing plant which is going to bring a rapid change in demand and stock level and finding faults. CPSS gives a customized production with the help of smart sensors technologies for monitoring and Analyzing [45]. It doesn't relocate to only production management but also the maintenance management.

Horizontal Integration: A new generation of global value chain networks has the horizontal Integration and real time networks for optimizing which enables integrated transparency and a huge flexibility for responding on huge problems easily and to facilitate the better optimization globally [46]. Basically, it means that the customized adaptions can be also made in Planning, development, ordering and distribution with better quality and low risk while production [6]. Engineering occurred for one of the main purposes of Industry 4.0 seamlessly for manufacturing, development and for the providing better service through producing better and updated products towards the customer [47]. Through engineering new advanced synergies are created as the data and information are always stored while the production cycle takes place and integrates the data via product stage and data modelling [11].

Exponential Technologies For acceleration: This accelerated and catalectic technologies are allowing cost savings, durability and customization in all industrial trial process and implementations [48].

A relevant example of industry 4.0 through Exponential technologies is 3D printing as it is mostly widely use across the world now. Mass customizations are provided by number of fields like healthcare, agriculture and aerospace industry. The most important tool for this manufacturing is Additive manufacturing as the leading manufacturers Like Eos, Stratasys uses this tool for better complexity [49]. AM provides a better customized products with improved performances.

Shape complexity: Any kind of virtual shapes can be found with AM with some complexity reduction techniques. Algebraic complexity will measure all the degree of polynomials which are required for representing the shapes match in its parametric and implicit form. Morphological Complexity analyze the feature size and smoothness whereas the combinational complexity Counts the Shapes as a vertex count with meshes hence the accuracy varies from this complexity to Representational Complexity for easy to use and provides the footprints of a data structure [50].

Material complexity: Manufacturing can be done easily with reducing the addition of complex material part and convert into low-cost production. Material Complexity has two focus areas like Disordered materials for achieving the targeted performance outcome, Design of defected and the guided synthesis for stabilizing the targeted concentration of degrees of disorder [Joint Center For energy storage Research].

3.2 Why Industry 4.0 is Important?

Industry 4.0 is an interface between physical world with digital counterpart and provides an efficient production system with the help of modern tools like cyber physical space, artificial intelligence, internet of things etc. [51–53]. It also improves firms' productivity in various fields starting from manufacturing to logistics and distribution chain. Analytics helps organizations to collect huge volume of customers' data, process and analyze it to understand the customers' characteristics, preference and taste. Looking into this, firms can design their marketing effort which is more customers centric; develop customized production and selling process. These I4.0 techniques help the industries to rethink their operations and develop capability to grab new competitive advantage [54].

Industry 4.0 is a composite of four important drivers like Cloud based manufacturing, smart manufacturing, Industrial Internet of Things and Artificial Intelligence which helps industry process more digitalized [55]. It reduces jungle of traditional manufacturing process and develops good network between suppliers, customers and company with better man machine interface [31].

3.3 Implementation of Industry 4.0 in Indian Industry

Industry 4.0 journey started from Germany in the year 2010. It is estimated by the expert that it can reach to $214 billion market by 2023. India has also observed significant changes and adaption of Industry 4.0 in past few years. In comparison to other country, performance of India is very less but make in India is starting point of revolution of Indian industry and India mostly focuses on SME to develop a better market proposition. In connection with I4.0, it provides greater opportunities of Indian manufacturing industries. Application of tools of Industry 4.0 like IoT, Additive manufacturing, Smart factories are in less percentage [56]. Major forces of Indian industries like MSME are not able to access the modern technology. "The Two critical enabling Industry 4.0 Technologies i.e., Big data, IoT which appears to be the most developing platform to make a root for small factories", comprise of 20% of the global AI market which will be commanded by India.

Most of Indian manufacturing industries adopted IoT and AI in their manufacturing process. It covers near about 60% of the total market in respect of Industry 4.0 tools. On the other side Big Data also captured a significant growth and is going to reach $1.6 billion by 2025 and the growing CAGR is 26% from $2 billion [57].

Digitalization in the manufacturing, service and other sectors are growing very fast in India. Most of the industries are adopting their process as per requirement of market and customers. It is noted that company related to marketing and services adopted digital culture and spreading most of the urban market. The automobile industries took a drastic step to implement Industry 4.0 by using robot technology in their manufacturing process. It is noted that they use 58 robots per 10,000 employees in comparison to other industries (3 robots per 10,000 employees) to make their manufacturing process more progressive and customized [58].

In respective of industries application of computer vision, artificial intelligence, robots, Government of India launched their IoT policy for skill development, use technology to upgrade their traditional infrastructure and making the IoT application-based product to meet the demand of the consumers and other market related requirements [59]. The government also designed National policy for advanced manufacturing for enhancing global competitiveness and to enhance contribution of manufacturing sector [60]. Samarth Udyog Bharat 4.0 is an initiative taken by ministry of heavy industries to promote the industry 4.0 in manufacturing, supply chain management, shipping, infrastructure and other allied industries with an aim to propagate technological solution of Indian manufacturing units by 2025 through various social initiatives program like awareness, training and demonstration center and customer solution center. The government is estimating to increase the manufacturing output to 25% by applying the industry 4.0 by Indian firms [61].

It is noted by various researchers that for growth and development of industry 4.0, IT industry play a major part. Due to this Indian IoT companies are gradually enhancing their way of operations to develop I4.0 capabilities through continuous R&D [62]. Government established center for product design and manufacturing for building smart factories, Bosch is continuously working for developing smart manufacturing, Robots and Collaborative robots used by automobile and FMCG industry to improve automation process, prevent accidents and reduce supervision and maintain effective workforce. Developing IoT hub to support private sector specially MSME. Indian companies also collaborating with software firms for machines-to-machines solution. Factory for the future lab has been set up of providing new age technology for creating a deep convergence of physical and cyber space [63].

Therefore, we may say that implementation and adaptation of industry 4.0 in India is in nascent stage but the good thing is that government and large corporations are trying to adopt this cutting-edge technology to develop a strong competitive market in the world. Make in India is an initiative by the government to reach to industry 4.0 destinations.

Like the top supply chain and logistics network built by Flipkart is banking a range on technology of Industry 4.0 by including the automation and robotics for creating a system which is robust, resilient and scalable to handle millions of products on a daily basis. Thus, the Sortation Robots are the next automation for flipkart. The worker just needs to place the package on the sorting Cobot and it will do the rest like they will drop the package to designated pin code bins for final delivery. Cobots has the ability to handle 4500 packages per hour as compared to 500 packages done by humans. when the energy is down, Cobots automatically goes to the charging point, hence Flipkart is anticipating 60% improvement in efficiency within the introduction of robots and will provide better service by fast delivery for the customer.

The Most Advanced Cancer Treatment—Proton Therapy by Apollo Healthcare is termed as the revolution of Health 4.0. In today's world there are mainly 3 types of treatment with cancer surgery like chemotherapy, radio therapy that uses photons for killing the tumor cells. Now with Health 4.0, Proton Therapy is an advanced form of treatment using protons [64]. These particles give Proton therapy more advantages like its accuracy and the ability to spare healthy tissues. Conventional X Rays are made of photons that passes through the body and deposits considerable energy before and after the tumor. Whereas protons have a particular physical property called Bragg Peak [65] as it has the ability to deposit less energy on their way to the tumor and do not affect any tissues beyond it. This probably allows to confine the radiation dose to the tumor and radically decreases the risk of injury to surrounding healthy tissues [66]. The protons originate the Ion source where the hydrogen atom is separate into electron and protons. The protons are injected into the System from where they are accelerated into high power protons then these high-power protons are sent through an energy selection system and the degrader that can adjust the proton's energy this proton arrives at treatment room via gantry. The process of photon therapy is shown in Fig. 2.

The gantry is a device which revolves 360° around the patient where the patient has been positioned so that the beam can be delivered at any angle. A high Accuracy of patient position is required. Apollo's Inference based therapy treatment room are equipped with the latest patient positioners alignment and imaginary system to ensure the patient's positioning. Today proton therapy has access to the state of art delivery mode called Pencil. The bean scanning process for detecting tumor is given in Fig. 3.

This allows doctors to match the shape of tumor through beam pixel by pixel. The proton dose can be maintained and can be changed according to the complexity and reduces the integral dose of radiation given to the patient. Tumor shape and size is recognized through Beam Pixel which is in the Fig. 4.

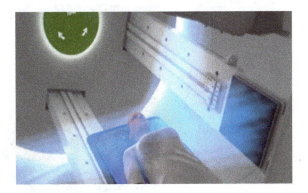

Fig. 2 Photon therapy process

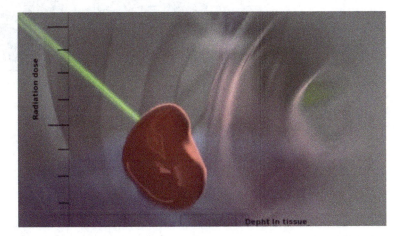

Fig. 3 Beam scanning

4 Proposed Work

The present study is an attempt taken by the researcher to analyze the impact of Industry 4.0 technology by the various companies. For the study we have considered Cognizant, Siemens and Tesla for discussion their performance level using computer vision, robotics and machine learning. In the second part we are going to review the challenges and opportunities in the management perspective. In this context we made certain assumption to address the issue and these are.

H1: Industry 4.0 is going to bring better insight into data and operations.
H2: Artificial intelligence is bringing change in the manufacturing system through effective workforce.
H3: Industry 4.0 is making the market more user-centric.

Fig. 4 Beam pixel

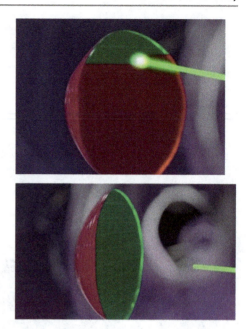

H4: Industry 4.0 is making supply chain more effective.
H5: Industry 4.0 is making sales effective.

4.1 Methodology

The present study is descriptive in nature and tries to explore the application and outcome of the industries using various methods of Industry 4.0. The study has been divided into two sections. For the objectives number one, information has been collected from website three reputed organization who adopted Industry 4.0. The information has been collected on computer vision, smart manufacturing technology, internet of things and robotics and this information helps us to know how these new technologies are going to make changes in the operational and financial performance of these selected firms. The study explains how Cognizant, Tesla and Siemens use various modern tools to enhance their industry output, productivity and their manufacturing process. This part assesses inbound part of industry using I4.0. The second part of the study conducted using focus group interview who are working with the new technology and responses have been noted using five assumptions, to get idea on better insight on data and operations, effectiveness of AI technology to manage better workforce, develop of market more customer centric, better supply chain system and high-end sales generation. This part help to understand the outbound part of industry using I4.0.

4.2 Discussion

The demand for high quality and highly customized manufacturing process is rising. Batch size 1 is the new paradigm, in mass production predictive maintenance was applied, however, for a batch size 1 it is not applicable mainly due to the large variation in the production environment [52]. For Instance, tools can be worn out which may impact the final product quality. Alien objects can damage the manufacturing tools and products or parts with low quality which can be used for final product, the problems are often detected too late that may impact the final quality and productivity and that result into revenue loss. Current problem monitoring system and providing solution for problem are much costlier and time taking. So new technology like neural network is used by manufacturer very often and develop a customer model with the help of neural network. This technology can recognize abnormal situation without intervention, operators receive notification within seconds, enabling immediate problem resolution. This signifies on reducing Non-performance cost & expensive tools breakdown, products are released more faster with highest quality and maximized profit margin.

To explain objective no-1, we have considered three industries as a case and discuss the impact of industry 4.0 technology in their overall performance.

Case-1 Computer vision and Machine vision in Smart manufacturing Industry: Computer vision is a new quality control system which requires a data management challenge towards a proper execution with an active training and testing models for providing more advanced non defect products to the valuable customers [67]. The goal is to create a smart machine which is going to work communicate and do exactly the same work as human performs and capable of [68]. Due to computer vision many Industries like Siemens benefitted from this technology as CV is mostly used for the purpose of Inventory and helps in saving accomplished metadata. (www.siemens.com/global/research-development).

- Why the companies like Siemens around the world are investing in AI and Computer vision (CV)

CV has proper time efficiency and automated systems which can run the manufacturing process 24/7 if needed. While the System is working for a longtime, the accuracy level mostly remains high which allows a quality control in the products and there will be less divergence for mistakes and less waste will be produced which can help reducing cost [39]. Anomaly detection is used to work on logical strategy to analyze new images and compare them for inspection with other images so that the defects in the products can be find by Statistical and model-based approach [69]. Computer vision has the ability to improve the structured and detailed management process and speed up the process of ordering when the barcodes and labeled scanner are available to scan very quickly and increases productivity. The mislabeled and misplaced items from the warehouse or in the manufacturing hub are not been found and misplaced can be harmful in cases of food items or medicines, so the computer vision helps companies by matching the databases and track those item [70]. Hence this system allows continuous repeatable cycles of training and testing, classifying, monitoring and operating in a large scale for an error implementation. It has observed significant changes of global market, in 2015–2020 it is $9.2 billion, and it is expected to reach $13.0 billion by 2025. Siemens also reported significant growth in their Digital Industries, in 2019, $3409 million to $3847 million in 2020 which is 13% change. But Smart Industries reduces their income like in 2019 it is around $1733 million to $1541 million and company expected that that is happening due to pandemic where market growth is in stuck but there is significant growth in Digital industries (www.siemens.com/iot).

Case-2: Industrial Internet of Things for evolving Manufacturing Firms: Smart manufacturing is powered by Industrial Internet of things as IoT is better defined as convergence of many several technologies like machine-to-machine automation and communication and sensors data [71]. It can significantly Improves the traditional human process like efficiencies can be identified and predicting maintenance and production issues early before the manufacture process starts repeatedly [72]. This shows that IIOT has sustainability practices, quality control and better supply chain efficiency. Thus, the outcome includes Predictive Maintenance, deployment of field technicians, asset tracking and energy management [73].

How cognizant is driving innovation in manufacturing with IIOT? Cognizant is leading the digital Revolution in Industrial IoT, transforming organizations and enhancing revenues with a new ecosystem and business models. Now it is the time for Cognizant to help the world get started on the path to improve efficiency, better customer services and increased growth (www.cognizant.com/iot, 2019). As they drive operations from R&D to production, sourcing to delivery and product monitoring for new services. Cognizant has successfully guided their clients for developing more efficient human-centered operating model. IOT connects real time data to real time outcomes. Cognizant can make the vision of customers into reality as their approach is always prioritizing customer's data and gathering it from multiple of sources to develop actionable performances. Hence this way will always lead to improve solutions which will allow manufactures to respond more quickly to their customers (www.cognizant.com/iot, 2019).

Cognizant worked with many leading packaging equipment manufacturers to help them monitoring thousands of machines and detect and prevent service bottlenecks. This made Cognizant's service parts revenue increased by 12% (www.cognizant.com/iot).

A global Firm equipment Manufacturer to visualize all the aspects of their sensors enabled production line. This has addressed problems in real time and helped accelerating outputs of new products per machine by nearly 50% [39]. Cognizant is a multinational leader to retrofit and IOT enable refrigeration equipment. By remotely predicting service needs, cognizant removes response times from 5 h to just 30 min and product waste was also reduced by 10%. From gearing up digital to ensure equipment uptime there was reduction of 5% in production downtime and by a digital remedy to harmonizing manufacturing system has saved more than 4000 person hours (www.cognizant.com) which shows that the empowered workers will increase their efficiency with the engineering environment and smartphones app [72].

Case Study-3 Robotics: Robots will be no longer a thing of dream or any science fiction of future but now they are here and driving a manufacturing revival that's creating jobs and making world's manufacturing Industry strong. Manufacturing today looks very different that it used to be before 10–20 years ago, Shop floors has been transformed in what's being called the industry 4.0 [39]. In 2019 according to Forbes, 38,000 Industrial robots were installed in us manufacturing facilities [74]. Robots offer precision speed and repeatability to make production faster and more flexible which makes the work more efficient [45]. They can reduce waste and cost as they are more likely to be a small manufacturer in the industry [75]. Rapid researchers and developers have given rise to a new generation of industrial robots which are more easily programmable and have initiate user interface which are equipped with sensors that allow them to be a better navigator [74]. Collaboration of robots and humans are quickly programmed in such a way for performing different tasks and works safely side by side with human co-workers and assist with unpleasant and dangerous tasks which further improves the manufacturing workplace experience. Robots will always have the capability to constantly optimize the process and adapt new accelerating technologies [75].

How Robots helped tesla for making battery replacement easier than refueling a vehicle? Over the last few years, Tesla has made and installed 160 Robots for assembling the vehicle. A robot in a Tesla Factory is mostly involved in bonding, welding, riveting and installing the component at the time of manufacturing shows that they are multi task performer where else other Robots typically does each job at a time [76]. Tesla has also started providing machine to machine connectivity for their S model vehicle in USA and in European Countries. This level of automation helped the company to achieve a profit margin at 25% in Q4 2013 as compared to ford which is 15.5% and Audi Motors with 12% with the collaboration with four types of robots in the industry. The manufacturing process system is given in Fig. 5.

In the second part of our discussion (objective no-2) is based on experts' opinion on application of Industry 4.0 on various operational field in industry. Expert mainly discuss on certain assumption and role of Artificial intelligence in changing business prospect. We interviewed two experts working with these tools for last one decade. They belong Aureus Tech Systems and Spancer Retail and they share their opinion in this regard with some assumptions.

H1: Industry 4.0 is going to bring better insight into data and operations: Industry is moving from human based business decision making to data driven approach for their operational decision making. It gives more accurate outcome of business problem and predicts the possible solution. To strengthen it more industries, apply artificial intelligence into their work system. Artificial intelligence makes structured data and helps it move towards better processing. It also helps to rectify their problem in advance manner which help them to redesign their business process.

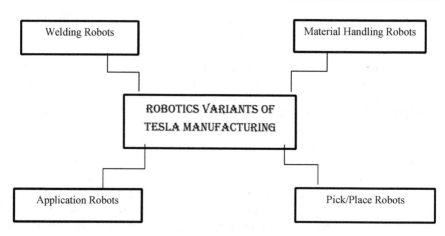

Fig. 5 Tesla manufacturing process

H2: Artificial Intelligence is bringing change in the manufacturing system through effective workforce: Every day, the market system and requirement are changing. It gets more customized in nature. Companies are facing a lot of difficulties to maintain high level of quality and to comply with quality regulation within short span of time. On the other hand, customer expect faultless product thereby pushing manufacturers to set up quality products. Quality is strongly associated with the companies' name and brand value. Industry 4.0 helps in developing quality products with help of artificial intelligence algorithms. It helps manufacturer to detect production fault and reduce the cause of product quality issues. It also helps manufacturer to find root cause of abnormalities of machine behavior and maintain good quality standard. Artificial intelligence works in market to collect product performance related data from market and also measures product performance. AI delivered information helps companies to design their strategic planning decision which facilitates product management teams in their way of operation. It can develop product structure and components as per market requirements.

H3: Industry 4.0 is making the market more user-centric: The major advantage of Industry 4.0 is to help marketer to make the market more user centric than marketer centric. Artificial intelligence is a major tool of Industry 4.0 with its wide operations. There are diversified approach and application of AI in industry and market. With the help of AI marketers design market structure, optimize their supply chain system, promote the product as per customer preference, and assess the market movement. Artificial intelligence facilitate marketer to estimate their demand pattern, demand supply mechanism in connection with geographical location, psychographic dimension and socioeconomic, macroeconomics factors and demographic factors [77].

H4: Industry 4.0 is making supply chain more effective: Industry 4.0 does make huge revolution in the supply chain network process. With the help of artificial intelligence, internet of things and big data analytics made the supply chain network more opportunities to reach the next horizon of operational effectiveness. Big data do better market prediction and find out the gaps in the market. AI and IoT help in sensitizing customer expectation and try to acquaint with the expected supply time. It allows the business to optimize their sourcing in terms of purchase and ordering system. It helps marketers to reduce their transportation, warehouse and supply chain administration cost. Use of Industry 4.0 facilitates marketers design the value chain network which helps to restructure better retailing and manufacturing strategies. With the help of AI, marketers do lot of development in supply chain domain like logistics, shipping, delivery date predictions, reduction of manual workload and informed decision making etc. [78].

H5: Industry 4.0 is making sales effective: Industry 4.0 is making huge contribution in sales effectiveness with the help of AI. Sales people are able to identify right prospect of their products and design their offering as per customer requirements. AI helps companies to discern in activities that lead in winning sales deals, identify buying and selling pattern and try to motivate customers looking into their previous buying history. There are certain areas of sales force effectives help by AI like CRM adaptation, improve productivity by focusing on engaging with prospects and customers, data driven sales coaching-identify right customers and approaches them with right products in right time. Improving forecasting with the help of real and current sales data, representative do better market prediction and design their policy. It creates unprecedented buying group visibility, identifying market challenges and tries to take precaution before processing to customers.

Creating upon the awareness of what's happening in the organization with understanding the root cause behind critical incidents. So, machine operators have all the possibility and changing the settings of the camera and look into the machine even in such areas where finding defects is not accessible during production time. It gives HD and 4K closure surveillance with slow motion and hour loop surveillance process which provides analysis of each and every picture's movement and results in visual aids. This achievement has shown a lot of improvements in finding the main reasons for recurring faults while the products are manufactured and redefining the optimized measures totally based on actual facts with reducing machinery breakdowns this results a low-price solution and easy installation for daily use, regular managerial function of the business firm in the current technological era.

5 Conclusion

Industry 4.0 is going to be new dimension for the industry for their operational efficiency. In our study, we observed that most of the industries are trying to adept the mechanism of Industry 4.0. In India, it is noted that automobile industries applying Robotics for smooth operation. Big data analytics are making the things easier for the industry to project the business forecasting. From the discussion, it is revealed that robotics, computer vision, IIOT and artificial intelligence improve the firm performance in terms of operations and financial growth. From the second part of our discussion indicates that use of tools of industry 4.0 firms could be able to manage their sales, workforce, identifying right prospect with right segment, effective use of supply chain networks in different geographic locations and consumer requirement. The present study also delineates the implementation of industry 4.0 facilitates firms for their better sustainability in the cutting-edge technology. The study will help researchers to study financial aspect and marketing aspect of the firm using Industry 4.0 technology. The study will help marketers to do a trial error approach on the application side of new technology. The study can help small firms in India to start new journey in the business.

Managerial implication: Industry 4.0 is in implementation phase in India; therefore, a thorough understanding of different techniques of Industrial 4.0 is important for managers who are dealing with current market and workforce. The study was trying to explain the impact of technologies used by large corporates and how it gives them benefit for smooth running of industry. Though, Indian market covers a large part on the performance of SME. Therefore, it is imperative for those industries to know about these new technologies and try to find out a way of implementation. This chapter will help managers to articulate their ideas for implementing the new path of journey and facilitate them to design better sustainability strategy. For the academician this paper will help them to understand basic concept and encourage them to conduct research to get better insights in these areas.

References

1. Qin, J., Liu, Y., & Grosvenor, R. (2016). A Categorical Framework of Manufacturing for Industry 4.0 and Beyond, Changeable, Agile, Reconfigurable & Virtual Production, Procedia CIRP 52 (2016) 173–178.
2. Neugebauer, R., Hippmann, S., Leis, M.J., & Landherr, M.H. (2016). Industries 4.0 - From the Perspective of Applied Research. *Procedia CIRP, 57*, 2–7.
3. Brettel, M., Friederichsen, N., Keller, M. (2014). How Virtualization, Decentralization and Network Building Change the Manufacturing Landscape: An Industry 4.0 Perspective, International Journal of Mechanical, Aerospace, Industrial, Mechatronic and Manufacturing Engineering, *8* (1), 37–36.
4. Lee, J., Kao, H.A., and Yang, S (2014). Service Innovation and Smart Analytics for Industry 4.0 and Big Data Environment, *Procedia CIRP*, Vol-16, 3–8, https://doi.org/10.1016/j.procir.2014.02.001.
5. Lobo, A F. (2015), The Industry 4.0 revolution and the future of Manufacturing Execution Systems (MES), *Journal of Innovation Management, 3 (4)* 16–21.
6. Rojko A. (2016). 'Industry 4.0 Concept; Background and Overview, *International Journal of Interactive Mobile Technologies*, 11(5), 67–76, https://doi.org/10.3991/ijim.v11i5.7072.
7. Agrifoglio, R., Cannavale, C., Laurenza, E., & Metallo, C. (2017). How emerging digital technologies affect management through co-creation'. Empirical evidence from the maritime industry. *Production Planning and Control, 28(*16), 1298–1306, https://doi.org/10.1080/09537287.2017.1375150.

8. Bogers, M., Hadar, R., and Bilberg, A. (2016). Additive manufacturing for consumer-centric business models: Implications for supply chains in consumer goods manufacturing, *Technological Forecasting and Social Change*, Volume *102*, Pages 225–239, https://doi.org/10.1016/j.techfore.2015.07.024.
9. Porter, M.E., & Heppelmann, J.E. How smart, connected products are transforming competition. *Harvard Business Review*, *92*(11), 64–88. 2014.
10. Chen, D., Heyer, S., Ibbotson, S., Salonitis, K., Steingrímsson, J. G., & Thiede, S. (2015), Direct digital manufacturing: Definition, evolution, and sustainability implications. *Journal of Cleaner Production*, *107*, 615–625, https://doi.org/10.1016/j.jclepro.2015.05.009.
11. Stock, T., Seliger, G. (2016). Opportunities of Sustainable Manufacturing in Industry 4.0, *Procedia CIRP*, Vol-40, 536–541. https://doi.org/10.1016/j.procir.2016.01.129.
12. Anderson, C. (2012). Makers: The new industrial revolution. Crown Business books, https://doi.org/10.1016/0016-3287(91)90079-H.
13. Goldhar, J.D., and Jelinek, M. (1983). Plan for Economies of Scope. *Harward Business Review*, 61 (6), 141–148.
14. Kagermann, H., Wahlster, W., and Helbig, J. (2013). Recommendations for Implementing the Strategic Initiative Industries 4.0, Final Report of the Industries 4.0 Working Group, Forschungs Union and Acatech. Retrieved from https://en.acatech.de/publication/recommendations-for-implementing-the-strategic-initiative-industrie-4-0-final-report-of-the-industrie-4-0-working-group.
15. Chesbrough, H.W. (2003). Open Innovation: The New Imperative for Creating and Profiting from Technology, Harvard Business School Press, ISBN: 9781578518371.
16. Charles, A. (2020). Fourth Industrial Revolution: How Latecomers and Laggards can Catch up, UNID, Vienna, February. 2020. Retrieved from https://www.unido.org/stories/fourth-industrial-revolution-how-latecomers-and-laggards-can-catch.
17. Bhat, T.P. (2020), 'India and Industry 4.0', Working paper, 218, Institute for studies in Industrial Development, New Delhi, January 2020.
18. Michael, E., & Wirtz, J. (2017). Unlocking value from machines: business models and the industrial internet of things, *Journal of Marketing Management*, 33:1–2, 111–130, https://doi.org/10.1080/0267257X.2016.1248041.
19. Hermann, M., Pentek, T., & Otto, B. (2016). Design Principles for Industrie 4.0 Scenarios. *2016 49th Hawaii International Conference on System Sciences (HICSS)*, 3928–3937.
20. Burmeister, C., Lüttgens, D., Piller, F. T. (2016). Business Model Innovation for Industries 4.0: Why the Industrial', (2016). Internet Mandates a New Perspective on Innovation, *Die Unternehm*, 70, 124–152. https://doi.org/10.5771/0042-059X-2016-2-124.
21. Nagy, J., Oláh, J., Erdei, E., Máté, D., & Popp, J. (2018). The Role and Impact of Industry 4.0 and the Internet of Things on the Business Strategy of the Value Chain—The Case of Hungary. *Sustainability*, *10*(10), 3491. https://doi.org/10.3390/su10103491.
22. Geissbauer, R., Vedso, J., Schrauf, S. (2016). Industry 4.0: Building the Digital Enterprise'. Global Industry 4.0Survey. What We Mean by Industry 4.0/Survey Key Findings/Blueprint for Digital Success. Retrieved from: https:www.pwc.com/gx/en/industries/industries-4.0/landing-page/industry-4.0-building-your-digital-enterprise-april-2016.pdf (accessed on 12 March 2018) 2016.
23. Ghadge, A., Er Kara, M., Moradlou, H. and Goswami, M. (2020). "The impact of Industry 4.0 implementation on supply chains", *Journal of Manufacturing Technology Management*, *31* (4), pp. 669–686. https://doi.org/10.1108/JMTM-10-2019-0368.
24. Andreas., F, Qaiser, F.H., Choudhary, A., & Reiner, G. (2020). The impact of Industry 4.0 on the reconciliation of dynamic capabilities: evidence from the European manufacturing industries, *Production Planning Control*, https://doi.org/10.1080/09537287.2020.1810765.
25. Ramos, C.M.Q. and Brito, I.S. (2020), "The Effects of Industry 4.0 in Tourism and Hospitality and Future Trends in Portugal", Hassan, A. and Sharma, A. (Ed.) *The Emerald Handbook of ICT in Tourism and Hospitality*, Emerald Publishing Limited, Bingley, pp. 367–378. https://doi.org/10.1108/978-1-83982-688-720201023.

26. Paul, S.; Riffat, M.; Yasir, A.; Mahim, M.N.; Sharnali, B.Y.; Naheen, I.T.; Rahman, A.; Kulkarni, A. (2021). Industry 4.0 Applications for Medical/Healthcare Services. *Journal of Sensors Actuator Network, 10*(43). https://doi.org/10.3390/jsan10030043.
27. Singhal. N (2021). An Empirical Investigation of Industry 4.0 Preparedness in India. *Vision, 25*(3), 300–311. https://doi.org/10.1177/0972262920950066.
28. Barzotto, M., De Propris, L. (2021). The value of firm linkages in the age of industry 4.0: a qualitative comparative analysis. The Annal *of Regional Science, 67*, 245–272. https://doi.org/10.1007/s00168-021-01047-0.
29. Chauhan, C., Singh, A., Luthra, S., (2021), Barriers to industry 4.0 adoption and its performance implications: An empirical investigation of emerging economy, *Journal of Cleaner Production, Vol-285*, 124809, https://doi.org/10.1016/j.jclepro.2020.124809.
30. Sima, V., Gheorghe, I. G., Subić, J., & Nancu, D. (2020). Influences of the Industry 4.0 Revolution on the Human Capital Development and Consumer Behavior: A Systematic Review. *Sustainability, 12*(10), 4035. https://doi.org/10.3390/su12104035.
31. Rüßmann, M., Lorenz, M., Gerbert, P., Waldner, M. (2015). Industry 4.0: The Future of Productivity and Growth in Manufacturing Industries, (April 09, 2015) 1–14., retrieved from https://www.bcg.com/enin/publications/2015/engineered_products_project_business_industry_4_future_productivity_growth_manufacturing_industries.
32. Tippayawong K., Šafár L., Sopko J., Dancaková D., Woschank M. (2021) General Assessment of Industry 4.0 Awareness in South India—A Precondition for Efficient Organization Models?. In: Matt D.T., Modrák V., Zsifkovits H. (eds) Implementing Industry 4.0 in SMEs. Palgrave Macmillan, Cham. https://doi.org/10.1007/978-3-030-70516-9_11.
33. Cioffi, R., Travaglioni, M., Piscitelli, G., Petrillo, A., & De Felice, F. (2020). Artificial Intelligence and Machine Learning Applications in Smart Production: Progress, Trends, and Directions. *Sustainability, 12*(2), 492. https://doi.org/10.3390/su12020492.
34. Verma, S., Sharma, R., Deb, S., Maitra, D. (2021). Artificial intelligence in marketing: Systematic review and future research direction, *International Journal of Information Management Data Insights*, 1(1), 100002, https://doi.org/10.1016/j.jjimei.2020.100002.
35. Chatterjee, S., Ghosh, S.K., Chaudhuri, R., Nguyen, B. (2019). Are CRM systems ready for AI integration? A conceptual framework of organizational readiness for effective AI-CRM integration, *The Bottom Line*, 32 (2019), pp. 144–157.
36. Nayyar, A., Rameshwar, R. U. D. R. A., & Solanki, A. (2020). Internet of Things (IoT) and the Digital Business Environment: A Standpoint Inclusive Cyber Space, Cyber Crimes, and Cybersecurity. The Evolution of Business in the Cyber Age, 10, 9780429276484-6.
37. Sha, N., Rajeswari, M. (2019). Creating a Brand Value and Consumer Satisfaction in E-Commerce Business Using Artificial Intelligence with the Help of Vosag Technology, *International Journal of Innovative Technology and Exploring Engineering*, 8 (8), pp. 1510–1515.
38. Nayyar, A., & Kumar, A. (Eds.). (2020). A roadmap to industry 4.0: Smart production, sharp business and sustainable development. Springer.
39. Mohd. J., Haleeem. A., Singh. RP. (2021).Significant applications of Big Data in industry 4.0, Journal of Industrial Integration and Management, vol 2, no-3, pp 18–27. https://doi.org/10.1142/S2424862221500135.
40. Vaidya. S., & Ambad, P., & Bhosle, S. (2018). Industry 4.0 – A Glimpse. 20. 233–238. https://doi.org/10.1016/j.promfg.2018.02.034.
41. Freeman, C. and L. Soete (1997), The Economics of Industrial Innovation, Psychology Press.
42. Büchi, G., Cugno, M., Castagnoli, R. (2020), Smart factory performance and Industry 4.0,*Technological Forecasting and Social Change*, Vol 150, 119790, https://doi.org/10.1016/j.techfore.2019.119790.
43. Piccarozzi, M., Aquilani, B., & Gatti, C. (2018). Industry 4.0 in Management Studies: A Systematic Literature Review. *Sustainability, 10*(10), 3821. https://doi.org/10.3390/su10103821.
44. PwC. Industry 4.0-Building the Digital Enterprise; PricewaterhouseCoopers LLP: Berlin, Germany, 2016; retrieved from https://www.google.com/search?q=PwC+%282016%29%3A+Industry+4.0+-+Building+the+digital+enterprise.

45. Elisis, M., Mahmoud, K., Lehtonen, M., & Darwish, M. M. F. (2021). Reliable Industry 4.0 Based on Machine Learning and IoT for Analyzing, Monitoring, and Securing Smart Meters. *Sensors, 21*(2), 487. https://doi.org/10.3390/s21020487.
46. Zhong, R.Y., Xu, X.W., Klotz, E., & Newman, S.T. (2017). Intelligent Manufacturing in the Context of Industry 4.0: A Review. *Engineering, 3*, 616–630.
47. Sommer, L. (2019). Industrial revolution - Industry 4.0: Are German manufacturing SMEs the first victims of this revolution?. Journal of Industrial Engineering and Management. 8. https://doi.org/10.3926/jiem.1470.
48. Zhou, R., & Cardinal, J. (2019). Exploring the Impacts of Industry 4.0 from a Macroscopic Perspective. Proceedings of the Design Society: International Conference on Engineering Design. 1. 2111–2120. https://doi.org/10.1017/dsi.2019.217.
49. Machado, C.G., Winroth, M.P., & Dener, E.H., and Silva, R. (2020). Sustainable manufacturing in Industry 4.0: an emerging research agenda, *International Journal of Production Research*, 58:5, 1462–1484, https://doi.org/10.1080/00207543.2019.1652777.
50. Rossignac, J. (2014). Shape Complexity, College of computing, GVU center and IRIS Cluster Georgia Institute of technology, Atlanta, Georgia.
51. Spath, D. (2013). Produktionsarbeit der Zukunft – Industries 4.0. IOA: Stuttgart. www.produktionsarbeit.de/content/dam/produktionsarbeit/de/documents/Fraunhofer-IAO-Studie_Produktionsarbeit_der_Zukunft-Industrie_4_0.pdf. retrieved on 12-3-2021.
52. Kusiak, A. (2018). Smart Manufacturing. *International Journal of Production Research 56* (1–2): 508–517. https://doi.org/10.1080/00207543.2018. 1351644.
53. Sjödin, D. R., Parida, V., Leksell, M., and Petrovic, A. (2018). Smart Factory Implementation and Process Innovation. *Research-Technology Management 61*(5): 22–31. https://doi.org/10.1080/08956308.2018.1471277.
54. Teece, D. J., Pisano, G., and Shuen. A. (1997) 'Dynamic Capabilities and Strategic Management'. *Strategic Management Journal, 18* (7):509–533. https://doi.org/10.1002/(SICI)1097-0266(199708)18:7.
55. Erol, S., Jäger, A., Hold, P., Ott, K., & Sihn, W. (2016). Tangible Industry 4.0: A Scenario-Based Approach to Learning for the Future of Production. *Procedia CIRP, 54*, 13–18.
56. Aulbur, W., and Singh, H.V. (2014), 'Next Generation Manufacturing: Industry 4.0: A Look at the Changing Landscapes in Manufacturing', CII and Roland Berger, New Delhi, September.
57. Nishimura, T. (2018), "Big Data Analytics Market – Future Scope in India", Silicon India, September, 12.
58. Roehricth, K. (2016). Study on emerging markets, with special focus on Asia, RockEU, 16 August 2016, https://www.eu-robotics.net/cms/upload/downloads/Rockeu1/2016-07-16_RockEU_Deliverable_412_v2.pdf, retrieved on 31-07-2021.
59. KPMG, Industry 4.0 Indian Inc, Gearing up for change, Internet of Things, Ministry of Electronics and Information Technology, accessed on 16 February 2018, https://resources.aima.in/presentations/AIMA-KPMG-industry-4-0-report.pdf, retrieved 02-08-2021.2018.
60. Jaitley heralds cyber physical systems mission to create new jobs, The Hindu https://www.thehindu.com/news/national/jaitley-heralds-cyber-physical-systems-mission-to-create-new-jobs, 2018. Retrieved 31-07-2021.
61. Kulshreshtha, M. (2021). Industry 4.0 Technology: The key game changer for Indian Manufacturing Sector, https://www.financialexpress.com/industry/industry-4-0-technology-the-key-game-changer-for-indian-manufacturing-sector, February 2021, retrieved 02-08-21.
62. Kumar, A., & Nayyar, A. (2020). SI 3-Industry: A Sustainable, Intelligent, Innovative, Internet-of-Things Industry. In A Roadmap to Industry 4.0: Smart Production, Sharp Business and Sustainable Development (pp. 1–21). Springer, Cham.
63. Nair, L. (2020). Industry 4.0-The Indian context, https://www.linkedin.com/pulse/industry-40-indian-context-lakshmi-nair, retrieved on 02-08-2021.
64. Lin. J., Carayon, P. (2021), Health Care 4.0: A vision for smart and connected health care, https://www.tandfonline.com/doi/full/10.1080/24725579.2021.1884627.
65. United states Environmental Protection agency, https://www.epa.gov/radon/health-risk-radon.

66. Mahal, A., Victoria., AK., Englgau, FM. (2013), The Economic Burden of Cancers on Indian Households https://journals.plos.org/plosone/article?id=10.1371/journal.pone.0071853.
67. Alonso, V., Dacal-Nieto, A., Barreto, L., Amaral, A., & Rivero, E. (2019). Industry 4.0 implications in machine vision metrology: an overview. *Procedia Manufacturing.* 41. 359–366. https://doi.org/10.1016/j.promfg.2019.09.020.
68. Ciora, R. & Simion, C.(2014). Industrial Applications of Image Processing. Acta Universitatis Cibiniensis. Technical Series, 64(1) 17–21. https://doi.org/10.2478/aucts-2014-0004.
69. Khan, T., Ashwin, D., Yadav, N., Roy, R.K. (2020). Automatic Detection and classification of Fabrics Using Computer vision, International Journal of Grid and Distributed Computing, vol 13, no-1. 600–612.
70. Kakani, V., Nguyen, V.H., Kumar, B.P., Kim, H., and Rao, V.P. (2020). A critical review on computer vision and artificial intelligence in food industry, *Journal of Agriculture and Food Research*, Volume 2,100033, https://doi.org/10.1016/j.jafr.2020.100033.
71. Georgios, L., Kerstin, S., & Theofylaktos, A. (2019), 'Internet of Things in the Context of Industry 4.0: An Overview'. International Journal of Entrepreneurial Knowledge. 7. 4–19. https://doi.org/10.2478/ijek-2019-0001. 2019.
72. Medhi, P.K. (2019). Is Academic Research in Industry 4.0 and IoT Aligned to the Industrial Needs - A Text Analytic Approach, https://doi.org/10.2139/ssrn.3450480.
73. Boyes, H., Bil, H.,Cunningham, J., Watson, T. (2018), The industrial internet of things (IIoT): An analysis framework, *Computers in Industry'*, *101*, 1–12, https://doi.org/10.1016/j.compind.2018.04.015.
74. Zhen, G., Tom, W., Singh, I., Gadhrri, A., and Schmidt, R. (2020). From Industry 4.0 to Robotics 4.0 - A Conceptual Framework for Collaborative and Intelligent Robotic Systems, *Procedia Manufacturing*, Vol 46, Pages 591–599, https://doi.org/10.1016/j.promfg.2020.03.085.
75. Pfeiffer, S. (2016), Robots, Industry 4.0 and Humans or Why Assembly work is more than Routine Work, *Procedia CIRP*, 40(Icc), 536–541, https://doi.org/10.1016/j.procir.2016.01.129.
76. Jerome B., Subrahmanyam KVJ (2017), Tesla Motors; A silicon Valley version of the Automotive Business Model, retrieved from https://www.capgemini.com/consulting/wpcontent/uploads/sites/30/2017/07/tesla_motors.pdf (2021).
77. Solanki, A., & Nayyar, A. (2019). Green internet of things (G-IoT): ICT technologies, principles, applications, projects, and challenges. In Handbook of Research on Big Data and the IoT (pp. 379–405). IGI Global.
78. Singh, K. K., Nayyar, A., Tanwar, S., & Abouhawwash, M. (2021). Emergence of Cyber Physical System and IoT in Smart Automation and Robotics: Computer Engineering in Automation. Springer Nature.

Foundation Concepts for Industry 4.0

Bhakti Parashar, Ravindra Sharma, Geeta Rana, and R. D. Balaji

1 Introduction

The fourth Industrial revolution is presently among top priority of many organizations, research centers and universities, but most academic experts believe that the term 'Industry 4.0' itself is not clear, and it is difficult for manufacturing companies to understand this phenomenon in a timely manner and to determine the necessary steps for the transition to Industry 4.0. In 2011, the German government used the phrase "Industry 4.0" as a concept and strategic strategy and other industrialized countries, such as the United States Advanced Manufacturing Partnership, China Made in China, the United Kingdom Smart Factory, and others, have taken comparable transformative measures towards fourth revolution [1].

It encompasses the use of the Internet of Things (IoT) and services in business operations [2]. Today, the real and virtual worlds are proliferating as the Internet of Things (IoT), which has inspired factories and governments to embark on the evolution journey towards Industry 4.0 [3]. It can also be said as a collection of ideas and technologies that distorts the distinctions between physical, digital, and biological activities [4]. Industry 4.0 (I4.0) provides new approaches to producing smart goods, as well as technical instruments to construct a new industrial concept: the smart factory. This new method, however, does not replace Lean, and the two may and should be combined [5].

B. Parashar
VIT-Bhopal University, Bhopal, India

R. Sharma · G. Rana
HSMS, Swami Rama Himalayan University, Dehradun, India

R. D. Balaji (✉)
University of Technology and Applied Sciences, Saadha, Oman
e-mail: balaji.majan@gmail.com

Although, it is technologically the most advanced, its implementation and validation are jeopardized without understanding of and judicious application of productive methods, but finding the appropriate approach to build competitive advantages in the age of Industry 4.0 is one of the most difficult challenges that businessmen today. It's a prerequisite for their long-term commercial existence [6, 7]. Industrialization began with steam and the first machines mechanized most of the jobs of our ancestors. Then we had electricity, assembly lines and the advent of mass production, then the third era of industry came with the advent of computers and the beginning of automation as robots and machines began to replace labor on these assembly lines.

This allows for integrated feedback to changes and integrates the relaunch plan into the management system [8] where industry 4.0 aspires to achieve a high degree of connectivity of all processes and products throughout the lifecycle. I4.0 is characterized by highly intelligent, interconnected systems that form a completely digital value chain. It is basically built on cyber physical production systems, which integrate communications, information technology, data, and physical elements to turn traditional plants into smart factories. With the use of digital technologies with the virtual and real-world of production, our production and industries can become global and master the endless increasing challenges in the dynamic and global markets.

This change and incorporated reviving planning of industry 4.0 into the management system is seen as a revolution in manufacturing by corporate companies in order to enhance productivity and shareholder value. While the promises of industry 4.0 have made many bold claims, one common misperception is that industry 4.0 is a panacea for all issues [9].

Today Information and communication technology (ICT) is advancing at a fast pace right now. Cloud computing, Internet of Things (IoT), big data analytics, and artificial intelligence are just some of the innovative technologies that have arisen [10]. Sensors are increasingly being utilized in equipment such as machine tools to allow them to perceive, act, and communicate with one another [11].

1.1 History and Evolution of Industrial Revolutions

The rise of the presumed Fourth Industrial Revolution, as well as the parallel growth of both the concept of Industry 4.0 and its attention in research area, signaled the beginning of the Fourth Industrial Revolution which is due to market development, globalization, and rising competitiveness. Industry 4.0 comes after three previous technological revolutions: steam power, which revolutionized the nineteenth century; electricity, which reinvented most of the twentieth century; and the computer era, which began in the 1970s [12]. The I4.0 idea was first introduced at the Hannover Fair in 2011, since it is indeed a relatively new phenomenon with a history of less than a decade. This may make it easier to identify the roots and origins of I4.0 than it is for many other management concepts and ideas, which can have decades-long histories and roots [13].

Fig. 1 Stages of industrial revolution

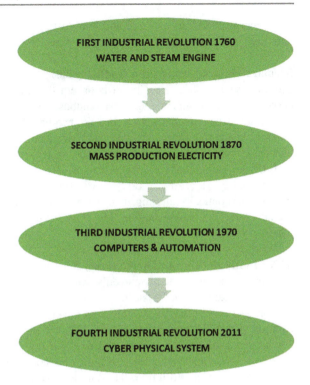

Food, clothing, housing, and weapons have all been made by hand or with the assistance of labor animals for ages. However, with the arrival of Industry 1.0 at the turn of the nineteenth century, manufacturing began to alter drastically, and operations quickly grew from there. Figure 1 elaborates the stages of Industrial revolution.

First Industrial Revolution (Industry 1.0)
Global industrialization began in the 18th century, before that the industries were limited to primary industrial sector, a secondary industry sector, and a tertiary industry sector only [14]. By the end of the 18th century, the industrial revolution finally arrived in Britain, taking with this the introduction of machines into production (1760–1840). This includes switching from manual manufacturing to steam-powered motors and the utilization of water as a source of energy. Since the first Industrial Revolution, manufacturing has evolved from water and steam powered machines to electrical and digital automated production, making the manufacturing process more complicated, automatic, and sustainable so that people can operate machines simply, efficiently, and consistently. From steam engines to automated electrical and digital production, succeeding revolutions have brought significant advances in manufacturing since the first industrial revolution which is also known as industry 1.0 (I1.0) [15]. Manufacturing processes have become

more complicated, mechanized, and long-lasting, making it easier, more efficient, and consistent for humans to operate equipment [16].

Second Industrial Revolution (Industry 2.0)
The Second Industrial Revolution is an era that incorporates new chemical processes, electricity, and the internal combustion engine, according to traditional definitions. These technological but science-based breakthroughs revolutionized industry in the late nineteenth and early twentieth centuries, starting in Germany and the United States and quickly spreading around the world [17]. It all started with the development of electricity and assembly line manufacturing in the nineteenth century. Henry Ford took some ideas and applied them to the automobile industry, dramatically altering it in the process. Previously, an entire vehicle was manufactured at a single station; now, cars are built in stages on a conveyor belt, which is faster and less expensive.

Third Industrial Revolution (Industry 3.0)
Industry 3.0 began in the Seventeen's, when ram devices and computers were first used to substantially automate tasks. One can now totally systematize a production process without requiring human involvement, thanks to the progress of these technologies. Well-known examples of this are robots that obey pre-programmed instructions without human intervention. "As industry turns digital, a third great transition is already gathering pace," the Economist magazine [18] wrote about the third industrial revolution. The circle is virtually complete, with mass production giving way to far more individualized manufacture. As a result, some of the employment that were lost to the emerging world could be returned to rich countries.

Fourth Industrial Revolution (Industry 4.0)
The Third Industrial Revolution began in the 1970s, when memory-programmable controllers and computers were introduced to partially automate activities. Thanks to technological advancements, we can now completely automate a manufacturing process without the need for human intervention. It's also characterized as a new level of organization and control over a product's whole value chain, with an emphasis on more customized services [19]. Industry 4.0s primary focused on meeting specific consumer needs, which impacts various areas such as order management, research and development, factory commissioning, delivery, and product utilization and recycling [20]. Production systems using computer technology are enhanced by a network link and, in a sense, have a digital twin on the Internet. These enable for communication with other systems as well as the production of data about themselves. This is the next phase in the automation of production.

Objectives of the Chapter
The main objective of this chapter is to understand the stages of industrial revolution how it started and how drastically and technically it has been changed. The following are the additional objectives of the chapter:

- To explain the fundamental aspects of Industry 4.0 along with its history and significance.
- To understand how India is ready to adopt Industry 4.0 to become globally competitive.
- To propose a conceptual model of industry 4.0 for the sustainable development.

Organization of the Chapter

This chapter is intended to explain the foundation concept of Industry 4.0, along with its history significance and readiness to adopt. The organization of this chapter is as follow. Section 2 explains the Industry 4.0 its key components and their meaning. Section 3 describes the significance of Industry 4.0 in global business such as technology, economy, society and so on and Section 4 give glimpse on how India is ready for adopting Industry 4.0 being technically backward country as compared to developed nations. Section 5 provides the conceptual model of Industry 4.0 for sustainable development where we discussed how all are technological pillars for the sustainable manufacturing process and leads to sustainable development. Section 7 the purpose of this section is to discuss the future scope of the present study for other researchers to extend. Section 6 concludes the paper with future scope.

2   Industry 4.0

The global recession has shifted the focus on the industrial sector in recent years, with a focus on the significant economic benefit it generates. Organizations who followed the recent movements of relocating operations in search of lower-cost labor are now working hard to reclaim their competitiveness. The German engineering approach played a significant part in that change, establishing measures to retain and promote the country's importance as a "forerunner" in the industry [21]. The buzzword "Industry 4.0" was adopted, and emerged with huge promises for addressing the most recent difficulties in industrial processes. Using its technology, Industry 4.0 is allowing and strengthening this trend, altering people's lifestyles, establishing new business models and production methods, and rejuvenating the manufacturing units for the pretended digital transformation. The term "Industry 4.0," combines the virtual and physical worlds with a focus on technical applications including robotics, digitization, and automation [8].

This global trend is driven by a plethora of possible technologies: Internet of Things (IoT), Big Data, Cloud Computing, Additive Manufacturing, Autonomous Robots, System Integration, Augmented Reality (AR), Cyber-Physical Systems (CPSs), 3D printing, electric vehicles and simulation. Here, the key components of I4.0 are given in Fig. 2.

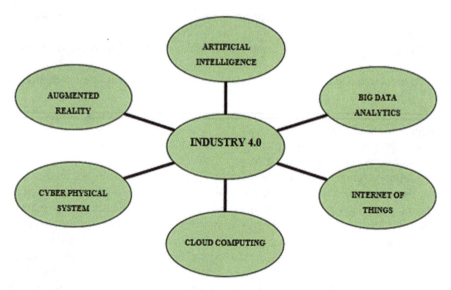

Fig. 2 Key components of industry 4.0

Artificial Intelligence

Digital automation has yet to make a significant impact on overall productivity gains [22, 23]. Apart from that, today's industries are confronted with new market requirements and competitiveness. They require a fundamental transformation known as Industry 4.0. AI integration with new disruptive innovations like the Industrial Internet of Things (IIoT) [24, 25].

It's worth noting that in the 1990s, John McCarthy [26], the father of artificial intelligence, described artificial intelligence as "the science and engineering of creating smart machines, especially intelligent system software." In general, when a computer replicate functions that humans identify with other human minds, such as learning and problem solving, the term "AI" is implemented.

Big Data Analytics

Although these companies are not strangers to holding huge information, practical and manageable insights from the data are frequently lacking. This is also known as a situation in which there is a lot of data but not much information. Big data analytics refers to an organization's ability to analyze large data collections in a systematic and computational manner, as defined by the 5Vs (volume, velocity, variety, veracity, and value addition) [27, 28]. In the future, big data analytics (BDA) could reshape and improve manufacturing and service operations.

This could help industries improve product estimations, performance management across multiple manufacturing and service units, product and service quality, operational visibility, customer preferences and buying patterns, real-time manufacturing process and asset condition monitoring, product design, and customer service. Furthermore, the availability of big data has exacerbated the complexity of supply chain management in the service and industrial industries. [29]. Big Data is booming, which is driven by particular more by rising capabilities of networked, distributed, and cloud computing systems [30, 31].

Internet of Things
The Internet of Things (IoT) could be a network of sensors, machines, and good nodes that may communicate with each other while not the necessity for human interference [32]. The materials square measure self-contained in their interactions with alternative objects. IoT nodes will deliver light-weight knowledge, access and authorize cloud-based resources for knowledge assortment and extraction, and create selections supported the information collected. People, services, sensors, and things square measure currently all connected due to the net of Things.

From good grids to health care and intelligent transportation systems, IoT devices square measure already getting used in a very wide selection of applications. In past few years, academic research on privacy and security challenges for IoT devices has shown promising results. Currently, the available strategies and security methods are predominantly based on traditional network security procedures. Due to the variety of the devices and protocols, as well as the scale or number of nodes in the system, implementing security methods in an IoT system is more difficult than in a typical network [33].

Cloud Computing
Cloud computing is a paradigm that allows users to access a shared pool of configurable computing resources on demand. IT-related capabilities are delivered as services in cloud computing, requiring little comprehensive understanding of the underlying technology and requiring minimal administration work [34]. However, the significant cost advantages offered by the cloud are outweighed by consumers' fears of security dangers. Cloud computing can also provide flexibility in resources, allowing for dynamic provisioning and scaling in responding to consumer demand. This method is designed to address both resource over-provisioning and resource under-provisioning, in which more resources are allocated than are required. As a result of the elastic management, the overall system resource utilization is improved, and the system efficiency improves [35].

IT and related smart technologies are enabling a substantial revolution in the manufacturing business. One of these clever technologies is cloud computing. The basic goal of cloud computing is to deliver on-demand computing services in a distributed environment with high dependability, scalability, and availability [36].

Cyber Physical System
With the evolving relevance of interactions between interconnected computing systems and the physical environment, the term cyber-physical systems (CPS) was coined in the United States in 2006 [37]. Computer security systems is a fast-evolving discipline that would soon have an impact on many aspects of lifestyle.

Semiconductors and the Internet improved and revolutionized our lives by reinventing and altering ways people interface with information, giving rise to digital technologies.

We are presently in the beginning of a new CPS paradigm, which will alter how we interact with and controlled system components [38].

CPS is a game-changer because it is a latest era of systems that blends computing and networking capacity with the motions of mechanical and physical systems. Because huge sums of money are being spent throughout the globe to make this technology, it is yet unclear how the CPS will influence the economy and social structure. On the other hand, the physical world is not totally predictable. As a result, we now have a large region accessible to researchers from many disciplines to investigate and exploit the challenges and vast research possibilities that exist in the Information security Interface field [39, 40].

Augmented Reality

AR is a branch of automation that involves superimposing software digital objects on the real environment in such a manner that the virtual content is aligned with physical reality and can be viewed and interacted with in perfect sync [41]. Researchers are concentrating on Mobile Augmented Reality (MAR). Due to the significant rise in smartphone performance and penetration. The main challenge in AR or MAR is 3D registration technology, and its performance has a direct impact on the performance of an AR system. For MAR, 3D registration is the tracking of the smartphone's position and pose in the real world in real time so that the virtual scene can be seamlessly merged into the real world utilizing the knowledge of position and pose. MAR's 3D registration technology, likewise AR's, has evolved from markers to natural features.

The need for using AR outside, as well as advancements in hardware, have influenced this transition. Currently, natural-feature-based 3D registration technology is a popular topic and the future development path. Although MAR's natural features-based 3D registration methods are developed from AR, some modifications are required due to the peculiarities of smartphones' limited computational power [42].

3 Significance of Industry 4.0 in Global Business

The fourth industrial revolution, promotes more industrial agility, as well as lean manufacturing, higher quality, and increased productivity. As a result, it enables businesses to meet the challenges of generating increasingly personalized products with shorter lead times and improved quality. Industry 4.0 places a high value on intelligent production. Typical resources are transformed into intelligent objects capable of sensing, acting, and behaving in a smart environment [43, 44]. The purpose of Industry 4.0 is to improve operational effectiveness and productivity while also increasing automation Industry 4.0 is beneficial the global business in many aspects, some of them are given below.

Technology

In the framework of Industry 4.0, manufacturing systems are upgraded to a higher degree. In order to meet a dynamic and worldwide market, smart manufacturing utilizes modern information and manufacturing technology to create flexible, smart, and adaptive production processes [45, 46]. It enables all physical processes and information flows across comprehensive manufacturing supply chains, various sectors, SMEs, and big companies to be available when and where they are required [47]. Intelligent manufacturing involves the use of specific core technologies that allow devices or machines to adapt their behavior in response to changing conditions and needs based on previous experiences and learning capabilities [48]. These advancements enable for beneficial collaboration with industrial systems, enabling for quick problem resolution and adaptive decision-making. Artificial intelligence (AI) is a technology characteristic that allows manufacturing systems to learn from their experiences in order to attain connected, intelligent, and ubiquitous industrial practices.

Economy

The transition to the fourth stage of industrialization, named as "Industry 4.0, is now shaping industrial value generation. Goods and services can be linked in a number of ways, including the internet and other network applications like block chains. Automated and self-optimized production of products and services, including distribution is feasible without human interaction. Decentralized control is used to manage value networks, while system parts (such as manufacturing plants or transportation vehicles) make autonomous decisions [49]. It's too early to say how the effects of Industry 4.0 will affect global and local economies. Only 7% of Industry 4.0 research is concerned with the issue of sustainability. The term "Industry 4.0" refers to the need for organizations' operational processes to change. However, the macroeconomic and microeconomic aspects of Industry 4.0 are still largely unexplored. There are studies concentrating on company innovation processes, capital replacement, and the repercussions in terms of rising unemployment rates and globalization [50].

Society

Sustainable Manufacturing (SMA) can be considered one of the most important issues to address for pursuing Sustainable Development, taking into account the social importance of industry and, in particular, manufacturing in our societies, as well as its huge impact on energy consumption, physical resource use, and emissions to the environment, which seems feasible with the application of I4.0 [51]. Furthermore, the enabling technologies of the fourth industrial revolution, commonly referred to as "Industry 4.0," can help provide the information needed for SMA, and they could become significant drivers in the quest of SM and asset lifecycle management. Customers are an important part of every company model, and Industry 4.0 offers them a number of benefits, such as improved communication throughout the value chain and improved customer experience [52]. It enables customers to order any product function, as well as an unlimited number of functions or commodities, even if only one of each is available. Customers could also amend their orders and ideas at any time during the production process, even at the last minute, without incurring any additional costs [9]. To increase sustainability, big data-driven analysis is required to evaluate trust, cultures, and behaviors across supply chains and cross-industry networks, which will help the economic as well as social development [53].

Manufacturing

Industrial production is the backbone of a country's economic growth and has a huge influence on people's lives. Emerging technologies have the ability to change industrial practices, methods, conceptions, and even whole companies. The three major advanced manufacturing technologies addressed in this section are intelligent manufacturing, IoT-enabled manufacturing, and cloud manufacturing. Additive production is a broad concept that attempts to make the most of sensors to be connected and manufacturing technologies to improve production and product relations [54].

It is defined as a new manufacturing model based on intelligent science and technology that significantly improves the design, production, management, and integration of a typical product's whole life cycle. Smart sensors, adaptive decision-making models, innovative materials, intelligent gadgets, and data analytics can help with the full product life cycle [55]. The intelligent manufacturing system (IMS) is a next-generation manufacturing system that is created by transforming a traditional manufacturing system into a smart system by using new models, forms, and processes. An IMS in the Industry 4.0 era provides collaborative, customizable, flexible, and reconfigurable services to end-users through the Internet, enabling a highly integrated human-machine manufacturing system [56].

Marketing

Data analytics, cyber security, the Industrial Internet of Things (IIoT), advanced robotics, and additive manufacturing (AM) are all used in the manufacturing industry. This industry has undergone significant transformations and will continue to do so in the future. A proper management system should be kept to manage all of the new emerging technologies [57]. These technologies can be used to their full potential with the right management mechanisms in place. Industrial management is more successful with the use of Industry 4.0 technologies. All of the expectations of the customers can be met in a shorter amount of time and at a lower cost with effective marketing.

Human Resource

Whereas the magnitude of prospective employment failures brought by modern technology is uncertain, Industry 4.0 will undoubtedly create future jobs by introducing new business models [58]. Advanced robots can execute repetitive tasks with far more efficiency and fewer faults than humans. Humans, on the other hand, have the unique ability to not only reproduce and transform knowledge, but also to generate new knowledge. In the era of IR 4.0, effective HRM practices could improve the working environment by encouraging employees' innovation [59]. As a result, it can be seen that, in terms of human resource issues, organizations are concerned with human resource practices in order to keep up with the IR 4.0 trend [60]. However, there is a dearth of discussion on the topic of human resource practices, with the majority of studies focusing on the technology side of IR 4.0 [61].

Finance
'FinTech' major focus remains mobile and digital payment methods. Furthermore, many technologies, for example, offer services in the following areas: Banking Web services, AI, Financial Planning, Sales, Financial assets, Commercial Investments, Network services, lending, Large-scale Finance, Wealth Management, Fund Transfers, Critical Data and Research, Financial Portals, Start-ups, Influence on the project, Block chain and Cryptocurrency Technology, Robo Assistants, and Next Geography are just a few of the topics covered. FinTech firms are flourishing all over the world. FinTech broadens options in a multitude of disciplines, including payment solutions, financial management, wealth management, and insurance; SMEs can profit from similar applications as well.

Customers' expectations are shifting as millenniums of individuals enter the workforce. In both production and consumption, the concept of the digital client is evolving, and offers must adapt accordingly. Digital platforms becoming more prevalent, digital platforms that allow business partners to contact with each other over a single Internet channel are becoming increasingly significant. Platform providers encourage service and product suppliers to establish internationally successful apps. FinTech broadens options in a range of fields, such as payment systems, credit solutions, asset management, and insurance; SMEs can also benefit from similar applications [62].

4 India Readiness for Industry 4.0

India as a newly industrialized countries have excellent current production bases, but future development may be difficult. India, for example, is the only country with its own Industry 4.0 initiative, branded as "Make in India." Despite being a freshly industrialized country, South Africa must demonstrate that it is ready for Industry 4.0 [63].

Many developed countries' industrial sectors are being transformed by robotics, AI, and IoT. These technologies can also be employed in India, a newly industrialized country, to help it bypass some development stages [64]. China's efforts to transform its manufacturing-intensive economy into an innovation-driven economy also show range of technological possibilities. Unlike developed countries, newly industrialized countries are less likely to have large infrastructure legacy difficulties, making them more adaptable to change. For newly industrialized countries, job losses due to population expansions and automation are an issue.

According to the United Nations' most recent world population forecasts, India, one of the newly industrialized countries, would overtake China as the world's most populated country in less than ten years [65]. The relatively close development of Industry 4.0, which is headed toward smart manufacturing, will not be fueled by this population explosion. Smart manufacturing usually necessitates fewer workers. Concerns include job dislocation, unemployment, energy shortages, and starvation.

The Indian automotive sector has begun to embrace Industry 4.0. Bajaj Auto, for example, began automating its operations in 2010.It now employs 100–120 "Cobots" (collaborative robots) in its manufacturing processes. Maruti Suzuki employs roughly 1700 robots to run seven process shops and five assembly lines. With the help of 437 robots, Ford manages to run its Sanand plant's 16 assembly lines and body shop. Hyundai has taken attempts to reduce labor costs by implementing 400 robots. Tata Motors' production lines for the Tata Nano include over 100 robots Renault is exploring on ways to prevent accidents through automation. All two-wheeler parts are designed utilizing additive manufacturing for fitment and functional testing by the large two-wheeler manufacturers. Companies have been able to shorten their time to market as a result of this. Textiles and packaging are two industries that have embraced Industry 4.0 technologies [66].

Eventually, India requires a new policy to encourage the adoption of digital technologies, develop the necessary digital ecosystem, boost competitiveness, and enable leapfrogging into the digital twenty-first century, all while filling skill gaps and ensuring jobs for millions of people entering the workforce. The government announced the launch of a new manufacturing policy aimed at increasing manufacturing's contribution of GDP to 25% by consolidating Make in India and embracing the Industrial Revolution 4.0.

5 Conceptual Model of Industry 4.0 for Sustainable Development

As a result of technical developments, the fourth industrial revolution is characterized by the widespread usage of smart devices in highly adaptable and networked production or manufacturing systems, but the interconnectivity of Industry 4.0 and sustainable development is not yet sure as this is a very new concept and have not being applied to many of the nations. There is very little research on the Industry 4.0 and sustainable development and the economic, environmental, and social ramifications of Industry 4.0 in terms of industrial digitization require further investigation. Here is a conceptual model which relates the fourth industrial revolution with sustainable development is mentioned in Fig. 3.

Foundation Concepts for Industry 4.0

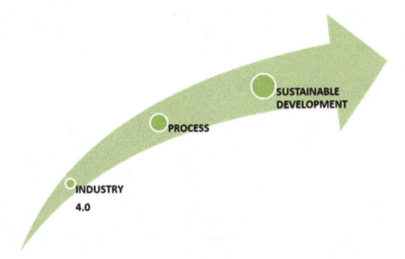

Fig. 3 Conceptual model of industry 4.0 for suitable development

Industry 4.0 includes Internet of Things, Big data analytics, cloud computing, Augmented Reality, Robotic System, Additive Manufacturing, Cybersecurity, Horizontal and Vertical System Integration, all are technological pillars for the sustainable manufacturing process and leads to sustainable development.

- **Internet of Things**: Helpful in real-time communication and interaction between devices with embedded computing.
- **Big data Analytics**: Useful in the collection and analysis of big data sets from many sources in order to assist real-time decision-making.
- **Cloud Computing**: Will be used for data-driven services and data sharing across several sites.
- **Augmented Reality**: Used to improve to improve work and maintenance operations as well as encourage virtual training.
- **Robotic System**: Use of robots in the production is evolving in their utility, increasing autonomy, flexibility, and interaction with humans and other robots.
- **Additive Manufacturing**: Will enable small batch production of customized and lighter products, also reducing logistics costs and stocks.
- **Cybersecurity**: Now a days Integrated networks necessitate protection for important industrial systems and manufacturing processes, as well as secure communication and data transfers.
- **Horizontal and Vertical System Integration**: Data-integration networks and internal cross-functions integration are created as a result of IT system integration across the full supply chain.

These technologies can be effective use for green design, product development and green manufacturing and supply chain and logistics, which will affect the economy in long run-in terms of environmental, economic and social sustainability.

6 Conclusion

Integration, automation, machine learning, and real-time data are all part of Industry 4.0, a new phase in the Industrial Revolution. Industry 4.0, also known as IIoT or smart manufacturing, combines physical manufacturing and operations with smart digital technologies, machine learning, and big data to create a more coherent and integrated system for manufacturing and supply chain management organizations. This research focuses on the development of conceptual analysis and model of fourth industrial revolution and implications for the development that is long-term of a newly industrialized country India. It has been analyzed that how each and every components of Industry 4.0 are useful for the economic as well as the sustainable development of countries with huge human capital and backward technological skills and development. We can conclude in the following ways:

- Sensors, actuators, controllers, and other equipment are connected through an IP backbone in automated building management systems, allowing for the monitoring of energy use from machines, lighting, HVAC, and fire alarm systems. This information may be coupled with larger datasets like weather predictions and real-time energy and other utility prices to provide a more comprehensive and informative perspective of building performance.
- Industry 4.0 is proven to be a significant driver of increased manufacturing plant process effectiveness. Intelligent IoTs, which are defined by smart information storage in goods and pallets, are an example of supply chain digitization that encourages the implementation of just-in-time production. Containers, pallets, and roll cages may be tracked and monitored before to delivery and while within the warehouse using RFID tags and sensors in conjunction with cellular 3G, LoRa WAN, NB IoT, Wi-Fi, and Bluetooth connectivity. This data may be used to expose information about location, inventory, and temperature, and it can also be used to build a foundation.
- Industrial resources may be networked together using sensors, software, and wired and wireless communication to offer an accurate overview of equipment performance at any given time. With the use of machine learning, this analysis may be advanced to real-time yield optimization in industrial environments, with outputs constantly re-calibrated to reach optimal performance based on a variety of variables. This guarantees that industrial assets are constantly operating at peak performance levels.

Thus, with the implementation of industry 4.0 manufacturing will become more decentralized, autonomous with more machine flexibility which is likely to become an exciting area of development.

References

1. Kumar, S.; Suhaib, M.; Asjad, M. Industry 4.0: Complex, disruptive, but inevitable. *Manag. Prod. Eng. Rev.* **2020**, *11*, 43–51.
2. Oberg, C. and Graham, G. (2016), "How smart cities will change supplychain management: a technical viewpoint", Production Planning & Control, Vol. 27 No. 6, pp. 529–538.
3. Shrouf, F., Ordieres, J. and Miragliotta, G. (2014), "Smart factories in Industry 4.0: a review of the concept and of energy management approached in production based on the Internet of Things paradigm", 2014 IEEE International Conference on Industrial Engineering and Engineering Management, IEEE, pp. 697–701.
4. A.A.F. Saldivar, Y. Li, W.N. Chen, Z.H. Zhan, J. Zhang, L.Y. Chen, 2015 21st Int. Conf. Autom. Comput. Autom. Comput. Manuf. New Econ. Growth, ICAC 2015 (2015) 11–12.
5. Nunes, M.L., Pereira, A.C. & Alves, A.C., 2017. Smart products development approaches for Industry 4.0. Procedia Manufacturing, 13, pp. 1215–1222.
6. RavindraSharma, Rakesh Saini, Chandra Prakash and Vinod Prasad (2021) Internet of Things and Businesses in a Disruptive Economy, 1st edn., New York: Nova Science Publishers.
7. Adamik, A.; Nowicki, M. Preparedness of Companies for Digital Transformation and Creating a Competitive Advantage in the Age of Industry 4.0. *Proc. Int. Conf. Bus. Excell.* **2018**, *12*, 10–24.
8. Kagermann, H., Helbig, J., Hellinger, A., & Wahlster, W. (2013), Recommendations for implementing the strategic initiative INDUSTRIE 4.0: Securing the future of German manufacturing industry; final report of the Industrie 4.0 Working Group. Forschungsunion.
9. Schumacher, A., Erol, S. and Sihn, W. (2016), "A maturity model forassesing Industry 4.0 readiness and maturity of manufacturing enterprises", Procedia CIRP, Vol. 52 No. 1, pp. 161–166.
10. RittinghouseJ W, Ransome J F. Cloud Computing: Implementation, Management, and Security. Boca Raton: CRC Press, 2016.
11. ZhangY F, Zhang G, Wang J Q, et al. Real-time information capturing and integration framework of the internet of manufacturing things. International Journal of Computer Integrated Manufacturing, 2015, 28(8): 811–822.
12. Cordes, F.; Stacey, N. Is UK Industry Ready for the Fourth IndustrialRevolution? The Boston Consulting Group: Boston, MA, USA, 2017.
13. Madsen, D.Ø. The emergence and rise of industry 4.0 viewed through the Lens of management fashion theory. *Adm. Sci.* **2019**, *9*, 71.
14. S.K. Babey, J. E. McFee, C. D. Anger, A. Moise, and S. B. Achal. Feasibility of optical detection of landmine tripwires. In A. C. Dubey et al., editors, Detection and Remediation Technologies for Mines and Minelike Targets V, pages 220–231, Seattle, WA, 2000. International Society for Optical Engineering.
15. Wahlster, W. (2012), From industry 1.0 to industry 4.0: Towards the 4th industrial revolution. In Forum Business meets Research.
16. J. Qin, Y. Liu, R. Grosvenor, A Categorical Framework of Manufacturing for Industry 4.0 and Beyond, Changeable, Agile, Reconfigurable & Virtual Production, Procedia CIRP 52 (2016) 173–178.
17. Hull J. The second industrial revolution and the staples frontier in Canada: rethinking knowledge and history. Sci Can 1994; 18(1):22e37 third industrial revolution. Economist, Special Report.

18. M. Rüßmann, M. Lorenz, P. Gerbert, M. Waldner, Industry 4.0: The Future of Productivity and Growth in Manufacturing Industries, (April 09, 2015) 1–14.
19. R. Neugebauer, S. Hippmann, M. Leis, M. Landherr, Industrie 4.0-Form the perspective of applied research, 49th CIRP conference on Manufacturing systems (CIRP-CMS 2016), 2–7.
20. E. Hofmann, M. Rüsch, Industry 4.0 and the current status as well as future prospects on logistics, Computers in Industry 89 (2017) 23–34, https://doi.org/10.1016/j.compind.2017.04.002. Fig. 26 Research gap between current manufacturing systems and I4.0. Adapted from [132].916V. Alcácer, V. Cruz-Machado/Engineering Science and Technology, an International Journal 22 (2019) 899–919.
21. Romeo L, Paolanti M, Bocchini G, Loncarski J, Frontoni E. An innovative design support system for industry 4.0 based on machine learning approaches. Proceedings of the 2018 5th International Symposium on Environment-Friendly Energies and Applications 2019.
22. Singh, R., Gehlot, A., Gupta, D., Rana, G., Sharma, R., & Agarwal, S. (2019). XBee and Internet of Robotic Things Based Worker Safety in Construction Sites.
23. Lee K. Artificial intelligence, automation, and the economy. The White. House Blog; 2016.
24. Da Xu L, He W, Li S. Internet of things in industries: A survey. IEEETrans Ind Inf 2014;10(4):2233–43.
25. Prakash, C., Saini, R., & Sharma, R. (2021). Role of Internet of Things (IoT) in Sustaining Disruptive Businesses. In R. Sharma, R. Saini & C. Prakash, V. Prashad, Role of Internet of Things (IoT) in Sustaining Disruptive Businesses (1st ed.). New York: Nova Science Publishers.
26. McCarthy, J.; Minsky, M.L.; Rochester, N.; Shannon, C.E. A Proposal forthe Dartmouth Summer Research Project on Artificial Intelligence. AI Mag. **2006**, 27, 12.
27. Moore, A. Carnegie Mellon Dean of Computer Science on the Future of AI. Available online: https://www.forbes.com/sites/peterhigh/2017/10/30/carnegie-mellon-dean-f-computer-science-on-the-future-of-ai/#3a283c652197 (accessed on 7 January 2020).
28. N. Shukla, M.K. Tiwari, G. Beydoun Next generation smart manufacturing and service systems using big data analytics Computers & Industrial Engineering, 128 (2019), pp. 905–910.
29. Ravindra Sharma, Ashwini Kumar Saini & Geeta Rana. (2021). Big Data Analytics and Businesses in Industry 4.0. Design Engineering, 2021(02), 238–252.
30. K. Govindan, T.C.E. Cheng, N. Mishra, N. Shukla Big data analytics and application for logistics and supply chain management Transportation Research Part E: Logistics and Transportation Review, 114 (2018), pp. 343–349.
31. Chifor B.-C., Bica I., Patriciu V.-V., Pop F.A security authorizationscheme for smart home Internet of Things devices Future Gener. Comput. Syst. (2017), https://doi.org/10.1016/j.future.2017.05.048.
32. Buyya, et al. Cloud computing and emerging IT platforms: vision, hype, and reality for delivering computing as the 5th utility Future Gener. Comput. Syst., 25 (6) (2009), pp. 599–616.
33. AmbrosinM., Anzanpour A., Conti M., Dargahi T., Moosavi S.R., Rahmani A.M., Liljeberg P. On the feasibility of attribute-based encryption on Internet of Things devices IEEE Micro, 36 (6) (2016), pp. 25–35.
34. Rana, G. & Sharma, R. (2019). Emerging human resource management practices in Industry 4.0. Strategic HR Review, 18(4):176–181.
35. M. b M Noor, W.H. Hassan Current research on internet of things (IoT) security: a survey Comput. Networks, 148 (2019), pp. 283–294.
36. C. Rong, S.T. Nguyen, M.G. Jaatun Beyond lightning: A survey on security challenges in cloud computing Comput. Elect. Eng., 39 (1) (2013), pp. 47–54.
37. X. Xu From cloud computing to cloud manufacturing Rob. Comput. Integr. Manuf., 28 (2012), pp. 75–86.
38. Lee, E. A. (2006). Cyber-physical systems—are computing foundations adequate? NSF Workshop on Cyber-Physical Systems: Research Motivation, Techniques and Roadmap, Austin, Texas.
39. Rho S., Vasilakos A.V., Chen W.F. Cyber physical systems technologiesand applications Future Gener. Comput. Syst., 56 (2016), pp. 436–437.

40. Jianhua Ma, Laurence T. Yang, Smart world: Physical-cyber-social-thinking converged world, 2015 Smart World Congress, Beijing, China, August 10–14, 2015.
41. Fei Hu, Yu Lu, Athanasios V. Vasilakos, Qi Hao, Rui Ma, Yogendra Patil,Ting Zhang, Jiang Lu, Xin Li, Neal N. Xiong, Robust cyber-physical systems: Concept, models, and implementation, Future Gener. Comput. Syst., 56 (2016) 449–475.
42. Hai Zhuge Cyber-Physical Society — The science and engineering for future society Future Gener. Comput. Syst., 32 (2014), pp. 180–186.
43. R.T. Azuma A survey of augmented reality Presence Teleoperators Virtual Environ., 6 (4) (1997), pp. 355–385.
44. McFarlane D., Sarma S., Chirn J.L., Wong C.Y., AshtonK. Auto ID systems and intelligent manufacturing control Eng Appl Artif Intel, 16 (4) (2003), pp. 365–376.
45. P. Chen, Z. Peng, D. Li, L. Yang An improved augmented reality system based on AndAR J. Vis. Commun. Image Represent., 37 (2016), pp. 63–69 (weakly supervised learning and its applications).
46. R.Y. Zhong, X. Xu, E. Klotz, S.T. Newman Intelligent Manufacturing in the Context of Industry 4.0: A Review Engineering, 3 (2017), pp. 616–630, https://doi.org/10.1016/J.ENG.2017.05.015.
47. ShenW.M., Norrie D.H. Agent-based systems for intelligent manufacturing: A state-of-the-art survey Knowl Inf Syst, 1 (2) (1999), pp. 129–156.
48. Wan J., Tang S., Li D., Wang S., Liu C., Abbas H., et al. A manufacturingbig data solution for active preventive maintenance IEEE Trans Ind Inform, 13 (4) (2017), pp. 2039–2047.
49. Wang S.Y., Wan J., Li D., Zhang C. Implementing smart factory of Industry 4.0: An outlook Int J Distrib Sens N, 2016 (2016), p. 3159805.
50. Rana, G., Sharma, R., Rana, S. (2017). The Use of ManagementControl Systems in the Pharmaceutical Industry. International Journal of Engineering Technology, Management and Applied Sciences. Volume 5, Issue 6, 12–23.
51. Hofmann,Erik, and Marco Rüsch. 2017. Industry 4.0 and the current status as well as future prospects on logistics. Computers in Industry International Journal 89: 23–34.
52. Antony, Jürgen. 2009. Capital/Labor substitution, capital deepening, andFDI. Journal of Macroeconomics 31:699–707.
53. Garetti, M., and Taisch, M. (2012). Sustainable manufacturing: trends andresearch challenges. Production Planning & Control, 23(2–3), 83–104.
54. Zhong, Ray Y., Xun Xu, Eberhard Klotz, and Stephen T. Newman. 2017.Intelligent Manufacturing in the Context of Industry 4.0: A Review. Engineering 3: 616–30.
55. M.L. Tseng, R.R. Tan, A.S. Chiu, C.F. Chien, T.C. Kuo Circular economy meets industry 4.0: can big data drive industrial symbiosis? Resour. Conserv. Recycl., 131 (2018), pp. 146–147.
56. M. Ammar, A. Haleem, M. Javaid, R. Walia, S. Bahl, Improving material quality management and manufacturing organizations system through Industry 4.0 technologies, Materials Today: Proceedings. 45 (P6) (2021) 5089–5096.
57. Kusiak A. Intelligent manufacturing systems Prentice Hall Press, Old Tappan (1990).
58. Li B., Hou B., Yu W., Lu X., Yang C. Applications of artificial intelligencein intelligent manufacturing: A review Front Inform Tech El, 18 (1) (2017), pp. 86–96.
59. Kolesnichenko, E. A., Radyukova, Y. Y., & Pakhomov, N. N. (2019). The Role and Importance of Knowledge Economy as a Platform for Formation of Industry 4.0 Industry 4.0: Industrial Revolution of the 21st Century (pp. 73–82): Springer.
60. Feeney A.B., Frechette S.P., Srinivasan V.A portrait of an ISO STEP tolerancing standard as an enabler of smart manufacturing system J Comput Inf Sci Eng, 15 (2) (2015), p. 021001.
61. Jen Ling Gan, Halimah Mohd Yusof, "Industrial Revolution 4.0: The Human Resource Practices" International Journal of Recent Technology and Engineering (IJRTE) 2277–3878, Volume-8, Issue-3S2, October 2019.
62. Nurazwa, A., Seman, N. A. A., & Shamsuddin. (2019). Industry 4.0 implications on human capital: A review. Journal for Studies in Management and Planning, 4(Special Issue-13), 221–235.

63. Sharma, R., & Rana, G. (2021). Revitalizing Talent Management Practicesthrough Technology Integration in industry 4.0. In R. Sharma, R. Saini & C. Prakash, Role of Internet of Things (IoT) in Sustaining Disruptive Businesses (1st ed.). New York USA: Nova Science Publishers.
64. Oktay Fırat, S. Ü. (2016). Determining What is Industry 4.0 Transformation and Expectations, Global Industrialists Magazine.
65. NITI Ayog (2018), "National Strategy for Artificial Intelligence," Government of India, Discussion Paper, June.
66. Singh R., Anita G., Capoor S., Rana G., Sharma R., Agarwal S. (2019)Internet of Things Enabled Robot Based Smart Room Automation and Localization System. In: Balas V., Solanki V., Kumar R., Khari M. (eds) Internet of Things and Big Data Analytics for Smart Generation. Intelligent Systems Reference Library, vol 154. Springer, Cham.

Challenges and Opportunities for Mutual Fund Investment and the Role of Industry 4.0 to Recommend the Individual for Speculation

Sanjay Kumar, Meenakshi Srivastava, and Vijay Prakash

1 Introduction

Industry 4.0 has added many professions to change. Humans are obligated to investigate new; normal duties but now are moreover forced to use hi-tech gadgets which might be fast turning into the maximum crucial issue of their operating life. In this chapter we discuss the impact of rapid change in industry 4.0 technology and its impact on the financial services and mutual fund industry in India. We discuss the impact of industry 4.0 technologies on India's existing mutual fund industry, the threats and challenges that investors and fund companies face when regulating disruptive new technologies. We also discuss the potential threats, challenges, and future prospects of upcoming technologies in the mutual fund industry. Based on the literature review, it is found that industry 4.0 have a positive and significant impact on the mutual fund industry in India. The AUM (Assets under Management) has seen a huge increase in the recent past and improves the customer experience with better access to the back office, even from remote locations. Additionally, the chapter also discuss the challenges faced by regulators who have not yet fully understood the impact of the rapidly changing technological environment around us. Ultimately, this chapter contributes to knowledge creation and

S. Kumar (✉) · M. Srivastava
AIIT, AmityUniversity, Lucknow, U.P, India
e-mail: k.sanjay123@gmail.com

M. Srivastava
e-mail: msrivastava@lko.amity.edu

V. Prakash
BBD University, Lucknow, U.P, India

© The Author(s), under exclusive license to Springer Nature Switzerland AG 2023
A. Nayyar et al. (eds.), *New Horizons for Industry 4.0 in Modern Business*, Contributions to Environmental Sciences & Innovative Business Technology,
https://doi.org/10.1007/978-3-031-20443-2_4

understanding of industry 4.0 and their impact on the mutual fund industry, challenges, and future prospects. In this chapter we also analyze the three mutual fund's Historical NAV for a period from 01-Jan-2020 to 01-Nov-2021. The first fund is Tata Infrastructure Fund-Direct Plan-Growth Option and the second fund is Aditya Birla Sun Life Infrastructure Fund-Growth and the last is LIC MF Infrastructure Fund-Direct Plan-Growth. We analyze the NAV historical data of these funds and show the analysis strategy with the help of graphical representation.

Objectives of the Chapters

The following are the objectives of the chapter:

1. To find the best tools and technology that can help to identify the right mutual fund to investors,
2. To find out the high-level safety and stability for the mutual fund investors by focusing on investment opportunities that offers the right selection of mutual fund and optimized return,
3. To understand how to utilize available resources and ensure to fulfill mutual funds investors needs and priorities,
4. And, to ensure that a fund manager makes decisions after thorough consideration of available mutual funds and potential risks in the future.

Organization of Chapter

Section 2 contains the Literature Review. Section 3 is all about bridging Industry 4.0 with mutual fund industry. Section 4 contains the introduction of mutual fund and the sub sections gives us the information about different types of mutual funds available in the industry, advantages and disadvantages are also defined in this section. Section 5 defines different parameters of mutual fund that are taken into consideration when we analyze the performance of any mutual fund. Section 6 defines the modern technologies available in industry 4.0 for different fields. Section 7 provides some tools and techniques that can be used in fund analysis. Section 8 gives scenarios of Case studies of Three Mutual Funds-Tata Infrastructure Fund, Aditya Birla Sun Life Insurance Fund and LIC MF Infrastructure Fund. Section 9 enlightens Fundings and Results. Section 10 concludes the chapter with future scope.

2 Literature Review

Antoney et al. [1] stated that humanity is currently in the age of the 4th industrial revolution, called the industry 4.0, which is characterized by the digital economy, new data, such as big data, cloud computing, business intelligence, and machine learning interplay, Robotics, Artificial Intelligence (AI), and Distributed Laser technology. Collectively, these technologies are again trying to transform the entire system of production, trade, commerce, banking, management, and administration and have begun to bring about significant changes in the ways of providing products and services in the industry. Patel and Khan [2] focussed on the major challenges facing the sales department of mutual funding companies and the challenges facing the world of VUCA in the 21st century. This paper offers numerous techniques focused on era involvement at diverse stages in solving digital income and distribution challenges. Modern-Day and palms-on discussions, supported by educational enter, ought to be useful to both students and practitioners. Kumar and Goel [3] analyzed how mutual price range has become critical for the reason that financial region won momentum in terms of globalization and liberalization. Mutual price ranges normally declare their own investment objectives and investors pick out the perfect mutual fund to spend money on as a part of their funding strategies. The overall performance of mutual fund products turns into more complex in phrases of both chance accounting and go back valuation in relation to funding objectives. Arun and Ankita [4] found that the level of participation and contextualization of these formal knowledge management systems was even lower due to low employee awareness. Based on this study, recommendations for improvement have been identified. Ruichan and Zhang [5] examined two forward-looking mutual fund ratings supplied through Morningstar the analyst ranking and the quantitative ranking based on the computer gaining knowledge of technique. The analyst ranking identifies outperforming funds, whilst the quantitative ranking fails to do so—much a difference is mostly due to the choice of analyst coverage. Moreover, the tone in the analyst record consists of incremental records in predicting fund performance. Finally, retail buyers do not follow analyst recommendations however, as an alternative chase the quantitative rating. Ordinary evidence highlights the importance of mutual fund analysts in fact manufacturing and implies a capital misallocation problem in mutual fund investment. David and Cleland [6] stated that project is an interdisciplinary activity that can be started in a timely manner as a multidisciplinary task. For large international projects, the number of supervisors and managers is high unless many reduce responsibility as a way to determine responsibility. The main purpose of the control is to see if the activity is achieving the desired result. Economic factors, cost factors for investment, demographic factors, the attraction of foreign investment, and various levels of environmental regulation are key characteristics of IP growth. Samrin and Ramanathan [7] analyzed that business of banking is credit and credit are the main basis on which the quality and performance of banks are centered. The failure of the banks was due to poor credit quality, meaning that the loans and advances made by them did not yield the expected returns which were not repaid to the lenders. Operational threat is the danger of damage from insufficient or failed internal processes, humans and structures, or some external event. Marketplace hazard is the risk of loss that may be amassed because of negative modifications in hobby costs, prices, and trade rates. Prabhavathi et al. [8] primary motive of the study was to understand the attitude, focus, and choices of mutual fund investors. Maximum respondents choose normal investment plans and gain their assets of information commonly from banks and economic advisors. Traders preferred mutual funds normally for expert fund control and higher returns and ranked funds usually on NAV and beyond overall performance. Saini et al. [9] analyzed mutual fund investments on the subject of investor behavior. An observation is done of investor perspectives together with the form of disputed fund scheme, the main purpose at the back of making an investment in a mutual fund scheme, the position of financial advisors and agents, investor opinion on factors that encourage to put money into mutual finances has been assets of statistics, expertise in offerings provided by mutual fund managers, challenges facing the Indian mutual fund enterprise, and many others. Suárez et al. [10] studied that, an artificial neural network modeling framework developed specially for use as lively data-driven robotic-consultant, as it can are expecting nowadays copper expenses five days before charge changes the usage of enter statistics, which mechanically can be fed into the version. This version tested inside the recent history of copper charges (from may also 2006 to September 2008 and from September 2008 to September 2010) the use of high volatility facts for two periods of return, suggests that this technique predicts in-sample and extra-sample charges resulting in rate modifications with high accuracy. How et al. [11] examined that the AI may be used via a user-pleasant human-targeted stochastic good judgment technique. The research made use of the 2018 Environmental performance Index (EPI) records of 180 nations, together with overall performance signs that cowl ecological fitness and atmosphere dynamism. AI-based totally predictive modeling techniques are carried out to the 2018 EPI information to uncover hidden tensions among the two fundamental dimensions of sustainable improvement: (1) environmental fitness. (2) The dynamism of the ecosystem. Mhlanga [12] analyzed conceptual and documentary evaluation from censored journals, reports, and different credible documents on AI and digital monetary inclusion to assess the impact of AI on digital financial inclusion. On this look at, AI has a sturdy effect on virtual a monetary inclusion in regions related to risk detection, measurement, and management, addressing information asymmetrical issues, and presenting customer service chat bots, fraud detection, and cyber security. Mohamad et al. [13] studied the adoption of financial technology (Fintech) in mutual fund/unit accept as true with investments a number of the people of Malaysia: Unified Theo-Ray of reputation and use of technology (UTAT). The cause of this study is to pick out the extent of monetary technical use in mutual price range/unit trusts amongst Malaysian investors. This takes a look at enables participants in financial services management to take advantage of the opportunities provided with the aid of fintech services. The implementation of Fintech will lead to carrier development and transformation for future investment control offerings. Calia et al. [14] demonstrated its disruptive capability in many fields, mainly in production, logistics, and engineering. Manufacturing procedures, business fashions, and IT infrastructures are predicted to be revolutionized: eventually, the supply chain will become an extra green, adaptable, and scalable workflow, ultimately driven by means of an actual-time and corresponding demand for brand spanking new or stepped forward services and products. Pantielieieva et al. [15] discussed the principal guidelines, demanding situations, and threats of digitization of the Ukrainian countrywide financial system. Author supplied ability and new possibilities to clear up public financial management troubles the use of blockchain technology. The development of FinTech, the modern-day financial technology, is diagnosed as a driving force for the digital transformation of economic services. It characterized the varieties of FinTech innovation, the intensifying opposition among FinTech groups and conventional financial intermediaries, and the trend of FinTech improvement in Ukraine. Das and Ali [16] mentioned the impact of FinTech on the present Indian financial gadget, as well as the threats and situations for regulators in regulating new disruptive technologies. The paper also mentioned the demanding situations faced by regulators who've not yet fully understood the impact of the swiftly converting the swiftly converting technological surroundings around us. Núñez et al. [17] analysis made it possible to propose a new classification of literature that identifies four research axes based on technology life systems: obsolete IDT in LSCM; Adult IDT in LSCM; Emerging IDT in LSCM and a general approach to information systems and IDT in LSCM. A series of implications are presented, which will be useful not only from an educational point of view but also from an administrative point of view, including recommendations for industry managers and policymakers. Bahrammirzaee [18] performed a comparative study of three popular techniques of artificial intelligence in the financial market, namely artificial neural networks, expert systems, and hybrid intelligent systems. The results show that the accuracy of artificial intelligence methods is better than traditional statistical methods for dealing with financial problems, especially with regard to online models. However, this success is not certain. Fethi and Pasiouras [19] provided a comprehensive overview of 196 studies using operational research (OR) and artificial intelligence (AI) methods to measure bank performance. Many important issues of literature have been highlighted. The document also points out several directions for future research. First, they discussed several applications of data coverage analysis; the most widely used operating technique in this area. Then, they discussed the use of other methods, such as neural networks, support vector machines, and multi sensor decision-making tools, which have in recent years been able to predict bank failures and reduce the bank's credibility and efficiency. QianZhu and Gee [20] presented the typography of cutting-edge social media services the use of the following categories: relationship, Self-Media, Collaboration, and innovative little. Author described in greater detail how each kind of social media meets fundamental human wishes and offers implications for social media marketing based on a lens of desires harmony. Kong et al. [21] contributed to the present literature to assess the performance of mutual fund using a hard and fast of extra state-of-the-art econometric models DELLO. Authors selected six consecutive ancient years to create a model for comparing the overall performance of mutual price range using the fast Adaptive Neural network Classifier (FNNC) and evaluating it with the results we've acquired with corrective resilience. Yang et al. [22] advised a method for commercializing the merits of monetary bailouts, solving present issues and troubles in the current R&D investment from the perspective of the perspective of monetary bailouts, and accomplishing fast boom. Brian et al. [23] used a method, that provide empirical evidence for the impact of IT funding declarations on the company's market cost for a sample of ninety-seven IT investments in the financial and manufacturing sectors from 1981 to 1988 at some stage in the announcement length, they did no longer get hold of any in addition returns for the total sample or for any samples from the industry. Jenderny et al. [24] developed a device for trying out paintings' situations 4.0 within the relevant dimensions of "era", "humans" and "organization". This report described the relevant criteria selected; consisting of feasible restrictions future work on this subject matter will include the introduction of profiles and similarly validation of requirements for 4.0 scenarios of labor. Sudheesh and Parimalakanthi [25] contributed that people are turning to existing scenario's, mutual fund is the maximum vivacious funding road amongst all of the pretty a number of handy options one could discover investment for mutual fund which in flip will provide them exposure to various asset schooling according with their decision on the inspiration of a variety of parameters like their age, financial position, danger tolerance, and return expectations. Panigrahi et al. [26] presented in this paper that a try and remember the performance of the top five ELSS schemes of specific mutual money in India the usage of some of tools like Beta, Sharpe ratio, Jensen ratio, and many others. The analysis shows that most of the people of price range have outperformed under Treynor's Ratio and Sharpe Ratio, giving consistent and substantial results at some degree inside the route. Jesus [27] exploited the model's go-sectional implication that the weakening of the hyperlink among float and previous standard overall performance is extra mentioned while information boundaries are better. Using a fund's age and retail clients as proxies for statistical costs, they obtain ample empirical aid for this prediction those findings are currently not due to numerous variables which can be correlated with the truthful rate and are impartial of fee-pushed modifications in fund manager behavior. Kancherla and Rao [28] concluded that fairness fund managers possess huge marketplace timing capacity and that institutional dollar managers are capable of time their investments, but, dealer-operated finances not show off market timing capability. Further, it has been determined empirically that fund managers are capable of time their investments with the conditions in the marketplace and possess brilliant timing capability. In this paper, author tried to focal point at the factors that inspire the traders to spend money on mutual price range. Anish and Majhi [29] proposed a feedback functional hyperlink synthetic neural community (FLANN) for the prediction of net asset value (NAV) of Indian Mutual fund range which includes less computational load and faster forecasting capability. Cuiyi [30] build an evaluation index set along with fund organizations, fund managers, threat degrees, and tiers of threat-adjusted income, and makes use of unbiased samples t-take a look at, linear regression, and KPCA and SVM mannequins to evaluate and classify securities funding budget. In line with empirical analysis, the mannequin is conceivable, and using the manner to choose out funds is simple, directly, and practical. Inchamnan and Anunpattana [31] focused on to illustrate the effect of high-quality feedback during sports on players' behavior. Gamification sports are designed to supply positive remarks via a know-how-primarily based system. This remarkable feedback would convince gamers to change their investment idea that is paintings research to examine the gamification workflow which inspires people to live in their lives with advanced generation. Yucan and Chen [32] examined the clustering effect of the open-ended fund as a shape of class of the included overall performance of the fund. To learn about the clustering effect of an open-ended fund, authors decided on the cumulative net fee of five signs to explain the consolidated running ability of cash throughout the length from January 1, 2008, to December 31, 2010, of one hundred stocks open-ended, with the resource of using cluster analysis, we received an image of the coins clustering impact outcomes. Veile et al. [33] developed a comprehensive framework that provides the necessary and effective actions for the implementation of future digitally connected industrial value creation. It also contributes to the study of technology management and innovation implementation in the dynamic digital age. Lee et al. [34] described trends in manufacturing service transformation in big data environments and the availability of intelligent predictive intelligence tools that manage big data for transparency and efficiency. Aridi and Querejazu [35] examined the current state of Industry 4.0 (I4.0) technology creation and adoption in the Czech Republic. The country has one of the strongest manufacturing sectors in Europe in terms of size, but productivity is falling and labor shortages are worsening. To stay competitive in the global economy and avoid the middle-income trap, the Czech Republic needs technologies that can displace the workforce and increase labor productivity through automation. Meenakshi and Yadav [36] advised that institutional investor, investment trusts play an important role in the proper functioning of the stock market. The research provided an analysis of the investment trust industry in India and a study of the three major companies in India. Iyer [37] conducted an investigation suggesting several important ways for emerging economies to move forward in Industry 4.0 to develop sustainable manufacturing processes around the world using India as a case study. Watzlawick [38] described specific data analyzes corresponding to their fields and categories, lessons, keywords and four research sub-questions and this result not only summarizes the current research activities but also suggested research programmers, and it also points out existing gaps and potential research directions. Muller et al. [39] identified and analyzed the determinants of smart manufacturing information and digital technology (IDT) implementation. The results show that the recognized benefits and management support of are two driving forces that serve as the foundation for implementing IDT smart manufacturing. Machado et al. [40] created sustainable value that can be solved by raising awareness for the benefit of the world. The ubiquitous application of modern information and communication technologies (ICTs) to form responsible global citizenship through knowledge transfer dramatically expands learning, education and productivity and the increased use of help with self-help. Nayyar et al. [41, 42] identified and analyzed to integrate smart sensors to work collaboratively without human intervention to provide the best applications and services. Das et al. [43] contributed to the literature on customer relationship management and service marketing by providing empirical support for customer loyalty and business relationships in the Indian context. Das et al. [16] analysis and discussion in the second part of the chapter broadly applied the previous framework, while the final part discussed important digital entertainment formats and options in engaging consumers.

3 Industry 4.0 Bridges with Mutual Fund

We are able to see that AI isn't hype however has the capability of remodeling the global financial system thru technological innovations; medical information and entrepreneurial sports. Elements are making AI the core generation responsible for excessive automation and connectivity for excessive automation and connectivity and accordingly, taking the arena closer to the sunrise ultra-modern the fourth industrial revolution. This could have intense influences on governments, communities, groups, and individuals. The innovation system and international Competitiveness is strengthening as a final result contemporary the adoption trendy various techniques by way of the company corporations (organizations and begin-ups) to come to be AI-companies. The actual purpose is to grow with the maximum superior generation modern AI and win the technological race. It has also furnished a better knowledge of the way AI can reshape the markets, transform and innovation approaches, the company contemporary studies & improvement, business techniques and the worldwide economy. With the structural liberalization guidelines absolute confidence Indian financial system is probable to go back to an excessive develop route in few years. Success modern day mutual fund but could vividly relying upon the implementation cutting-edge hints. Kumar and Nayyar [42] discussed that the Industry 4.0 involves as many industries as possible and adapts and improve existing technologies to better meet the requirements of digital manufacturing. Industry 4.0 includes concepts, tools and applications that complete an intelligent integrated system of machines capable of communicating with each other and with people and performing autonomous tasks in industrial production processes [51, 53, 55, 56]. Core tools include cyber-physical systems (CPS), Internet of Things (IoT), big data and cloud computing, autonomous robots, simulation and visualization models, and additive manufacturing. Automation and robotics are perceived as key components, arms and legs, of Industry 4.0 cameras and other sensors are perceived as senses; data and connectivity are compared with the nervous system; and artificial intelligence (AI) is the brain. Artificial intelligence improves industrial processes by enabling synergistic collaboration between humans and robots in smart factories for mass customization. The major technological advances of the Fourth Industrial Revolution, or Industry 4.0, are revolutionizing industrial manufacturing. The first industrial revolution was based on mechanization and steam engines; the second on the use of electricity and mass production; and the third on electronics and information technology, which translates into a high level of automation in production. Industry 4.0 takes the automation of production processes to a higher level with intelligent autonomous systems capable of self-recognition, self-optimization and self-personalization. Industry 4.0 will affect all sectors and disciplines, causing a structural transformation of the global economy and leading to a new division of labor, which will have a major impact on developing countries [48, 49, 52, 62–64]. A new wave of outsourcing and outsourcing will be unleashed, with new technologies like additive manufacturing using innovations like 3D printing, making outsourcing superfluous. In shoring could become a new trend in industrialized countries, depriving developing countries of job opportunities. Many predict that Industry will cause a polarization of the workforce, with an increasing share of high- and low-wage jobs and a declining share of middle-wage jobs. There is likely to be a concentration of low-paid jobs in developing countries, as well-paid jobs will require higher digital skills and developing education systems are weak.

4 Introduction to Mutual Fund

These days, any proper investment layout tends to encompass a plan to invest in mutual funds. While investing in a mutual fund is the correct idea, it, extra often than not, leaves viable traders scratching their heads about what this whole mutual fund aspect is all about. Mutual fund is noticeably easy to understand. The basics of a mutual fund are that you invest money in a fund alongside a load of different people. The organization that offers the fund then invests the cash and you get the returns. Geetha and Ramesh [44] stated that there are many investment options and one should choose the most suitable one. The person managing the project must know all the different options for investment and how to choose it for the purpose of achieving the overall objectives. Walia and Kiran [45] identified serious gaps in existing mutual fund frameworks. It is important to understand the importance of redesigning existing mutual fund services through the recognition of the investor-centered quality of service agreements (IOSQA) to guide investor behavior and understand and introduce financial innovations. Today's mutual fund and exchange-traded fund (ETF) income teams are drowning in data. Whether it pertains to products' positions, performance, flow, environmental, social, and governance (ESG) criteria, meetings and calls, macro and micro, history, estimates or predictions—there is an information overload. Ostensibly, this data exists to help income teams discover and qualify prospects, but that's less difficult stated than done. One client, a publicly-traded mutual fund manager, informed us that qualifying a single lead used to take a group of workers members greater than forty minutes [60].

Here's what the step-by-step process looked like this:

1. Log in to the Broad ridge Economic Answer for a listing of registered investment guide (RIG) possibilities in a precise territory.
2. Collect asset flow for these potentialities from Market Metrics market brain software.
3. Collect additional brain about the prospects from the client's Registered Investment Advisor database monetary facts and marketing solution.

4. Cross-reference the facts in opposition to a patron relationship administration customer relationship management gadget for information on and consequences from past meetings.
5. Assemble all the facts in Microsoft Excel.
6. Rank the possibilities based on formulation and judgment.
7. Review the top precedence prospect's funding philosophy through its internet site to decide the most useful pitch.
8. Schedule a call or set up a meeting.

The consumer set up record feeds with all of its vendors, aggregated its inside and third-party statistics into a fact's lake, and packaged them for use through consumer interfaces. These included a self-service interface for salespeople and a greater superior one for the Business intelligence (BI) team. Before the transformation, the sales group had to drill down one client at a time. The self-service interface helps with group habits evaluation throughout a couple of clients. This has opened the door to a wide variety of fruitful data-driven conversations. For example, the merchandise the firm had before prioritized for the sales group became out to be neither the best-performing nor the most sellable. Report Library once the income and advertising records used to be centralized and integrated, the purchaser developed a library of reports to drill down into the data. The goal was once no longer to reproduce existing reports, summarize the pipeline, or describe "how matters are going" Rather, the motive used to be to force choices about who to name and what to pitch. The aggregate of area expertise in the industry and business with digital technologies proved critical here. The client's excellent and most senior salespeople had the experience, skill, and instinct to identify patterns of probable shoppers and probable pitches. The library document codified this information and made it reachable to the complete team.

4.1 Types of Mutual Funds

There are two wide types of mutual fund available in India. They are open-ended and closed-ended mutual funds. Figure 1 highlights different types of Mutual Funds.

4.1.1 Open-Ended

Open-ended mutual dollars are these the places investors can indulge in the buying and selling of gadgets at any time. There is no maturity duration or funding duration for these funds. They can additionally be similarly categorized into four types, which are:

a. **Debt/Income**
   Investments are made in bonds and treasury bills in debt earnings mutual cash money invested in such a fund can be put into monthly earning plans, short period plans and flexible maturity plans etc. These investments are perfect for

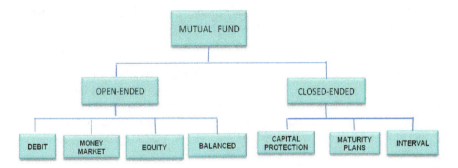

Fig. 1 Types of mutual funds

those who are searching for safer surroundings for their cash to grow in because they offer a very low threat elements and low returns.

b. **Money Market/Liquid**
Money market or liquid mutual money are those the place investments are made in treasury payments and constant earnings securities among different devices like short time period bank certificates of deposits. The purpose of this money is to grant traders liquidity consequently they come with short maturity durations of about 90 days.

c. **Equity/Growth**
Equity and growth mutual funds are where the buyer's money is invested in equity shares with the idea of either generating a profit or capital gain. Sometimes they can be investments made with the cause of generating both gains and income.

d. **Balanced**
Balanced funds, because the call indicates, make investments the cash in balanced a way among steady income securities and equity bucks in an effort to furnish traders with the chance to invest aggressively but with caution.

4.1.2 Closed-Ended Mutual Funds

Closed-ended mutual budget are in which the fund is available in with set maturity duration and also permit for investments to take vicinity solely in the initial tiers of the fund. It has two types, the capital safety fund, and the constant maturity plans fund.

a. **Capital Protection**
The capital safety mutual fund invests in each regular income securities plans however and fairness the funding in equity is marginal thinking about the aim of the scheme is to defend the primary while though getting returns.

b. **Fixed Maturity Plans**
These plans, unlike most other plans, can also come with the lowest costs for the scheme due to the fact they are no longer managed actively like other funds. In

a fixed maturity plan, funding is made primarily from debt devices that mature alongside an equal timeline as this fund in view comes with a fixed maturity period.

c. **Interval**

Sometimes you will note that a mutual fund will only allow funding at precise periods. It does not allow funding at any time, but it still permits investments to take place an awful lot later into the scheme's period too. This is so due to the c programming money functions as a mixture of language mutual every open and close-ended fund where investments may be made at particular durations.

4.2 Benefits and Disadvantages of Mutual Fund

Benefits:

There are some of advantages to mutual funds some of those are:

- For the reason that investments are definitely by means of experts, investors aren't required to have an expert perception of the markets and the way they characteristic.
- The funding in a mutual fund may be completed in a lump sum or in installments.
- Investments made in tax saver mutual money are exempt from tax beneath phase 80 C.
- Investors can pick out the risk degree from the low, medium, and excessive threat coins based totally on their appetite for chance.
- Dangers are mitigated by means of manner of funding the money in one-of-a-kind shares and bonds.
- Mutual fund that does not have lock-in intervals can provide liquidity whilst wanted.
- There's no need for massive sums of money to begin investing in mutual price range.
- Invests also can be made in SIPs, (systematic funding plans) where a specific quantity may be paid toward the mutual fund each month.

Disadvantages of Mutual Funds

The fundamental disadvantage that a mutual fund may have been that there is a component of chance involved in contrast to constant deposits. The boom in a mutual fund depends on market performance and can grant returns or even reason for the investor to insight a loss while constant deposits supply constant and safe growth.

4.3 Making Investment in Mutual Funds

To determine if the investment will be achieved in a lump sum or via SIP is the very first step to invest in mutual cash. Once that is resolved, you want to discern your urge for food. In general excessive risk funds will also provide the best returns, but if you want less return in safer surroundings instead of searching for high returns, then you can go in for a medium or low chance mutual funds. To use preliminary socio-environmental data is necessary for analysis to inform policy makers, decision—making about sustainability in development. However, it is not easy to use artificial intelligence for the people who are not trained in computer science but AI based methods are useful for analyzing data. The exposition of how AI can be used through a user-friendly human centered stochastic logical approach in the significance and originality of this paper. The analysts who are not computer scientist can use AI to analyze stability related API data using this method.

5 Mutual Fund Parameters

There are many key investment risk indicators used when analyzing mutual fund performance. These are alpha, beta, R-square, standard deviation and Sharpe ratio and many more. Modern portfolio theory (MPT) uses these parameters when analyzing risk with respect to past performance of mutual funds. Modern portfolio theory (MPT) is a well knows method used to compute production by estimating mutual fund investments with respect to market benchmarks. In this chapter we discussing and studying these parameters with respect to benchmark index such as the S&P 500 and Nifty 50.

Some most important indicators are discussed below:

5.1 Alpha

Alpha is measures' the risk level of a mutual fund with comparison to benchmark index. Volatility is the main factor that is calculated on portfolio with respect to reference index. Excess return of mutual fund is alpha against the benchmark return. For example, if the value of alpha is 5% and the benchmark return is 20% in a year then total aggregate return is 25% so fund is surpassed its benchmark. Suppose that an alpha below than 0 of 2% and benchmark return is 10% in a year then total aggregate return is 8% so fund has underperformed. The formula to calculate the alpha is:

$$F.R = R.F.R + Beta \times (B.R - R.F.R)$$

F.R-Fund Return, R.F.R-Risk free rate, B.R-Benchmark return

5.2 Beta

Beta is calculated using regression analysis and constitutes the return-on-investment gains that acknowledge to market movements. Beta is the measures of sensitivity of mutual fund. It shows the volatility and investors can judge the performance of asset on the basis of risk factor. It can be calculated as follows:

$$\text{Beta} = (F.R - R.F.R) \div (B.R - R.F.R)$$

F.R-Fund return, R.F.R-Risk free rates, B.R-Benchmark return

5.3 R-Squared

Relation between portfolio and benchmark index is called r-squared. The value of r-squared is a range for 1–100 it is not a performance indicator of mutual fund. Low value of this parameter indicates that portfolio is well organized and return good wealth. Funds with a value of 70 or less do not usually function as indexes. High r-squared value indicates that asset is more correlated with its benchmark. R-squared 0 values indicate that asset has no relation with its benchmark.

5.4 Standard Deviation

Standard deviation is measures of dispersion of data and can be calculated using statistical tools. Essentially, the more information is spread, the bigger the variance than the norm SD is described as the volatility of past mutual fund returns by the investors. More increased volatility is the indication of a greater sd. that indicates performance of mutual funds swing high and low from the average. So, the terms volatility and SD convertibly used by most of the investors.

5.5 The Sharpe Ratio

William Sharp is the inventor of Sharpe ratio it reflects the adjusted performance of the Sharp Index Risk. Sharpe ratio is quite helpful in measuring the risk adjusted returns capacity of a mutual fund. So, a greater Sharpe ratio specifies better return yielding potential of a fund for any more risk taken by it. It is an alibi for underlying volatility of the fund. Moreover, you may use Sharpe ratio to distinguish the fund. The high proportion of investors is told whether the return on investment is due to sound investment decisions or excessive risk.

6 Industry 4.0 Technologies

During the early 20th century industry 2.0 started. Electrical power was used to operate machines in this revolution. Electric power is more environments friendly and economical compared to water and stream based power to operate the machines. Using electric power the efficiency and production power of machine also dramatically also increase during this period. Different production management procedures such as at the last-minute manufacturing and rudimentary process enhanced in standard and output. Optimization process perfectly works on labors allocation, plan of action and allotment of asset. The second industrial revolution was a time of great economic growth, with increases in productivity, but also an increase in unemployment as many factory workers were replaced by machines.

Industry 3.0 was the era of industrial insurrection in this period the advanced techniques are came into picture. This resolution was also called digital resolution. Various new and advanced devices were discovered like transistor and integrated circuits which resulted the less effort, increase efficiency and precision. In few areas this inventions replace the human power and provide different more accurate operation using less efforts. The use of electric equipment and machines in manufacturing systems increases the software development rapidly. Use of software increases the Software Development Industry. Now different management process heavily uses the software tools for example resource planning, inventory management, logistics etc. It is important to mention that Industry 3.0 still exists, with most factories at this level of development.

Figure 2 elaborates technologies used in Industry 4.0 [57–59, 61].

1. **IIoT**: IIoT stands for Industrial Internet of Things and is a concept that shows the connections between people, data, and machines related to manufacturing [46, 47, 54].

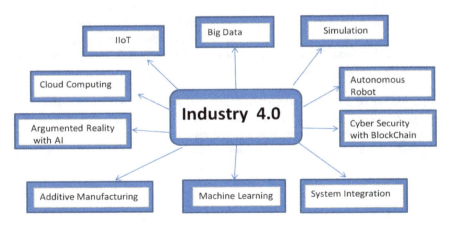

Fig. 2 Technologies used in industry 4.0

2. **Big Data**: Big data is a large set of structured or unstructured data that can be compiled, stored, organized, and analyzed to reveal patterns, trends, relevance, and potential.
3. **Artificial intelligence (AI)**: Artificial intelligence is a concept that refers to the ability of a computer to perform tasks and make decisions that previously required some intelligence.
4. **Machine Learning**: Machine learning refers to the ability of a computer to learn and improve independently through artificial intelligence, without being explicitly requested or programmed.
5. **Cloud Computing**: Cloud computing refers to the storage, management, and processing of information using interconnected remote servers hosted on the Internet.
6. **Additive Manufacturing**: Industry 4.0 facilitates the integration of intelligent manufacturing technologies and systems. Among them, the multilayer model plays an important role in meeting some of the most important requirements of the Fourth Industrial Revolution.
7. **System integration**: System integration is a process commonly performed in the fields of engineering and information technology. This includes combining many different computer systems and software packages to create larger systems. It is the driving force behind the optimal functioning of Industry 4.0.
8. **Cyber-Security with Block-Chain**: The fourth industrial revolution poses new operational risks for smart manufacturers and digital supply networks. In the age of Industry 4.0, cyber security strategies must be secure, vigilant, resilient and fully integrated into the organization's strategy from the start.
9. **Simulation**: Simulation is an important technology for developing planning and discovery models to optimize decision making, design and operation of complex and intelligent manufacturing systems.
10. **Augmented Reality with AI**: AR is a technology that gives the real world a big dimension by superimposing information such as text, images and voice on the world we see. In this way, we can create an interactive experience of the real environment which is enhanced by inserting computer-generated images into it. Much of the AI hype in manufacturing revolves around industrial automation, but it's only one aspect of the smart factory revolution. AI also brings to the production desk is the ability to open up a whole new avenue in business.
11. **Automated Robots**: Industry 4.0, also known as Connected Industry, integrates connectivity and uses collaborative robots in industrial processes to create a workspace where humans and robots can collaborate securely, share information, and optimize processes. Therefore, better decisions are possible.

7 Tools for Analyzing Mutual Funds

The following are the tools for Analyzing Mutual Funds. Figure 3 also gives detailed picture of these tools.

7.1 Artificial Intelligence

Finance is in the midst of a revolution. Processes are being digitized. Decisions are turning into increasingly more data-driven approaches from the backside up. Artificial intelligence (AI) is taking care of business while we work from home [50]. The revolution has affected every market, firm, and branch—besides for product distribution. How et al. [11] examined that sustainable development is important for humanity it's miles essential to use preliminary socio-environmental facts for analysis to inform policymakers' selection-making about sustainability in improvement. Artificial intelligence (AI) primarily based techniques are useful for analyzing statistics. However, it is not clean for folks that aren't trained in laptop technology to use AI. The significance and novelty of this paper is that it suggests how AI can be used via a user-pleasant human-focused stochastic good judgment approach. The use of this method, analysts who aren't laptop scientists can use AI to investigate stability-associated EPI records. Lee et al. [46] evaluated that AI emerges from science fiction to become a world-converting generation front, the

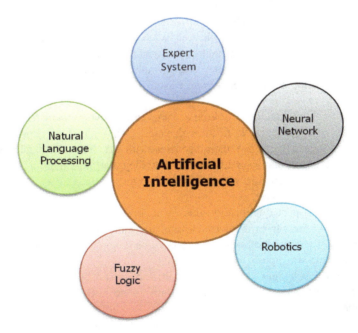

Fig. 3 Areas of artificial intelligence

systematic development and implementation of AI is urgently needed to see its actual effect on the following generation of commercial structures and industry 4.0.

There are many areas where the Artificial Intelligence can use some of them are discussed below.

a. **Expert System**
   In artificial intelligence (AI), expert systems are computerized decision-making systems designed to solve complex problems. Expert systems were one of the first successful forms of artificial intelligence.
b. **Neural Network**
   Neural networks are a set of algorithms that try to recognize the underlying relationships of a set of data through a process that mimics the functioning of the human brain. In this sense, a neural network is essentially an organic or artificial nervous system.
c. **Fuzzy Logic**
   In general, it is a method of reasoning. But it sounds like human inference. In addition, it has a human decision-making approach because they mean all the intermediate possibilities between the numbers YES and NO.
d. **Robotics**
   Robotics is an interdisciplinary field that integrates computer science and engineering. Robotics includes the design, construction, operation and use of robots. The goal of robotics is to design machines that can help and assist humans.
e. **Natural Language Processing**
   Natural Language Processing (NLP) allows computers to understand human languages. Behind the scenes, NLP analyzes the grammatical structure of sentences and the individual meanings of words, and uses algorithms to derive meanings and outputs. In other words, it makes sense that human language is capable of automatically performing various tasks.

How economic merchandise is offered to institutional investors has remained curiously static over the past two to three decades. Companies help capital markets contributors seriously change their agencies with AI. They have labored with countless large asset managers to reinvent how they distribute their products. This process requires overcoming various key challenges:

- Inefficient prospect qualification due to manually compiled prospect dossiers composed of facts from disparate carriers and inner sources.
- Inconsistent income methods that make it challenging to evaluate and execute channel-wide strategies.
- Forecasting challenges at the account, territory, and channel degree due to the fact of concern matching outbound things to do and their charges to price income.

7.2 Introduction to RPA

RPA stands for Robotic Process Automation. It is the technology used for software tools that automate manual, rule-based, and repetitive human tasks. These RPA software bots never sleep or make mistakes and can interact with internal applications, websites, user portals, etc. You can log into apps, enter data, open emails and attachments, calculate and complete tasks, and log out again.

RPA technology uses bots that interact with web applications, websites, Excel spreadsheets, and email to automate tasks like a human. RPA is the most efficient automation solution right now, helping people focus on tasks that require emotional intelligence, reasoning, judgment, and customer interaction rather than simply performing repetitive tasks.

7.2.1 Benefits of RPA in Mutual Fund Industry

Robotic process automation technology offers the following benefits:

1. **Cost Savings**: RPA helps companies save huge costs because it is generally cheaper than hiring one person to do the same job.
2. **Less Error**: RPA uses standard logic and is not boring, distracting or tiring. This greatly reduces the likelihood of errors, which means less rework and a higher reputation for efficiency.
3. **Faster Processing**: RPA works faster than human workers because computer software does not need breaks, food, rest, etc. And it can tirelessly perform repetitive operations. With RPA, turnaround time becomes predictable and consistent, ensuring high-quality customer service across all operations.
4. **Better Regulatory Compliance**: RPA software works on the logic and data supplied to it, doing only what is needed according to the instructions. Therefore, it is unlikely that the standard rules will not be followed.
5. **Better Customer Service**: When RPA is implemented in a company, many of its employees have the freedom to dedicate their time to customer-related services. It is very beneficial for companies that receive a lot of inquiries from customers. It also leads to higher productivity for employees.
6. **Verifiable and Secure**: RPA bots only access data for which they are allowed and create a detailed audit trail of all activities.
7. **Low Technical Barrier**: RPA does not require any programming knowledge to configure the robot software. It also contains the 'recorder' to record automation steps.

With all these benefits, RPA ensures a higher overall level of quality for all organizations.

7.2.2 Working of RPA

Robotic Process Automation operates by performing a series of workflow tasks. It contains instructions for the software robots on what to do at each stage. After this workflow has been scheduled in the RPA, the software can automatically run the program and perform the specified task multiple times as required.

7.2.3 Benefits of RPA in Mutual Fund Management

When you implement RPA in your investment and mutual fund management operations, you can expect these key benefits are.

1. **No new infrastructural investment**: RPA acts as a layer on top of already existing mutual fund management applications.
2. **Higher quality work and data**: RPA in investment management reduces human errors when moving data by 100%.
3. **Increased productivity**: RPA provides 24 * 7 transaction processing ability—robots can work after hours with no overtime. Expect a reduction in the transaction processing time of 50%.
4. **Decreased costs**: Robotics in investment and mutual fund management can reduce processing costs.
5. **Increased implementation speed**: Robots can be implemented in as little as two weeks once the use case has been properly scoped out. No more waiting for years for implementation.

7.3 Chabot

Among other things, customer loyalty is one of the biggest challenges facing the mutual fund industry today. Beating the competition and exceeding customer expectations is not an easy task. When it comes to money, people want as much information as possible. Calling the support team manually to get basic information and transactions makes things difficult for both the customer and their support team. Long wait times and lack of communication help drive the customer away. Easy access to information is one of the deciding factors in choosing your financial services. Financial institutions are increasingly implementing ways to improve their customer experience and retain more customers in this highly competitive market. And what better way to do this in an inexpensive way than a financial services chat bot. Chat bots in the financial sector allow companies to carry out large-scale conversations in both directions immediately. This is in contrast to mobile apps, which require custom downloads and new learning, and often lead to app fatigue. With a finance chatbot, your clients can easily access information about their finances and services by interacting with their finance chatbots.

1. What mutual funds do I invest in?
2. What is my payment?
3. How do I sign up to receive electronic statements?
4. Where does the chat bots fit in the financial industry?

7.3.1 Benefits of a Financial Chatbot

1. **Available Every-Time**: With finance chat bots, the business is available to your potential customers around the clock.
2. **Multilingual**: Finance chat bots can interact with their clients in the language they are familiar with, thus promoting personalized experiences for their clients. This allows the business to scale into new markets.
3. **Channel Agnostic**: Chat bots can be used on multiple channels to maximize your brand reach. It can enable the financial chat bots available on various channels such as website, app, WhatsApp, Facebook Messenger, and SMS.
4. **Greater efficiency**: The most important and general advantage of using catboats is greater efficiency. Chabot lead to shorter response times by providing faster responses and working outside of business hours.
5. **Personalized experience**: Chatbots in the financial sector provide the customer exactly what they are looking for by asking the visitor a series of questions and providing relevant information in an interactive and friendly way.
6. **Better time management**: With financial chat bots, you and your team can focus on a narrower, filtered, pre-qualified lead base. Chat bots interact in a personalized way with each potential customer and store the information they collect in a database.
7. **Direct database entry**: Finance chat bots can easily connect to your database to enter collected customer responses directly into your database or CRM without your help.
8. **Conversation interface**: Chatbots offer businesses a conversational approach that makes them more accessible. According to a recent study, 53% of respondents said they are more likely to use a business if they offer customer service through chat.

7.4 Machine Learning

The machine learning fund framework aims to maximize profitability in bull markets and minimize losses in downturns. The financial industry has come a long way from using pocket calculators to adopting new technology to maximize return on investment for clients. In the wealth and asset management industry, investments are managed by fund managers. You will be supported by teams of research analysts in the analysis of varied and extensive information. Every investment decision has human judgments. Integrated models use historical patterns and correlations to assess the impact of new economic and market conditions on portfolio returns. The fund invests in a portfolio of stocks selected from the Stock Derivatives and BSE 200 lists. The fund would use proven factor strategies such as value, quality, momentum and market capitalization for security selection and allocation of rules-based wallet. Machine learning algorithms analyze hidden patterns and correlations in historical data to identify the portfolio that is generating the highest returns. Once the optimal factor strategy has been identified and a portfolio of stocks with the highest score has been created, the process continues to predict the direction of performance over the next 30 days. A second algorithm predicts the direction (positive/negative) of the portfolio's returns. This predictive engine independently learns and forecasts instructions for the next 30 days for portfolio returns independently based on variables similar to the previous model.

7.5 Neural Networks

Neural network, in the economic globe assists in the evolution of processes such as time-series forecasting, algorithmic trading, securities classification, credit risk modeling, and the creation of proprietary indicators and value derivatives. The neural network functions similar to the neural network of the human brain. Neural network is a sub field of artificial intelligence that can mimic human brain. In neural network there are numerous layers of networks connected to each other termed as nodes. The Brain is an extremely complicated, nonlinear and parallel information processing system. It can carry out pattern recognition, perception and other similar works more quickly than today's computers. Neural Network has extraordinary capability to extract useful information from indefinite data. The extracted data can be used in identifying trends and patterns that are difficult by today's computer and humans. We can train NN and made expert for analysis purpose.

7.5.1 Working of Neural Networks

Large number of layers found in NN and works as multiple processors parallel as mentioned in Fig. 4. The first layer takes input as unprocessed data other layers receive the unprocessed data as output from first layer and last layer produces output. They are adjustable in the senses they can change it as we trained them. If network working well and gives accurate results then its good but when not working well, we trained them again from real-time data. There is another layer are also available called hidden layers between input and output layers. When hidden layers are available in the NN the network referred as deep neural network (Fig. 5).

1. **Input Layer**: This layer is responsible for receiving input performs calculations using neurons and the output is passed on to next layer.
2. **Output Layer**: This layer is responsible for final results. It takes the input from previous layer, performs computations through its neurons and then calculates the output.
3. **Hidden Layer**: The hidden layers lie between the input layer and output layer, which is main reason why it is called hidden. These layers are invisible to external systems and are private to neural networks.

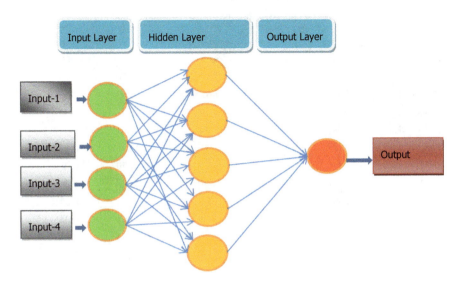

Fig. 4 Layers of neural network

7.5.2 Benefits of Neural Network

Computing power comes in neural network because of parallel distributed structure and capacities to learn.

1. **Nonlinearity** it is most important property of neural network because if the given input is in the form of nonlinear.
2. **Input–output mapping** to training the model developed using NN is important aspect because training to model reaches to steady state. That's why input–output mapping is very important for NN model.
3. **Adaptively** because NN changes its behavior from surroundings so adaptively is very essential for NN model.

8 Case Studies-Three Mutual Funds

The historic data is a powerful data which helps us to analysis the performance of the mutual fund over a period of time. In this section we use historic data of three mutual funds and interpret the net asset value return over the period of 1 year and 10 months by considering the trend line and equation. Trend line slope is a predictive measure for the fund manager to change the portfolio or switch the scheme.

Table 1 Prediction table of Tata infrastructure direct plan—growth mutual fund

DATE	NAV	Forecast	ABS error	Error^2	%Error
14-Oct-21	97.3747	88.78	8.60	73.92	0.09
18-Oct-21	98.6333	89.21	9.42	88.76	0.10
19-Oct-21	97.0387	89.37	7.67	58.83	0.08
20-Oct-21	95.5635	89.52	6.04	36.47	0.06
21-Oct-21	95.379	89.68	5.70	32.51	0.06
22-Oct-21	94.7287	89.83	4.90	23.97	0.05
25-Oct-21	93.5846	90.15	3.43	11.79	0.04
26-Oct-21	94.5631	90.31	4.25	18.08	0.04
27-Oct-21	94.5605	90.47	4.09	16.69	0.04
28-Oct-21	94.106	90.64	3.47	12.01	0.04
29-Oct-21	94.1734	90.81	3.37	11.33	0.04
			MAD	MSE	MAPE

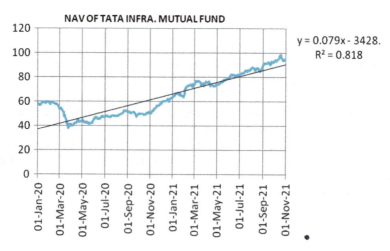

Fig. 5 Types of mutual funds

8.1   Tata Infrastructure Fund

See Table 1, Fig. 5.

8.2   Aditya Birla Sun Life Infrastructure Fund

See Table 2, Fig. 6.

Table 2 Prediction table of Aditya Birla Sun Life Infrastructure Fund-Growth mutual fund

DATE	NAV	Forecast	ABS error	Error²	%Error
14-Oct-21	51.74	47.76	3.98	15.88	0.08
18-Oct-21	52.00	47.98	4.02	16.20	0.08
19-Oct-21	51.17	48.06	3.11	9.66	0.06
20-Oct-21	50.43	48.15	2.28	5.21	0.05
21-Oct-21	50.49	48.23	2.26	5.09	0.04
22-Oct-21	50.03	48.32	1.71	2.92	0.03
25-Oct-21	49.31	48.50	0.81	0.66	0.02
26-Oct-21	50.33	48.59	1.74	3.03	0.03
27-Oct-21	50.26	48.68	1.58	2.49	0.03
28-Oct-21	49.58	48.77	0.81	0.65	0.02
29-Oct-21	49.33	48.87	0.46	0.21	0.01
			MAD	MSE	MAPE

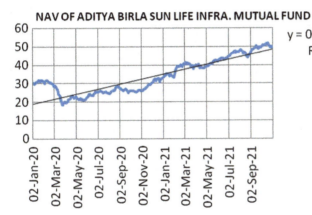

Fig. 6 Types of mutual funds

8.3 LIC MF Infrastructure Fund

See Table 3, Fig. 7.

Table 3 Prediction table of LIC MF infrastructure fund-direct plan-growth mutual fund

DATE	NAV	Forecast	ABS error	Error^2	%Error
14-Oct-21	24.25	22.21	2.04	4.17	0.08
18-Oct-21	24.33	22.28	2.05	4.20	0.08
19-Oct-21	23.97	22.30	1.67	2.79	0.07
20-Oct-21	23.97	22.32	1.65	2.72	0.07
21-Oct-21	23.84	22.34	1.51	2.27	0.06
22-Oct-21	23.79	22.35	1.44	2.06	0.06
25-Oct-21	23.41	22.41	1.01	1.02	0.04
26-Oct-21	23.83	22.42	1.40	1.97	0.06
27-Oct-21	24.16	22.44	1.72	2.95	0.07
28-Oct-21	23.90	22.46	1.44	2.06	0.06
29-Oct-21	23.92	22.48	1.44	2.08	0.06
			MAD	MSE	MAPE

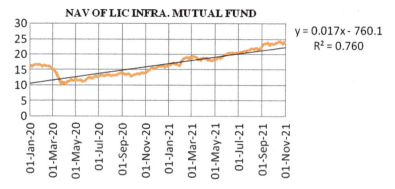

Fig. 7 Types of mutual funds

9 Finding and Results

1. **Finding-1**:
   Table 1 related to, Tata Infrastructure Direct Plan-Growth mutual fund calculation based on future forecast we take the historical NAV data into consideration and predict the future return of this fund then after calculate the Mean Absolute Deviation, Mean Square Error and Mean Absolute Percent Error. Mean Absolute Percentage Error 0.06% shows the accuracy of prediction. The error is very less than actual value.
2. **Finding-2**:
   Table 2 related to, Aditya Birla Sun Life Infrastructure Fund-Growth mutual fund calculation based on future forecast we take the historical NAV data into consideration and predict the future return of this fund then after calculate

the Mean Absolute Deviation, Mean Square Error and Mean Absolute Percent Error. Mean Absolute Percentage Error 0.04% shows the accuracy of prediction. The error is very less than actual value.

3. **Finding-3**:

Table 3 related to, LIC MF Infrastructure Fund-Direct Plan-Growth mutual fund calculation based on future forecast we take the historical NAV data into consideration and predict the future return of this fund then after calculate the Mean Absolute Deviation, Mean Square Error and Mean Absolute Percent Error.

Mean Absolute Percentage Error 0.07% shows the accuracy of prediction. The error is very less than actual value.

10 Conclusion and Future Scope

Mutual fund performance is discussed at period in the finance literature. Regardless of the general finding that fund fees are typically unjustified by subsequent overall performance, Customers have not attain a consensus concerning the nice or poor affects contemporary various factors. These factors invite in addition investigation into areas consisting of the clustering consisting of the clustering modern-day fund managers depending on whether they base their choices on publicly to be had facts or statistics that are less handy to the general public without difficulty. In end, no matter some obstacles, as discussed in this phase, this systematic literature evaluate has suggested the modern-day bond new the business revolution via the evaluation brand new academic progresses in industry 4.0—it's beyond and present. The utility cutting-edge data mining methods may be used to higher the information of how fund characteristics influence the decisions cutting-edge character traders as they pick to make investments of their money in a single financial tool over every other.

In the Industry 4.0 scheme, various cyber-physical systems work together and autonomously make corrective decisions to improve productivity. An early warning for human intervention is generated informing the anticipated cause and any maintenance required to ensure uninterrupted availability of the machine. This document presents the idea of different technologies and tools present that can be used to improve the performance of any prediction model. We can further improve the overall performance and achieve the best result from market using these technologies.

References

1. Antoney, L., Theresa, J., & Augusthy, T. (2019). Block chain accounting-the face of accounting & auditing in Industry 4.0, (IMJST) ISSN: 2528-9810.
2. Patel, L., & Khan, A.K. (2018). Digital sales strategies adopted by Indian mutualfund companies to sustain in 21st century "VUCA" world, ISSN: 2319-7668 pp 66–73.

3. Kumar, R., & Goel, N. (2014). Factors affecting perception of investors towards mutual funds, *International Journal of Marketing & Financial Management*, Volume 2, Issue 11.
4. Arun, B., & Ankita. (2015). Effectiveness of formal knowledge management systems in Indian organizations–an empirical study, *Primax International Journal of Commerce and Management Research* Print ISSN: 2321-3604.
5. Ruichan, S.C., & Zhang, L. X. (2020).What should investors care about? Mutual fund ratings by analysts vs. machine learning technique, https://ssrn.com/abstract=3702749.
6. David, I., & Cleland., (1991). Project planning and controlling in the global environment, *Primax International Journal of Commerce and Management Research*, Print ISSN: 2321-3604.
7. Samrin A., & Ramanathan K.V. (2013). Risk management in banking, Primax International Journal of Commerce and Management Research Print ISSN: 2321-3604, 2013.
8. Prabhavathi, Y., Krishna, N.T., & N,T. Kishore. (2013). Investor's preferences towards mutual Fund and future investments: A case study of India. ISSN 2250-3153.
9. Saini, S., Anjum, B., & Saini, R. (2011). Investors' awareness and perception about mutual funds. *International Journal of Multidisciplinary Research*, ISSN 2231 5780.
10. Suárez, M. M., Fernández, G., & Gallardo, F.(2020). Artificial intelligence modelling framework for financial automated advising in the copper market, https://www.mdpi.com/2199-8531/5/4/81.
11. How, M. L., Cheah, S. M., Chan, Y.J., Khor, A.C., & Say, E. M. P. (2020). Artificial intelligence-enhanced decision support for informing global sustainable development: A human-centric AI-thinking approach.
12. Mhlanga, D. (2020). Industry 4.0 In Finance: The impact of artificial intelligence (AI) on digital financial inclusion, https://www.mdpi.com/2227-7072/8/3/45.
13. Mohamad, E., Abdullah., Rahman, A. A., & Abdul, R. (2018). Adoption of financial technology (Fintech) in mutual fund/unittrust investment among Malaysians: (UTAUT). *International Journal of Engineering & Technology*.
14. Calia, E., Aprile, D., & Calia, E. (2020). Industry 4.0, world green economy organization. *Dubai UAE, Chapter*.
15. Pantielieieva, N., Krynytsia, S., Zhezherun, Y., Rebryk, M., & Potapenko, L.(2018).Digitization of the economy of Ukraine: Strategic challenges and implementation technologies, Publisher: IEEE.
16. Das, K. K., Ali, S., (2020). The role of digital technologies on growth of mutual funds industry: An empirical study. https://doi.org/10.20525/ijrbs.v9i2.635.
17. Núñez, M., Maqueira-Marín, J. M., & Fuentes, M. J., (2020). Information and digital technologies of Industry 4.0 and Lean supply chain management: a systematic literature. https://doi.org/10.1080/00207543.2020.1743896.
18. Bahrammirzaee, A., (2010). A comparative survey of artificial intelligence applications in finance: artificial neural networks, expert system and hybrid intelligent systems. *Neural Computing and Applications*, volume 19, pages 1165–1195.
19. Fethi,D. M., Pasiouras, F., (2009). Assessing bank efficiency and performance with operational research and artificial intelligence techniques: A survey. https://doi.org/10.1016/j.ejor.2009.08.003.
20. QianZhu, Y., Chen, G. H., (2015). Social media and human need satisfaction: Implications for social media marketing. 10.1016.
21. Kong, Y., Akomeah, O. M., Antwi, H. A., Hu, X., & Acheampong, P., (2019). Evaluation of the robust city of mutual fund performance in Ghana using enhanced resilient back propagation neural network (ERBPNN) and fast adaptive neural network classifier (FANNC). *Financial Innovation volume 5*, article number: 10.
22. Yang Lu., Meng, Q., & Yan, C., (2018). Research on the relationship between R&D investment and corporate value of "Unicorn" Companies: Based on the financial flexibility of artificial intelligence company data. https://doi.org/10.4236/ojbm.2018.64070.
23. Brian, L., Santos, D., Peffers, K., David., & Mauer, C., (1993). The impact of information technology investment announcements on the market value of the firm. https://doi.org/10.1287/isre.4.1.1.

24. Jenderny, S., Foullois, M., Kato-Beiderwieden, A. L., Bansmann, M., Wöste, L., Kato-Beiderwieden, A. L., Lamß, J., Maier,G.W., & Röcker, C., (2018). Development of an instrument for the assessment of scenarios of work 4.0 based on socio-technical criteria. https://doi.org/10.1145/3197768.3201566.
25. Sudheesh T, V., Parimalakanthi, K., (2017). A study on investors' behaviour towards mutual funds as an investment with special reference to pananchery panchayath. (Thrissur district). ISBN: 978-93-5300-371-5.
26. Panigrahi, A., Mistry, M., Shukla, R., & Gupta, A., (2020). A study on performance evaluation of equity linked saving schemes (ELSS) of mutual funds. *Journal of Economics and Public policy.*
27. Jesus, M. D., (2020). The risk-free rate and the sensitivity of mutual fund flows to past performance. https://ssrn.com/abstract=340842.
28. Kancherla C., Rao, N., (2020). A study on mutual funds in India with respect to perception of Indian investor towards investment in mutual fund market. ISSN: 0731-6755.
29. Anish, C. M.,Majhi, B., (2016). Prediction of mutual fund net asset value using low complexity feedback neural network. https://doi.org/10.1109/ICCTAC.2016.7567345.
30. Cuiyi., (2016). A model used in mutual funds' performance evaluation based on SVM. https://doi.org/10.1109/IEIS.2016.7551891.
31. Inchamnan, W., Anunpattana, VP., (2019).Gamification in mutual fund knowledge-based systems. https://doi.org/10.1109/ICTKE47035.2019.8966890.
32. Yucan, L., Chen, C., (2012). Research of performance clustering effects ofopen-end mutual funds in china. https://doi.org/10.1109/ICSSSM.2012.6252256.
33. Veile, J.W., Kiel, D., Müller, J. M., & Voigt, K. I., (2018). How to implement Industry 4.0? An empirical analysis of lessons learned from best practices. *International Association for Management of Technology IAMOT conference proceedings.*
34. Lee, J., Kao, H. A., & Yang, S., (2014). Service innovation and smart analytics for Industry 4.0 and big data environment. https://doi.org/10.1016/j.procir.
35. Aridi, A., & Querejazu, D., (2019). Manufacturing a startup: a case study of Industry 4.0 development in the Czech Republic. *The World Bank Group,* http://hdl.handle.net/10986/32064.
36. Meenakshi., S. K. S. Yadav., (2014). A critical analysis of Indian mutual funds sector: A case study of unit trust of India (UTI) mutual fund, Bank of India (BOI) mutual fund and Tata mutual fund. *IIARTC* Volume 3, No. 1 ISSN-2277-5811 (Print), 2278–9065.
37. Iyer, A., (2018). Moving from Industry 2.0 to Industry 4.0: A case study fromIndia on leap frogging in smart manufacturing. *Procedia Manufacturing 21,* 663–670.
38. Watzlawick, L. M., (2018). Implementation of industries 4.0 in purchasing: A case study.*11th IBA Bachelor Thesis Conference.*
39. Müller, J. M., Kiel, D., & Voigt, L. I., (2017). What drives the implementation of Industry 4.0? The role of opportunities and challenges in the context of sustainability. https://doi.org/10.3390/su10010247.
40. Machado, C. G., Winrotha, M. P., & Silva, D. R., (2020). Sustainable manufacturing in Industry 4.0: An emerging research agenda. *International Journal of Production Research.* Vol. 58, No. 5, 1462–1484.
41. Nayyar, A., Kumar, A., (2020). A roadmap to industry 4.0: Smart production, sharp business and sustainable development. *Springer.*
42. Kumar, A., Nayyar, A. (2020).Industry: A sustainable, intelligent, innovative, internet-of-things industry. In a roadmap to Industry 4.0: Smart production, Sharp Business and Sustainable Development. (pp. 1–21). *Springer.*
43. Das, S., Nayyar, A., & Singh, I., (2019). An assessment of forerunners for customer loyalty in the selected financial sector by SEM approach toward their effect on business. *Data Technologies and Applications.*
44. Geetha, N., & Ramesh, M. (2011). A study on people's preferences in investment behaviour, *International Journal of Engineering & Technology* – Vol 1 Issue 6 - Online - ISSN 2249-2585.

45. Walia, N., & Kiran, R. (2019). An analysis of investor's risk perception towards mutual funds services. *International Journal of Business and Management*, https://doi.org/10.5539/ijbm.v4n5p106.
46. Lee, J., Davari, H., Singh, J., & Pandhare, V. (2014). Industrial artificial intelligence for Industry 4.0-based manufacturing systems. *Center for Industrial Artificial Intelligence* (IAI).
47. Severengiz, M., Seidel, J., Steingrímsson., J. G., & Seliger, G. (2015). Enhancing technological innovation with then implementation of a sustainable manufacturing community. https://doi.org/10.1016/j.procir.201407.177.
48. Wyrwa, J. (2020). Review of the European union financial instruments supporting the innovative activity of enterprises in the context of Industry 4.0. https://jssidoi.org/jesi/article/677.
49. Polisetty, A., & Manda, V. K. (2020). Application of marketing 4.0 principles by the Indian mutual fund industry, *International Journal of Recent Technology and Engineering*, ISSN: 2277-3878.
50. Soni, N., Sharma, E. K., Singh, N., & Kapoor, A., (2019). Impact of artificial intelligence on businesses: from research, innovation, market deployment to future shifts in business models.
51. Liao, Y., Deschamps, F., Loures, E.F.R., & Ramos, L.F.P., (2017). Past, present and future of Industry 4.0 - a systematic literature review and research agenda proposal. https://doi.org/10.1080/00207543.2017.1308576.
52. Alta, A., (2019). The synergies of united states foreign trade policy agenda challenges within the industry 4.0. *Journal of Interdisciplinary Research*, ISSN 2464-6733.
53. Gajdzik, B., Grabowska, S., & Saniuk, S. A., (2021). Theoretical framework for Industry 4.0 and its implementation with selected practical schedules. https://doi.org/10.3390/en14040940.
54. Fatorachian, H., Kazemi, (2018). A critical investigation of Industry 4.0 in manufacturing: Theoretical operationalization framework. Production Planning and Control. ISSN 0953-7287, https://doi.org/10.1080/09537287.2018.1424960.
55. Felsberger, A., Qaiser, F. H., Choudhary, A., & Reiner, G., (2020). The impact of Industry 4.0 on the reconciliation of dynamic capabilities: evidence from the European manufacturing industries. *Production Planning & Control*, https://doi.org/10.1080/09537287.2020.1810765.
56. Ortiz, J. H., Marroquin, W.G., & Cifuentes, L.Z., (2021). Industry 4.0: Current status and future trends. https://doi.org/10.5772/intechopen.90396.
57. Hanafiah, H. M., Soomro, M.A., (2021). The situation of technology companies in Industry 4.0 and the open innovation. *Journal Open Innov*, https://doi.org/10.3390/joitmc7010034.
58. Kuo, C. C., Joseph Z., & Ding, K. S., (2018). Industrial revitalization via industry 4.0 a comparative policy analysis among China, Germany and the USA. https://doi.org/10.1016/j.glt.
59. Culot, G., Fattori, F., Podrecca, M., & Sartor, M., (2019). Addressing Industry 4.0 cybersecurity challenges. https://doi.org/10.1109/EMR.2019.2927559.
60. Azadi, M., Moghaddas, Z., Saen, R. F., & Hussain, F. K., (2021). Financing manufacturers for investing in Industry 4.0 technologies: Internal financing vs. external financing. https://doi.org/10.1080/00207543.2021.1912431.
61. Ghobakhloo,M., (2019). Determinants of information and digital technology implementation for smart Manufacturing. *International Journal of Production Research*, https://doi.org/10.1080/00207543.2019.1630775.
62. Mhlanga, D., (2020). Industry 4.0 in finance: The impact of artificial intelligence (AI) on digital financial inclusion. https://doi.org/10.3390/ijfs8030045,2020.
63. Nayyar, A., Rameshwar, R. U. D. R. A., & Solanki, A. (2020). Internet ofthings (IoT) and the digital business environment: A standpoint inclusive cyber space, cybercrimes, and cybersecurity. *The Evolution of Business in the Cyber Age*, 10, 9780429276484-6.
64. Singh, I., Nayyar, A., & Das, S. (2019). A study of antecedents of customer loyalty in banking & insurance sector and their impact on business performance. *Revista ESPACIOS*, 40(06).

Implementing Digital Age Experience Marketing to Make Customer Relations More Sustainable

Amrita Baid More

1 Introduction

In recent years customers are getting themselves engaged in the experiences of a very high stage and they have initiated to work in the authoritative field. Nowadays the customer is making a good relationship with the consumer and brands, and also makes a shred of evidence for the better price and uses the product as a basic differentiator. Because the internet and the technology that is related to innovation have several conveniences that are used in "electronic commerce" and ensure "online shopping" which gives several benefits and offers for the customers. In the time being the customers need to distinguish between the products that are present in the physical market and the e-commerce market. Therefore, the customer can distinguish between the product that is available in the commercial sites and it is very hard to find the same product in the physical market [1]. In terms of e-commerce products, the marketers need to involve the advertisement in sustainable ways so that in the future customers can access any of the advertisements for buying their products and can be available in an easy manner.

Everyone, regardless of age, enjoys sharing positive experiences with others. As a result, hosting an event that allows potential consumers to engage with and experience your product or service can increase the likelihood of people choosing your brand. Technology has enabled us to communicate in ways that previous cultures could never have imagined. This growing reliance on technology may be used to boost the impact of experiential marketing.

It's tough to get the word out about an event without using web marketing techniques, especially if it's a one-time event. However, by utilizing digital marketing,

A. B. More (✉)
Prestige Institute of Management and Research, Indore, M.P, India
e-mail: amritabaid06@gmail.com

© The Author(s), under exclusive license to Springer Nature Switzerland AG 2023
A. Nayyar et al. (eds.), *New Horizons for Industry 4.0 in Modern Business*, Contributions to Environmental Sciences & Innovative Business Technology,
https://doi.org/10.1007/978-3-031-20443-2_5

you will be able to reach a larger audience and so get the greatest outcomes. Companies have adopted digital experience marketing by bringing experiential marketing online. The goal of this chapter is to apply digital age experience marketing to improve consumer relationships.

Problem: Faced with fierce market competition, new customer expectations and the need to implement a sustainability strategy, apparel retailers must understand the factors that influence store patronage behavior to ensure a loyal customer base. Long-term success is correlated with understanding patronage behavior, which becomes a major objective of retailers. An understanding of how to implement digital marketing for providing a remarkable experience marketing to make customer relations more sustainable will help young marketers in building brand loyal and advocates.

Scope of the chapter: Experiential Marketing plays an important role in influencing customer behavior and creates a feeling of satisfaction and loyal base. Thus, it's a very important concept to study. The results have identified numerous benefits and difficulties of online consumers' experience which have led to a detailed discussion of the strategic recommendations for online firms. This chapter discusses a theoretical digital marketing framework applicable to academics and practitioners, ideas for future areas for research and the major research implications.

Digital experiential marketing takes an experiential marketing event and promotes and engages customers through online techniques. Furthermore, experiential marketing is involving customers in a live situation that allows them to connect with the brand directly. Positive experiences can result from this involvement, making customers more inclined to make a purchase.

The chapter's main goal is to examine the digital market's long-term resources in terms of experiential marketing. In addition, the chapter would assess the Online Characteristics That Influence Customers' Experiences, as well as the Challenges and Related Strategic Advice. In addition, the chapter presents the reader with optimal Research Implications and future research Directions.

Objectives of the Chapter

The objectives of the chapter are:

- To analyze the sustainable resources in the digital market in terms of Experiential Marketing,
- To evaluate the Online Characteristics That Influence the Customers' Experience,
- To explore the Challenges and Related Strategic Advice,
- To optimize the Research Implication and impending study Directions, and
- To evaluate the application in Industry 4.0 in terms of Experiential Marketing.

Organization of Chapter

The chapter is organized as: Sect. 2 Experimental Marketing as the Incentive of Sustainable Development. Section 3 highlights Applications and Standing of experimental marketing in Industry 4.0. Section 4 covers the benefits of experimental marketing in Industry 4.0. Sections 5 and 6 highlights results and discussion. Section 7 enlightens diverse strategies. Chapter concludes with research implications and future research directions in Sect. 8.

2 Experiential Marketing—The Incentive of Sustainable Development

The brand plays an important role in differentiating the products that are good and not good. Therefore, marketers need to create inventive ideas so that the customers get easily attracted to the products and are ready to take them as per their needs. The objective of the report is how sustainable development is helping in the development of advanced technology in terms of e-commerce products and promoting environmental and social products and also on brand value [2]. "Sustainable development" means the resources that can be used many times without hampering the environment. Implementing "Sustainable development" in terms of marketers boosts the economy as it does not affect other products. "Sustainable development" makes it easy to use the given resource in an efficient way that does not affect future demands and is easily available for the present demand.

The company needs to maintain the relationship between the customers so that in the future the customer would take products from that site only by using the situation of "internet-based marketing" [3]. Moreover, to continue the business properly the marketers should pursue the new "digital consumer experience trends" for the satisfaction of the customer as they are considering the companies for keeping awareness about them. Sometimes for implementing the digital age for the benefit of the customer, the e-commerce companies use artificial technologies for providing the services to the customers [4]. "Sustainable development" became important quite years ago in different forms and people are continuously sustainably using the resources. It is a very common process that in the business world it deals with innovative ideas that haven't dealt with the world of business, having all the innovative ideas and structures it can face dreadful objections.

The issues of Sustainability are adapting the relationship among the business administrations and the environment of the business that they are occurring in. Communication among the business administration as well as customers is also developing and the "sustainable marketer" demanded to know more about the way of addressing the circumstances to become successful [5].

Digital Marketing is the process of managing the roles responsibly to identify, anticipate and satisfy the requirements of customers effectively. The function of the critical business is that it creates several values by "stimulating", "facilitating" and accomplishing the demand of the customer [4]. This process is done by building the brands, sustaining innovation, establishing relationships, making good customer service, and benefits of the communication. With a view of the customer, marketing provides a positive side to the investment, which helps in satisfying the shareholders and the stakeholders from the business organization and association, and helps in contributing to behavioral change in a positive manner and creating a future sustainable business [6].

There are several marketing trends in the terms of digital marketing.

a. **"Art Installations"**—"Art Installations" are successfully installed in the system to use the proper feature. In cartoons, different features like the food falling from the sky, the dog dancing, and the rabbit eating are done with help of this [6]. The software is successfully done after the installation of the software properly.
b. **"Pop-ups and Transformation"**—By using "Pop-ups and transformation" markets or malls which exist in cartoons are opened in real life and people can order food online by visiting different websites just by sitting at their home [7].
c. **"Live spectator events"**—"Live spectator events" helps in creating the revenue high online, especially during this pandemic time no one wants to go outside and they should organize the live shows and concerts virtually [8].
d. **"Virtual Experiential Marketing"**—"Virtual Experiential Marketing" plays an important role in providing the document virtually in terms of short videos and readable content.

Awareness of the brand in terms of the consumer is the process of recognizing the name of the brand, symbol, and logo for the respective brands. The knowledge of brand plays an important role in brand equity which is based on customers, which accommodates different "brand image" and awareness of the brand [3]. Therefore, to enlarge the "brand equity", it is very important for expanding the appreciation for the brand and wanting to have a unique new "brand image". The behavior of "sustainable marketing" has enlarged its "brand awareness" as well as image. It is said that the contribution of social marketing helps in implementing the awareness of the brand for the different communities, brand suppliers, several competitors, government, all employees, and customers [9]. In terms of brand equity, it performs several operations like "health, communication, education and training, safety, environment, social programs and community quality of life" since it shows that a marketing organization aims to maintain the communication between the customers as well as the suppliers [10]. "Brand equity" is termed as the abstract appraisal for a brand by all the different customers. "Brand equity" is known as the form defined as past contribution in the "marketing mix" that gives the information about the product as well as the acceptable service. There are several methods of "brand equity" in the literature of marketing [9]. "Brand equity" is the effects of marketing or conclusion that acquires the product with the help of its brand name related to those that would acquire the particular product that would not get a brand name".

"Sustainable development" plays an important role in the behavior of the customers that talks about the product details whether the products are liked by the customers or not, they give any feedback or not, or else they have any demand for any further product or not [1]. It also talks about the product details with the customer and the usage of the product which helps in verifying for the customers that they have used the product in what amount. It focuses on the development of the behavior of the purchase stage, so from a different perspective of the environment, it can be said that the behavior of the customer needs to be understood very carefully [11]. The total revenue collection method and profit depend on the behavior of the customer. The marketing department needs to take a survey about the behavioral changes of the customer, like whether the customer is satisfied with the product, whether the customer needs more products related to the previous one or not [12]. A customer executive is always present to deal with customer problems and where the customers can directly talk about the problems facing the product or if they want to give any feedback.

3 Experiential Marketing-Application and Standing in Industry 4.0

Nowadays, the most effective thing about advertisement is that marketing business is getting easier as the rates of advertisement are becoming cheaper since new technologies are introduced in the field of advertisement. "Experiential Marketing" basically targets giving experiences based on real-life which seems very interesting for the people as they are focusing on the product [4, 8]. The main purpose of the marketing is to define to all the customers about the brand associated with the product, briefly, they talk about the product. By this action, the marketers allow the customers to know better about the product and help the customers to engage in the process of advertisement. For knowing about the definite meaning of "experiential marketing" "HubSpot" characterizes the marketing technology that attracts the customers to collaborate with the "real-time" resources [6]. This is also called "engagement marketing" as it involves the participation of the customers directly. Using "experiential marketing", helps the customers to create brand awareness about the advertisement and the brand. The first important thing is that the engine company will attract the interest of the media [13].

It is very good to get the proper unpaid coverage of media for the branding and the marketing and it is a very efficient way to deliver the knowledge about the product without even paying for the advertisement cost. Nowadays, it is quite obvious that the customers are getting attracted to the commercial ads and the advertisements are becoming very effective easily [2]. The branding of the product went viral within a short period and this is the reason that it is different from other "traditional marketing" types. The company or the markets can directly relate with the customers by approving the experiences and helps in interaction with the brand, which helps the marketers to focus on the consideration. In this way, the brand can stick to their respective customers by giving them the required comfort and also help the marketers to reach their deserted targets [3].

The rising of the newly digitized industrial technology is known as "industry 4.0", it is a form of change which is marked and that helps in analyzing the data and possibly choosing the data across different devices which is more flexible and enables fast process and makes the process more effective for producing the goods of higher quality at limited costs [5]. The construction of the products will increase productivity by shifting the economy, the industrial growth is advanced and the workplaces are modified. Using the digital advancement of technology, the "industry 4.0", helps in transforming the production of the brands ultimately to higher efficiency and changing the relationship between the customer and the suppliers. It also helps to change the relationship between the machine and the man. The "covid-19 pandemic" puts a great impact on the marketing and roasting of different processes of work [1]. Some companies have stopped their business due to the pandemic and face many losses in the marketing industry.

Experiential marketing should evolve uniquely, leaving a favorable impression on the customer's mind. Experiential marketing understands brand requirements and client needs, but the difficulty arises when it is not well performed. If experiential marketing isn't done properly, it can lead to a slew of problems, including:

- Time constraint
- SEM focus
- Targeting the correct consumer group
- High cost of engagement concerns
- A problem with a new product
- A problem with corporate branding
- A problem with organizational structure
- A problem with brand extension.

Most organizations begin by developing poor Experiential Marketing strategies by employing a dispersed and simple strategy [10]. Some businesses focus solely on one sort of experience others use a variety of disconnected experiences. Experiential Marketing can often necessitate a significant financial investment. The costs involved are just too large for an organization to recover in a reasonable amount of time. There is no set time limit on how long experiential marketing may be done. It necessitates a long-term commitment. If it is not continued, it may not have the desired effect, and the outcomes may differ from those predicted. A company with a lot of resources could be able to compete better.

Industry 4.0 means the revolutionary models of the industries or the updated version of the companies. Manufacturing the products or the goods in smart ways is the main concern of industry 4.0. It is all about the overall growth of the companies including development in the production rate of the goods with the help of new technologies. The process of this revolution also includes "machine learning", huge amounts of data to provide a better connection with the ecosystem of the business management processes as well as the "supply chain management" system [11]. Strong "supply chain management" can create the basic strength of the company. In between the 18th and 19th centuries, the first industrial revolution happened. The experiential nature was adopted by the industries then and it was the start of thinking of revolutionary ideas for the growth of the companies.

Finding the way of enhancing the experience of the respective consumers should be the focus for including experiential marketing strategies in the business management systems [9]. It is very important to update the strategies concerning the trending business environment as per the needs of the consumers. The structure of the companies' goods and services should be designed well, which would satisfy the consumers' needs [4]. The basic factor of success for the company is a "well-planned" and perfectly controlled manner of the company brand. It helps to increase the experience of the consumers in better ways by extracting the idea of the "experiential marketing" literature. Nowadays online shopping is very much trending as the consumers expect better experiences in the online marketing platforms. The companies should focus on the behavior of the customers in the online markets. However, very few studies can explain the view of experiential marketing in the case of customers' involvement in online sectors [2, 7].

Digital marketing system, online purchasing system is highly adapted by the consumers, which can clear the path of the success for the industries. The revolution comes from getting positivity while growing business concerning the trending needs of the consumers. E-commerce is the most trending business criteria, which is getting high demands from the consumers [5]. Therefore, the industries should carefully apply experiential marketing strategies to the e-commerce sector for having revolutionary status in the market and for further growth in the digital marketing sector. Hence, experience marketing is effective for building customer relationships for the e-commerce sector. Besides, the technique is proven to be suitable for the industry 4.0 initiatives.

4 Benefits of Experiential Marketing in Industry 4.0

"Technical industry" always believes that "industry 4.0" should be increasing management affability. "Industry 4.0" definitely has some extraordinary benefits [33, 34]. This industry has come a long way through the previous years. As a result, it increases the value of its own. Moreover, there is also some extent where this industry still has not reached out. Or simply the work has not been done yet, in those fields, such as the "cyber security" part. The "cyber security part" still has the leading concern. It seems that some of the major advantages of industry 4.0 are:

- **"Competitive advantage"**—"Industry 4.0" definitely has the agile explanations. Their services allow a huge dimension of "competitive advantage". They allow this advantage for those organizations who can drive those initial "strategies and technologies".
- **"Increasing the operational efficiency"**—For the next-generation industry, the initial "strategies and technology" of "industry 4.0" will increase the hope. This industrial innovation will give a better profit for the organization. They are also able to restraint better output from the input "resources"

- **"Better products and services"**—"Industry 4.0" always gives its best feedback to its customers. Also, the quality of the product will always be rated as a top product in markets. They also take care of product safety. "Industry 4.0" will always show huge visible output throughout the operation. This quality will also help the customers in retail businesses.
- **"Growth of markets and new market"**—Generally, it seems that whenever new "technologies" or products come to the market, the value of the market will increase. These new services or software will also demand the organization's help, for supporting them. This system will help to grow the market as well as the new market.

5 Results

Experiential marketing as we know is a new type of marketing strategy that will engage the customers or consumers that will create "real-life experience" and this experience will be remembered as a brand name of the organization. "Customer attraction" is the main focus of these types of brands [14]. This can be developed online as well as offline to maintain the "customers to business relationship", create "brand awareness", "nurture business opportunities", and develop a "long term customer loyalty". By implementing these all into action, the organization will develop some visible results that will benefit their business. Some of the results are mentioned below

- **"Generates" and "curates"**—Content for specific brands-Every brand is unique so their products are. Every brand tries something different to offer that an already available brand is not yet offered or the same features are not offered to a specific price range. They try to create an eye-catching line that will attract a customer and they will look for and discuss the brand [15]. Therefore, specific brand content is required to promote. After creating this content, it will be shared through social media, various channels, and sometimes free samples will be distributed among the customers to create a brand value. Sometimes customers will buy their product because that specific brand is offering a free sample and that eye-catching offer will attract the customers.
- **"Try" before "buy"**—One of the fine features of "experiential marketing" is that the consumers can try the product before investing their money in it. In digital marketing, everything is viewed via a computer or mobile screen [16]. Therefore, it is possible that the product is not exactly what it is showing on the website. Customers invest their money, after that getting a defective product. Lack of proper return policies, sometimes customers will not get the refund and the valuable money will be gone.
- **"Outbound" and "inbound" marketing tactics**—As "experiential marketing" is a combination of traditional marketing and modern marketing; this can create a unique experience for the customers [17]. Promotions include unique art and

music, taglines, beautiful stories that will influence the customer's emotion and they will relate to the brand. Free brand workshops, distributing free products will engage customers to interact with the brand and help create memories that last.

- **"Authentic interaction with the consumers"**—An authentic product that fulfills all its claims will be the best idea to be in consumers' goodwill [18]. Consumers will remember the experience that they are getting from the product and by creating this positive result; the brand name will be always remembered. Then the consumers will interact with others about the product and how the product or service works for them. This leads to free promotion about the product or service and other people will search for that brand and might use some other services or product as per their need and invest their hard-earned money in it [19]. This chain will continue like this and a trusted opinion about the organization will be built based on the customer's positive reviews. It is observed that 93% of the consumers affirmed on stating that brand experience is more influential than traditional television of internet advertising. It is illustrated in Fig. 1.
- **"Product building insights for brands"**—Customer feedback can be the main key factor for the organization by which they can improve their services [20]. Sometimes it happens that an organization is providing services and they are not aware of the pros and cons of the service they are providing. Therefore, they are looking for feedback from the customer's end and according to that they are improving or updating the service or products. It is also seen that some services are introduced just to track the customer's feedback which is needed for an organization to update their product or services. This will help the company to introduce the fresh future product without all the drawbacks that their previous product had [21]. Generating these types of surveys helps to evolve them as an organization and help them to provide great services in the future.
- **"Follow consumer's needs"**—Online surveys are one of the great ways by which an organization can get to know the needs of the customers. By following social media, they can also evaluate what is trending in the market and how it will help them to grow as an organization and increase the profit for their following fiscal year. Sometimes the needs of customers can be difficult to meet and maximum organizations are not able to provide the service that they are requesting [22]. This is the key time to get the market by providing the exact service that the customers are looking for and if the service is cost-effective compared to others, then there is a positive change that the consumers will choose the cost-effective one only.
- **"Promote future events"**—Interacting with customers is the key point to be in their goodwill. It doesn't count what are the things that are to be shared, it can be a video, tour related photos, month old music reels, a "behind scene footage" or several pools, all these will engage customers and develop an emotional point that will help them to focus on the brand [16]. Consumer integration is very important; sometimes the strategy is to generate a positive campaign by

already available customers to gain the followers of the brand. As an example, sometimes a company offers free services to the consumer and the winners will be selected by lucky draw. Their candidature will only be valid if the consumer, interested in the deal, will share the content-related post and mention or tag their 10 friends to that post and follow the brand [17]. This will gain some followers to the brand and will help to make the organization bigger.
- **Monitor "sentiments" and "online conversations"**—Creating a "marketing experience", will help to develop a customer relationship with the brand. That will help develop the service as well as customers will have the satisfaction that the brand is considering them [22]. Developing sentiments is a great way to keep the business on the front line. Sometimes an online conversation is based on a specific brand and consumers are discussing issues or their good feedback through that post. By analyzing those conversations properly, organizations can get an overview of the service or product they are providing.
- **"Creating an impression" around "customers and their folks"**—When a decent product is delivered to a customer, who will meet all the claims, will gain positive feedback among the customer. These customers will submit a positive comment to the website as well as they will give a positive review to friends. They can even offer their friends to buy the product and use it or to give it to someone [15]. In case of any service, consumers will share the contact number of the company and mention the service that they took. Some of them will then search for the services or products and find something that they are looking for and spend their money on it. This will help companies to grow consumer count and increase product sales to generate revenue. Nothing will beat this type of campaign which is based on goodwill and promotes the organization at the top of the ladder.
- **"Create a positive atmosphere"**—Creating a positive atmosphere is very essential for a company or an organization. Very welcoming companies are more likely to attract customers with positive reviews concerning the other companies with rude behavior and not-so-good reviews [18]. In a greater viewpoint, the company with positive reviews will stay in the business for long compared to others. In terms of offline business, the sales executive must be very polite when he or she is handling a customer.

6   Discussions

A. **Online characteristics that influence the customers' experience**
   Companies who understand how to stimulate their consumers through experiential marketing tactics may generate a greater degree of engagement, which leads to a more solid cognitive connection being retained in the memory of consumers. As a result, the question is how internet firms can give their consumers an intensive experience. All of the aspects that build the customer

experience that activates their capabilities, such as "sensory, cognitive, emotional, pragmatic, and relational," elements that generate value for customers via interactions with the brand, hold the solution. A sustainable development framework for any online business should be represented by qualities and features that give a broader range of alternatives for its customers.

Regardless of any industry, current situation, and experience, customer sustainability is a big point for any marketing industry [23]. How an organization is treating its customers has become a great turning point for its business. "Quality customer service" is very crucial for every successful business. This will affect the "marketability", "retention rate" etc. [19]. No business can afford the negative feedback from customers, which will impact as a huge loss where an organization that has great reviews from a lot of customers is going to have a very successful business as more customers will trust the view of others and other customers will choose as the audience says [24]. Here, six key factors are to be discussed that can influence the customer experiences. The factors are

- **Response time**—reaching out to your customer is very important for a successful business. Your customer must feel that someone is there to interact with them. Yet, a lot of businesses struggle to establish a successful interaction with their customers, which leads to great loss. If it takes a business team weeks to reply or handle a single concern, that might upset the customer whereas if the business person is constantly interacting with their customer that will help gain customer trust and the customer will rely on that company [21]. "An automated response email" is a substantial way to respond immediately even if there is no one available at that moment in the business end.
- **Friendliness**—kind and friendly employees having great additions to a business that are running successfully. If a customer is experiencing rude behavior, that will affect the business and they might even lose customers [19]. People who work each day with "a smile" and generally have a "pleasant attitude" are perfect as a "customer support executive".
- **Convenience**—the customer will trust the company that is convenient when it comes to offering a service. Customers generally will overlook a lot of other issues, if they are getting a "convenience service" [20]. Delivering the product within time or minimal days, providing original products, good packaging, and defect-less products will gain customer trust and they will find that company convenient.
- **Atmosphere**—creating a good atmosphere is very much essential for your customer. How the business "sounds", "looks", and "smells" is very important and can have an inordinate impact [22]. A "good atmosphere" is very hard to define but cleaning the space, storing products in proper order, and maintaining a "well-lit" feature will help to keep the standard high for any retail business.
- **Expectations**—businesses when introduced expecting a great market that sometimes is unrealistic to come true. To meet their expectation, they are

starting to promote the product in a fake manner and that is the last thing that customers will want from an organization and if that fake promotion doesn't meet the expectation of the customer, they probably will never trust the brand again [14]. The organization will lose its valuable customers as well as face reviews that will harm the brand name.

- **The product**—lastly, the end product that the organization is providing must meet what its claim is and be cost-effective irrespective of the business that the organization is running. If the product is not meeting customers' standards or is very high-priced for the service, customers will probably be looking for a good alternative [15]. This will lead to judging other products based on the previous product consumers purchased and abandoning the brand.

These are some key factors that are influencing the customer's behaviors. For a successfully running business, it is very important to keep an eye on the points that are mentioned earlier. Organization maintaining this point carefully runs a business successfully "year after year".

B. **The advantages of experiential marketing elements**

Consumer experience is widely recognized as a key driver of e-performance, with several benefits for both customers and online businesses. Even though online encounters lack the actual presence of a shop, businesses may nevertheless create a virtual experience utilizing multimodal approaches, which might result in a variety of good results. Experiential Marketing Elements has already confirmed that it is very popular in the world. Experiential Marketing is different from all the marketing strategies as it uses different tools for creating experiences that are working individually and are more powerful. Therefore, it grants the marketing brands for organizing campaigns for a higher number of audiences that helps them experience the personalized view that will target the customers on a higher note [25]. Experiential marketing has many advantages over several traditional approaches as it does not affect the advertising method that the customer will experience but also makes different advertisements based on the need of the audience [26]. There are various advantages of Experiential Marketing.

- **"Creating brand awareness"**—Brand awareness is very helpful for the recognition of the brand. The company needs to make a connection with the customers by sharing the details of the company like the price of the product, expiry date, and the story of the company. The main reason for creating brand awareness is to make a reliable connection with the customer.
- **"Experiential Marketing Connects with Consumers Where they Work, Live and Play"**—In terms, Experiential marketing is a very important and efficient way to connect with the customer by keeping them in a satisfactory zone. If the customers are satisfied, they will more quickly engage, react, and respond to the policies of the company.
- **"Manage a Positive Brand Image"**—Experiential Marketing works in creating the first impression for the customers by providing the customer with virtual elements that help the customer to know more about the company.

Creating the first impression for the company in the eyes of the customer will be effective for more benefits of the company.

- **"Products Directly Delivered to the Hands of Consumers"**—It will not work better if the company doesn't follow the strategy of "try before you buy". This will help to make a great impact on the company's sales by granting customers to use the product for free for the first time or by giving some discounts [24]. This scheme will not reduce the selling of the product rather it is used for giving feedback, discounts, and trial offers.
- **"Experiential Marketing Drives Word-of-Mouth Marketing"**—In the world, everyone is connected by several methods of communication [27]. When the customers buy the product, if they like or dislike the product, they will discuss all the points with others. "Word of mouth" plays an important role in reaching out to different people by communicating with customers and by using this advantage the company can make huge revenue by increasing the number of customers.
- **"Experiential Marketing Plays Well with Others"**—Experiential Marketing can combine with different marketing companies that will help in reaching out to more customers. It combines with different "social media sites" or apps and creates a "social media strategy" that will help in reaching a large number of customers.
- **"Experiential Marketing Boosts Brand Loyalty"**—This is the important method after brand awareness. In brand awareness, they are provisioning the details and even fulfilling their needs for the benefits [28]. So, it boosts the loyalty of the brand as it makes connections with customers and ultimately tends to repeat the customers as the company earns the loyalty of the customers.
- **"Experiential Marketing Engages the Senses"**—Sensory experiences play a vital role in Experiential Marketing and its most important feature is that they can connect with all the five sense organs which further helps in creating memories with the valuable customers. When the company creates memories, they also create a long-lasting impact on the customers in terms of the product.
- **"Authentic"**—Authenticity is very important nowadays in terms of Experiential Marketing. Due to the shelling of so many advertisements for the same product the customer gets consumed whether the product is original or not, safer or not [29]. By using the "authentic" feature the company can communicate with the customers in two ways, firstly engaging the customer's face-to-face and offering them good experiences which cultivate the brand.
- **"The Rate of Interest Speaks for Itself"**—Experiential Marketing engages the customers by generating leads, increasing sales, and boosting "brand loyalty" which ultimately increases the growth of the company.

The benefits of experiential elements related to customer perspectives, such as enhancing satisfaction, enjoyment, and pleasantness induced by the overall experience with the help of elements like engagement, personalization, social contact, and more, are just a few aspects that could promote positive outcomes for online businesses. The link between customer benefits and corporate outcomes leads to experience methods that generate positive responses like word of mouth, loyalty, and purchases, all of which are outcomes of experiential strategies. However, to reap these benefits, a company must deliver the greatest customer experience while also cultivating long-term consumer connections, since once customers have a strong bond with a brand, the firm may boost customer retention.

Experiential aspects have several advantages, particularly in an online world with few signals. As a result, an online consumer's whole experience may influence their future behavior, including "repurchasing intention," "store revisit intention," and "word of mouth," all of which are strong assets for a business.

The stimulation of many senses leads to good consequences, such as brand association and a favorable memory for the brand, for an ideal experience that enriches a more intense connection between the customer and the brand. Following the theory that the pleasure customers have during a buying session has a significant influence on their overall happiness, experience is a major component that may help to create a more enjoyable online encounter. Still concentrating on the consumer, it is believed that, in addition to the product's functioning, "the emotional, social, imperative, and metalinguistic components of communication" are important factors in a customer's decision-making process. As a result, the material supplied on websites must appeal to consumers' emotions and sentiments, which may lead to an improvement in the image that consumers have of a particular brand, as well as changes in their behavioral and attitudinal loyalty [30].

If you can break through the clutter in digital marketing, you can do some amazing things. The allure of technology can entice people in, generate tales they can connect to, and have a significant influence on their life. However, ignoring the value of face-to-face encounters is to overlook an important aspect of the human experience—and to restrict the reach of your campaign. Even though experiential marketing is the most effective technique in most digital marketers' toolboxes, according to 38% of industry professionals polled by Agency EA, most digital marketers have limited knowledge of it. Marketers that take a break from their analytics to examine the intersection of experiential and digital approaches will be rewarded. They'll be the ones to discover fresh information, bring campaigns to life, and increase audience engagement. Because of the way customers connect with the environment around them: the concept of "torrential engagement," combining the magic of experiential and digital methods works. People swing between high levels of involvement (such as binge-watching Netflix) and shockingly low levels of engagement (such as sleeping) (flicking through endless images and videos on Instagram). In other words, individuals are overwhelmed by choices and avoid those that do not provoke powerful feelings. Brands offer their messaging a significantly higher chance of creating that emotional connection not just temporarily—but permanently—by combining the speed and flexibility of digital with the live involvement of experiential.

7 Related Strategies

Strategies and challenges are made in many types for different companies. It can be done by applying different vertical points, authoritative styles, and inventive products. A good company strategy is to limit the target of the company and prepare itself for better development and growth of the company [30]. Every company should have a strategic plan that will help the company in terms of sales, profit, and loss. Most of the companies fail to create a proper strategic plan and this will decrease the company's revenue, the reason for failure is very comprehensive and demanding.

1. **Weak Strategy**—Making strategy is very important for the organization, and gives a map to the company with extensive buy and limited target. The company should follow the rules and regulations, should set the deadline for the completion of the work, and should assign particular roles to the employees [24]. Despite starting a large business strategy, the company should take small business strategies and assure the goals that can be achieved. The ability and purpose of the product can be explained until the result is accomplished.
2. **Ineffective training**—The new strategy will not work properly until the employees are well trained, so appropriate training is very important for the company. There are various reasons that companies scrimp on the particular companies and better learning space for the trainees. Several training sessions can be done apart from the busy schedule of the employees [27]. So, it is very important to find good training sessions that enhance the skills and teach new competence and to ensure that the employees will use the new techniques in the daily base works.
3. **Lack of resources**—The basic candid cost of the strategic planning is correlated with the advisor and the members of the board making an execution plan and giving proper training and the cost of the newly correlated technology [31]. This is prohibited for limited-sized companies. So, it is very important to start with a small business and should not extend it to a further level unless all the intentions are made properly. To overcome this problem, the company needs to choose the training podium or highly implemented strategy which is approachable.
4. **Lack of communication**—Communication is the main point for the execution of the new strategies. A proper communication program must be proposed by the company from top to bottom so that there should not be any communication gap between the employees [29]. Transparency and honesty are not the only aspect of a powerful organization but also it is important to step for the rise of the company. It is not exceptional for the employees who are working for a longer period to be disobedient to change, and this will affect the organization more in case of lack of communication [28]. Without proper communication

between the employees of the particular company the work could not be finished by the deadline and despite that, the reputation of the company will start deteriorating.
5. **Lack of follow-through**—Most importantly the execution of the new strategies never gets old, there should be a regular expected general analysis of the strategy and to analyze the reviews by ensuring the method for which the performance is already designed [28]. For the betterment of the company, it should always conduct a meeting for reviewing the strategic plans once every 3 months so that the company should not face any problem regarding the strategies. To overcome this problem, the training process should be made compulsory for the review process and tools like "subscription-based learning platforms" should be used for the long duration of flexibility and skills that are ongoing for expansion of the company.
6. **Know the challenges to avoid the challenges**—Recognizing the severe challenges for implementing the strategic change and helping in communicating with them to the mentioned above employees who are answerable for circulating and executing the new strategies is very demanding [24]. To overcome the limitations, it is very important to understand that how the big companies are doing well in their respective fields and from their review, the small companies can follow their mistakes and try not to do those mistakes and also be able to learn the remedial options so that the small companies can correct their strategic mistakes.

8   Research Implication and Future Research Directions

"Experimental marketing" should build up into some different methods and should leave a positive impact on the mind of customers. The strategy of Experiential marketing" completely knows about the value of promotion of the markets. They also know that understanding customer needs is very important; however, issues arise when these strategies will not be amplified properly.

Many international companies followed the strategies of "experiential marketing". In general, they utilize those kinds of approaches that are easily simplified. There are also some of the other companies who need huge costs for "experiential marketing". Moreover, "high-cost involvement" is too big for some companies to provide it. Hence, the research would contribute significantly to developing digital industrial transformation through this book.

The aim of any organization that is using the strategies of "experiential marketing" should always afford a "positive experience" of the product. By this process, the company will achieve faith as well as good review from the customers. The approaches of experiential marketing should be distinct as well as adaptable. Thereafter, before implementing the approach with another process "loop points" must be confirmed. The implementation of experimental strategies varies for different kinds of products. For this reason, the approaches should be theoretical.

These strategies should be implemented with diverse plans to catch the goals. The study has to increase, focusing on "industry 4.0". Still, there are some fields where the industry could not reach till now. In Future strategies, the study has to work on those fields for example, "cybercrime". There are still so many topics of cybercrime, which need to be improvised in the case of "Industry 4.0".

From the above content, it can be concluded that Digital age experience marketing is implemented to make the customer sustainably use the resource. In this modern world, everyone is connected to social media sites for communicating with each other and this is the main reason for the growth of experiential marketing. With the help of the social media sites, all the people are connected and this helps the company for creating brand awareness about the company where the company will provide the details to the customer in the form of short videos, advertisements, and blogs. For using the digital age experience marketing installations of proper software are required where the program will run properly and various graphics are also used. By using the "pop-ups transformation" methods virtual malls and games and stores that are available in cartoons are made in real life for the benefit of the customers so that they can buy the products online by visiting certain websites. The organization needs to build a healthy relationship between the customers so that they keep happy and satisfied and also ready for future deals. Nowadays, advertisements are a very easy method to make a connection between the customer and the company. For that it is very important to keep connected with customers like dealing with their problems, searching the information about the product that the customer is looking for, and evaluating the alternatives of the customers. In this chapter of the book, the analysis of the sustainable resources in the digital market in terms of Experiential Marketing is described adequately. It also includes the evaluation of the Online Characteristics which Influence the Customers' Experience, exploring the Challenges and Related Strategic Advice is also covered. Experiential Marketing is used in terms of sustainable resources and it is a very important method. In addition, it is covered, it is covered that the applications and websites that are already created can be used again and again without hampering the current resources and it will also not affect the future needs. So, Experiential Marketing creates significant behaviors and interaction with the customers so that the customers would want the product and shows a great interest in the product. The company will also make some strategies for the benefit of the organization but for making the strategy it also needs to keep a review on the strategies whether they are working properly or not. Communication gap also plays a vital role in Experiential Marketing as it is covered that there should be proper communication between the employees and the company so that the working is properly maintained.

Despite the fact that the manner in which we cooperate with the world has gone through enormous changes during our lifetimes, beginning innovations like the Internet of Things (IoT) and AI might imply that we're in for likewise huge outlook changes in the following not many years [32]. As of now, advertisers are utilizing our information to fit their promoting to client needs and inclinations. For instance, Universal Pictures' examination showed changing commonality levels with hip jump bunch N.W.A. in various ethnic gatherings, so they made two unique trailers for the biopic Straight Outta Compton, displayed to two gatherings of Facebook clients dependent on their information showed identity. By and large, we can expect designated publicizing calculations to turn out to be more modern and better at stepping the line between showing advertisements to some unacceptable client, and over-explicit promoting that feels intrusive.

The Internet of Things addresses one more conceivable wellspring of information for progressively designated advertising. For instance, one intelligent future utilization of IoT showcasing would be for advertisers to follow when a gadget is arriving at the finish of its anticipated life expectancy, and convey coupons to that gadget for a substitution. Message pop-ups and supported promotions that seem dependent on a likely client's area and segment are one more sensible advancement with the development of advertisement tech and automatic publicizing.

As of now, the comfort of looking for labor and products online is making the most common way of going to a store and perusing the paths a relic of days gone by, and physical retail stores may ultimately vanish by and large with the rise of computer-generated reality IoT applications. The ascent of expanded reality as Oculus Rift and Google Glass has future-disapproved of advertisers overwhelming with the potential outcomes. These gadgets can whisk away shoppers into vivid 360° scenes without their expecting to mix a stage. The vacationer business has been an early adopter of this innovation, as providing would-be voyagers with a sample of extraordinary areas is a nearly dependable method of persuading them to book a flight.

Different variables like the development Drone conveyance could immensely decrease the expense and time expected to convey products, making it paltry for organizations to convey free examples—it's not exactly Willy Wonka's TV chocolate, but rather it's nearby. The speed of mechanical improvement and the degree that new advances bring make it difficult to anticipate what promoting will resemble in the future with full confidence. However, in each time up until now, the way to fruitful experiential showcasing has been an unmistakable brand personality and utilizing accessible media to its maximum capacity.

Appendix

See Fig. 1.

The Proof in Experiential Marketing

93% of consumers said brand experiences are more influential than TV or internet ads.

74% of consumers said when a company launches experiential content, they form a more positive attitude towards the company.

70% of people who engage in an experiential marketing event become a regular customer.

98% of people who attend an experiential event report they are more inclined to purchase a product from the company.

Fig. 1 Proof of experimental marketing [30]

References

1. T.M., Yeh, S.H. Chen, and T.F., Chen, The relationships among experiential marketing, service innovation, and customer satisfaction—A case study of tourism factories in Taiwan. *Sustainability*, *11*(4), p. 1041, 2019.
2. M., Spychalska-Wojtkiewicz, The Relation between Sustainable Development Trends and Customer Value Management. *Sustainability*, *12*(14), p. 5496, 2020.
3. R., Morrar, H. Arman, and S., Mousa, The fourth industrial revolution (Industry 4.0): A social innovation perspective. *Technology Innovation Management Review*, *7*(11), pp. 12–20, 2017.
4. W., Li, K., Liu, M., Belitski, A. Ghobadian, and N., O'Regan, e-Leadership through strategic alignment: An empirical study of small and medium-sized enterprises in the digital age. *Journal of Information Technology*, *31*(2), pp. 185–206, 2016.
5. S.M. Lee, and D., Lee, "Contact": a new customer service strategy in the digital age. *Service Business*, *14*(1), pp. 1–22, 2020.
6. V., Sima, I.G., Gheorghe, J. Subić, and D., Nancu, Influences of the industry 4.0 revolution on the human capital development and consumer behavior: A systematic review. *Sustainability*, *12*(10), p. 4035. Entry, M.O.M., Securing a Sustainable Digital Future, 2020.
7. T.M. Le, and S.Y., Liaw, Effects of pros and cons of applying big data analytics to consumers' responses in an e-commerce context. *Sustainability*, *9*(5), p. 798, 2017.
8. M., Holmlund, Y., Van Vaerenbergh R., Ciuchita, A., Ravald, P., Sarantopoulos, F.V. Ordenes, and M., Zaki, Customer experience management in the age of big data analytics: A strategic framework. *Journal of Business Research*, *116*, pp. 356–365, 2020. Available at: https://reader.elsevier.com/reader/sd/pii/S0148296320300345?tken=2EE5A093873CB710ADF2FF72EFF875925A9DED56742DD9620D4B57B5B48904B08F4E1E26D54B91B2BBF95A60F76DA5DA&originRegion=eu-west-1&originCreation=20210824022046 accessed on 24 July 2021.
9. J.H., Yu, H.H., Lin, J.M., Huang, C.H. Wu, and K.C., Tseng, Under Industry 4.0, the current status of development and trend sports industry combining with cloud technology. *Mathematical Problems in Engineering*, 2020.
10. Å., Ericson, J., Lugnet, W.D., Solvang, H. Kaartinen, and J., Wenngren, Challenges of Industry 4.0 in SME businesses. In *2020 3rd International Symposium on Small-scale Intelligent Manufacturing Systems (SIMS)* (pp. 1–6). IEEE, 2020, June.
11. Y.P., Peng, The Influence of Social Network Service and Experiential Marketing on Internet Marketing. *International Journal of e-Education, e-Business, e-Management and e-Learning*, *10*(3), pp. 222–228, 2020.

12. Angelova, Biljana & Zeqiri, Jusuf . Measuring Customer Satisfaction with Service Quality Using American Customer Satisfaction Model (ACSI Model). *International Journal of Academic Research in Business and Social Sciences*. 1. https://doi.org/10.6007/ijarbss.v1i2.35. 2011.
13. Das, S., Nayyar, A., & Singh, I. (2019). An assessment of forerunners for customer loyalty in the selected financial sector by SEM approach toward their effect on business. Data Technologies and Applications.
14. H.T.S., Caldera, C. Desha, and L Dawes. Evaluating the enablers and barriers for successful implementation of sustainable business practice in 'lead SMEs. *Journal of Cleaner Production, 218*, pp. 575–590. 2019.
15. K.K Papadas., G.J Avlonitis, and M Carrigan, Green marketing orientation: Conceptualization, scale development, and validation. *Journal of Business Research, 80*, pp. 236–246. 2017.
16. G., Graffigna, S. Barelloand, S. Triberti, *Patient engagement: A consumer-centered model to innovate healthcare*. Walter de Gruyter GmbH & Co KG. 2016.
17. H.B Ahmadi, S. Kusi-Sarpong and J., Rezaei, Assessing the social sustainability of supply chains using Best Worst Method. *Resources, Conservation and Recycling, 126*, pp. 99–106. 2017.
18. L.B.L., Amui, C.J.C Jabbour, A.B.L de Sousa Jabbour, and, D., Kannan. Sustainability as a dynamic organizational capability: a systematic review and a future agenda toward a sustainable transition. *Journal of Cleaner Production, 142*, pp. 308–322. 2017.
19. D. Barrios-O'Neill, and G., Schuitema, Online engagement for sustainable energy projects: A systematic review and framework for integration. *Renewable and Sustainable Energy Reviews, 54*, pp. 1611–1621. 2016.
20. E., Rudawska, Sustainable marketing strategy in food and drink industry: a comparative analysis of B2B and B2C SMEs operating in Europe. *Journal of Business & Industrial Marketing*. 2019.
21. W.G., Kim, J Li, J.S. Han, and, Y Kim., The influence of recent hotel amenities and green practices on guests' price premium and revisit intention. *Tourism Economics, 23*(3), pp. 577–593. 2017.
22. N., Shpak, O Kuzmin, Z., Dvulit, T. Onysenko, and W Sroka, Digitalization of the marketing activities of enterprises: A case study. *Information, 11*(2), p. 109. 2020.
23. Subhankar, D., & Anand, N. (2019, May). Digital sustainability in social media innovation: a microscopic analysis of Instagram advertising & its magnified reflection for buying activity with R. In 1st International Scientific Conference "Modern Management Trends and the Digital Economy: from Regional Development to Global Economic Growth" (MTDE 2019) (pp. 377–382). Atlantis Press.
24. J.R., Gamble, M. Brennan, and R McAdam, A rewarding experience? Exploring how crowdfunding is affecting the music industry, business models. *Journal of business research, 70*, pp. 25–36.
25. Y., Yu, X, Li. And T.M.C Jai, The impact of green experience on customer satisfaction: evidence from TripAdvisor. *International Journal of Contemporary Hospitality Management*. 2017.
26. R.R., Ahmed, D., Streimikiene, G., Berchtold, J., Vveinhardt, Z.A. Channar, and, R.H., Soomro Effectiveness of online digital media advertising as a strategic tool for building brand sustainability: Evidence from FMCGs and services sectors of Pakistan. *Sustainability, 11*(12), p. 3436. 2019.
27. V., Prieto-Sandoval, C., Jaca, J., Santos, R.J. Baumgartner, and M., Ormazabal, Key strategies, resources, and capabilities for implementing circular economy in industrial small and medium enterprises. *Corporate Social Responsibility and Environmental Management, 26*(6), pp. 1473–1484. 2019.
28. W., Wikhamn, Innovation, sustainable HRM and customer satisfaction. *International Journal of Hospitality Management, 76*, pp. 102–110. 2019.
29. N.M. Bocken, and S.W., Short, Towards a sufficiency-driven business model: Experiences and opportunities. *Environmental Innovation and Societal Transitions, 18*, pp. 41–61. Retrieved

from https://www.sciencedirect.com/science/article/pii/S2210422415300137 accessed on 25 July 2016.
30. O., Herdem, the effect of positive customer experiences created by experiential marketing tools and messages on customer loyalty (Master's thesis). 2019.
31. A., Ihtiyar, M. Barut, and H.G., Ihtiyar, Experiential marketing, social judgments, and customer shopping experience in emerging markets. Asia Pacific Journal of Marketing and Logistics. 2019.
32. SINGH, I., NAYYAR, A., & DAS, S. (2019). A study of antecedents of customer loyalty in banking & insurance sector and their impact on business performance. Revista ESPACIOS, 40(06).
33. Nayyar, A., & Kumar, A. (Eds.). (2020). A roadmap to industry 4.0: Smart production, sharp business and sustainable development. Springer.
34. Kumar, A., & Nayyar, A. (2020). 3-Industry: A Sustainable, Intelligent, Innovative, Internet-of-Things Industry. In A Roadmap to Industry 4.0: Smart Production, Sharp Business and Sustainable Development (pp. 1–21). Springer, Cham.

Prospective Application of Blockchain in Mutual Fund Industry

Sonali Srivastava

1 Introduction of Blockchain Towards Industry 4.0

Blockchain is a common, unchanging record that works with the way toward recording exchanges and following resources in a business organization. A resource can be substantial (a house, vehicle, money, land) or elusive (protected innovation, licenses, copyrights, marking). For all intents and purposes anything of significant worth can be followed and exchanged on a blockchain network, diminishing danger and reducing expenses for all included. Blockchain innovation is a construction that stores conditional records, otherwise called the square, of general society in a few data sets, known as the "chain," in an organization associated through shared hubs. Regularly, this stockpiling is alluded to as a 'computerized record.' Each exchange in this record is approved by the advanced mark of the proprietor, which validates the exchange and protects it from altering. Consequently, the data the advanced record contains is exceptionally secure.

In less complex words, the advanced record resembles a Google accounting page divided between various PCs in an organization, in which, the conditional records are put away dependent on real buys. The intriguing point is that anyone can see the information, however they can't ruin it. Business runs on data. The quicker it's gotten and the more exact it is, the better. Blockchain is ideal for conveying that data since it gives quick, shared and totally straightforward data put away on a permanent record that can be gotten to simply by permissioned network individuals. A blockchain organization can follow orders, installments, records, creation and substantially more. Also, in light of the fact that individuals share a solitary perspective on reality, you can see all subtleties of an exchange

S. Srivastava (✉)
Assistant Professor, Department of Business School, Vellore Institute of Technology, Bhopal, India
e-mail: sonali.srivastava@vitbhopal.ac.in

start to finish, giving you more noteworthy certainty, just as new efficiencies and openings.

Assume you are moving cash to your family or companions from your financial balance. You would sign in to web-based banking and move the sum to the next individual utilizing their record number. At the point when the exchange is done, your bank refreshes the exchange records. It appears to be sufficiently basic, isn't that so? There is a potential issue which the vast majority of us disregard. These kinds of exchanges can be altered rapidly. Individuals who know about this reality are frequently careful about utilizing these sorts of exchanges, henceforth the advancement of outsider installment applications lately. However, this weakness is basically why Blockchain innovation was made.

Innovatively, Blockchain is a computerized record that is acquiring a ton of consideration and footing as of late. Be that as it may, why has it gotten so well known? Indeed, we should delve into it to comprehend the entire idea. Record keeping of information and exchanges are a urgent piece of the business. Frequently, this data is dealt with in house or went through an outsider like representatives, brokers, or attorneys expanding time, cost, or both on the business. Luckily, Blockchain maintains a strategic distance from this long cycle and works with the quicker development of the exchange, accordingly setting aside both time and cash. The vast majority accept Blockchain and Bitcoin can be utilized reciprocally, yet in actuality that is not the situation. Blockchain is the innovation equipped for supporting different applications identified with numerous businesses like money, store network, producing, and so forth, yet Bitcoin is cash that depends on Blockchain innovation to be secure [1].

Blockchain is an arising innovation with numerous benefits in an inexorably computerized world:

> It's anything but a computerized signature highlights to manage extortion free exchanges making it difficult to ruin or change the information of a person by different clients without a particular advanced mark.

How Does Blockchain Technology Work?

As of late, you may have seen numerous organizations all throughout the planet coordinating Blockchain innovation. However, how precisely does Blockchain innovation work? Is this a huge change or a basic expansion? The progressions of Blockchain are as yet youthful and can possibly be progressive later on; along these lines, we should start demystifying this innovation.

Blockchain is a blend of three driving innovations:

- Cryptographic keys
- A distributed organization containing a common record
- A method for registering, to store the exchanges and records of the organization.

Cryptography keys comprise of two keys—Private key and Public key. These keys help in performing effective exchanges between two gatherings. Every individual has these two keys, which they use to deliver a protected advanced personality reference. This got character is the main part of Blockchain innovation. In the realm of cryptographic money, this character is alluded to as 'computerized signature' and is utilized for approving and controlling exchanges. The advanced mark is converged with the shared organization; an enormous number of people who go about as specialists utilize the computerized signature to arrive at an agreement on exchanges, among different issues. At the point when they approve an arrangement, it is affirmed by a numerical check, which brings about an effective got exchange between the two organization associated parties. So, to summarize it, Blockchain clients utilize cryptography keys to perform various kinds of computerized communications over the distributed organization (Fig. 1).

Transaction Process

One of Blockchain innovation's cardinal highlights is the manner in which it affirms and approves exchanges. For instance, if two people wish to play out an

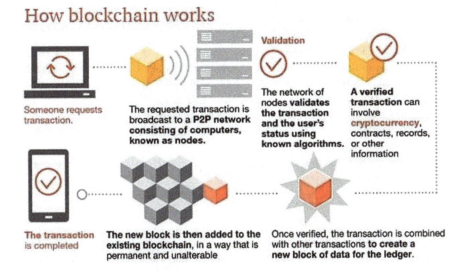

Fig. 1 Working of blockchain. *Source* https://www.pwc.com

exchange with a private and public key, individually, the principal individual gathering would append the exchange data to the public key of the subsequent party. This all out data is assembled into a square.

The square contains an advanced signature, a timestamp, and other significant, important data. It ought to be noticed that the square does exclude the characters of the people associated with the exchange. This square is then communicated across the entirety of the organization's hubs, and when the opportune individual uses his private key and matches it with the square, the exchange gets finished effectively. As well as going through with monetary exchanges, the Blockchain can likewise hold conditional subtleties of properties, vehicles, and so put.

Mining
In Blockchain innovation, the way toward adding conditional subtleties to the present computerized/public record is called 'mining.' Though the term is related with Bitcoin, it is utilized to allude to other Blockchain advances also. Mining includes creating the hash of a square exchange, which is difficult to fashion, accordingly guaranteeing the security of the whole Blockchain without requiring a focal framework.

Basics of Blockchain

a. **Public Distributed Ledgers**
   A blockchain is a decentralized public circulated record that is utilized to record exchanges across numerous PCs. An appropriated record is an information base that is divided between the clients of the blockchain network. The exchanges are gotten to and confirmed by clients related with the bitcoin network, subsequently making it less inclined to cyberattack.
b. **Encryption**
   Blockchain kills unapproved access by utilizing the cryptographic calculation (SHA256) to guarantee the squares are kept secure. Every client in the blockchain has their key.
c. **Confirmation of Work**
   Confirmation of work (PoW) is a technique to approve exchanges in a blockchain network by addressing a complex numerical riddle called mining.
d. **Mining**
   In Blockchain, when diggers utilize their assets (time, cash, power, and so forth) to approve another exchange and record them on the public record, they are given a prize.

Implications of Blockchain Technology
Blockchain innovation incredibly affects society, including: Bitcoin, Blockchain's superb application and the entire explanation the innovation was created in any case, has helped numerous individuals through monetary administrations like

advanced wallets. It has given microloans and permitted micropayments to individuals in under ideal financial conditions, in this manner presenting new life on the planet economy.

The following significant effect is in the idea of TRUST, particularly inside the circle of worldwide exchanges. Beforehand, legal advisors were employed to connect the trust hole between two distinct gatherings, yet it burned-through additional time and cash. In any case, the presentation of Cryptocurrency has drastically changed the trust condition. Numerous associations are situated in regions where assets are scant, and defilement is far and wide. In such cases, Blockchain renders a critical benefit to these influenced individuals and associations, permitting them to get away from the stunts of untrustworthy outsider middle people.

Blockchain innovation can make a decentralized shared organization for associations or applications like Airbnb and Uber. It permits individuals to pay for things like cost charges, stopping, and so forth Blockchain innovation can be utilized as a safe stage for the medical care industry for the motivations behind putting away touchy patient information. Wellbeing related associations can make an incorporated data set with the innovation and offer the data with just the fittingly approved individuals. In the private customer world, Blockchain innovation can be utilized by two gatherings who wish to manage a private exchange. Notwithstanding, these sorts of exchanges have subtleties that should be worked out before the two players can continue:

What is the T&C of the trade? Are generally the terms clear? When does the trade begin?

When will it wrap up?

When is it out of line to end the trade?

Luckily, since Blockchain innovation utilizes a common record, disseminated record, or some other decentralized organization, the gatherings can rapidly acquire answers to these trade connection inquiries. Likewise, exchanges or data on a Blockchain stage can be followed from flight to the objective point by the entirety of the clients in the inventory network.

Objectives of Chapter
The Objectives of the chapter are:

(a) To understand the importance of blockchain in Mutual Fund Industry,
(b) To understand blockchain towards the industry 4.0, and
(c) To study blockchain exchange traded fund.

Organization of Chapter.
The chapter is organized as: Sect. 2 gives the overview of Mutual Funds and Blockchain importance in current industry 4.0 scenario. Section 3 illustrates Industry 4.0 and importance of Industry 4.0 in Asset Management Industry. Section 4

highlights the role and impact of Blockchain technology on Mutual Fund Industry. Section 5 enlightens blockchain exchange traded fund. Section 6 stresses on challenges of blockchain in mutual fund industry. Section 7 concludes the chapter with future scope.

2 Organization Overview

Blockchain can be utilized to oversee securities and serve clients from various perspectives that would likewise save money on handling times and expenses. The execution of blockchain stages might actually save India's common asset industry a lot of cash, for the two firms and clients the same. Also, with common asset industry zeroed in on abundance age, blockchain innovation will empower store chiefs to get to ongoing client information. The information—made accessible from a solitary, secure area—will permit supervisors to settle on better-educated choices in an opportune way, to choose the best speculations for customers. The Indian shared asset industry has likewise seen issues, like fluid assets, where awful advances have brought down the worth of fluid assets and minimized enormous organizations [5, 6, 18, 19, 60, 77, 78, 80].

Considering these turns of events, blockchain might be the best answer for monitor information in the common asset framework, as the innovation could be depended upon to give early alerts and distinguish blemishes in speculation measures. Whenever executed, blockchain innovation could significantly disturb the manner in which common assets work, like its impact on the money and banking industry. Perhaps the greatest test of the common asset industry is the unified climate. This includes causing significant expenses of keeping up computerized framework. The current organization of shared assets depends on various outsiders and cycles. The appropriation specialists are the outsiders and they are not associated with the assets. They cooperate to interface financial backers with the asset. Move specialists, for example, banks or other monetary organizations are doled out to track financial backer records. They will likewise be administering the instalment of profits and giving articulations to financial backers. As such countless gatherings are engaged with joining another financial backer, the exchange times are long and thus it takes three to four working days from the place of membership to the mark of settlement. This makes the contributing cycle moderate. This is the place where blockchain innovation comes into the image.

A cross breed blockchain will be utilized in this proposed framework to such an extent that the permissioned private portion gives data, (for example, financial backer exchange data) just to and by approved gatherings while the public fragment gives data, (for example, day-end NAV information, factsheets and so on) to all clients with no limitations. Savvy agreements will be utilized for the data (and asset) stream between the partners as and when important. The UI and connection with the blockchain organization will be through a Decentralized Application [67].

Shared asset administrators, Investment Management and Resource Management have set up another program that will test blockchain in an exchanging climate, with the goal of decreasing working expenses. It is the first run through resource supervisors have met up to test with blockchain. Private value fills in as a decent use case. For instance, a financial backer may have put resources into a private value framework reserve. To work with the private value venture, numerous gatherings are engaged with passing manual documentation to and fro. Blockchain can possibly mechanize the sharing of documentation and lessen working expenses for the asset. Be that as it may, all together for the innovation to be extraordinary, numerous individuals should be associated with a solitary exchange at one time. This sort of communitarian innovation will likewise permit various gatherings to add data without changing the record or requiring a focal position. Because of blockchain's underlying encryption, the data can be shared immediately and safely, which will increment trust between counterparties just as lessen the need for a focal record. This is essential to controllers as they keep on searching for customer securities around speculation straightforwardness with regards to conditional developments, a blockchain-based framework may take into consideration better following when both money and protections change hands. In expansion to disentanglement, it will likewise lessen the opportunities for human mistake or altering. At last, it might permit shared assets to decrease their operational expenses. Fintech has been a central participant in expanding financial development; the COVID-19 pandemic has been at a defining moment for the more noteworthy reception of advanced loaning, online speculations, banking, miniature administrations, robot/AI administrations, and that's just the beginning. Thusly, there is one more part of monetary innovation, which is perhaps the most recent pattern in the monetary business sectors called 'Blockchain Technology'.

As of late, Invesco Mutual Fund documented papers with SEBI to dispatch another asset offer (NFO). This will be an open-finished asset of asset conspire, it will put resources into Invesco Elwood International Blockchain UCITS ETF, an abroad trade exchanged asset recorded on worldwide financial exchange [15, 16, 17]. The asset house means to present a shared asset conspire that will offer Indian financial backers the openness to worldwide organizations that are associated with blockchain innovation. Financial backers can hope to make gains from the development of organizations advancing with the blockchain innovation. The fundamental plan, for example Invesco Elwood International Blockchain UCITS ETF, will give openness across organizations universally in created and developing business sectors that partake and can possibly include in the blockchain environment. For financial backers, there are two wide regions that attention on blockchain innovation, one is the unadulterated digital money speculations; and the second is interest in a shared asset conspire that by implication puts resources into organizations focusing on the development capability of blockchain innovation. The chance to put resources into blockchain innovation allows financial backers the opportunity to use the potential presented by this progressive innovation. Prior to putting resources into plans counts on the future development possibilities of blockchain innovation, you should evaluate your danger resistance, venture objective and skyline. The blockchain innovation has a low openness in India and numerous financial backers are ignorant of the nitty-gritty. Try not to make a speculation impacted by a market pattern; set aside the effort to acquire an intensive comprehension of the asset and its basic plan [8–11].

3 Industry 4.0-Introduction

Industry 4.0 is utilized conversely with the fourth modern upheaval and addresses another stage in the association and control of the mechanical worth chain. Digital actual frameworks structure the premise of Industry 4.0 (e.g., 'brilliant machines'). They utilize present day control frameworks, have inserted programming frameworks and discard an Internet address to interface and be tended to through IoT (the Internet of Things). Industry 4.0 alludes to the smart systems administration of machines and cycles for industry with the assistance of data and correspondence innovation (Platform Industry 4.0) [4, 12, 49, 63–65, 68, 79].

Along these lines, items and methods for creation get organized and can 'convey', empowering better approaches for creation, esteem creation, and continuous streamlining. Digital actual frameworks make the capacities required for brilliant production lines. These are similar abilities we know from the Industrial Internet of Things like far off observing or track and follow, to make reference to two. Industry 4.0 has been characterized as "a name for the latest thing of mechanization and information trade in assembling advancements, including digital actual frameworks, the Internet of things, distributed computing and intellectual registering and making the keen plant".

Industry 4.0 is frequently utilized conversely with the idea of the fourth modern upheaval. It is described by, among others,

(1) much more computerization than in the third mechanical transformation,
(2) the spanning of the physical and advanced world through digital actual frameworks, empowered by Industrial IoT,
(3) a shift from a focal modern control framework to one where brilliant item characterize the creation steps,
(4) shut circle information models and control frameworks and
(5) personalization/customization of items.

The expression "Industry 4.0" is utilized to imply the start of the fourth modern insurgency—the past three being mechanical creation, large scale manufacturing, and afterward the advanced transformation. It very well may be contended that Industry 4.0 is basically a combination of the three past times in assembling, however Industry 4.0 is ready to be considerably more significant than that.

Industry 4.0 includes "new innovations that consolidate the physical, advanced and natural universes, affecting all orders, economies and ventures. These advances can possibly keep on associating billions additional individuals to the web and definitely improve the productivity of business and associations." In its application and widespread comprehension of Industry 4.0,

The development of computerization and information innovations controlled by the web of things (IoT), the cloud, progressed PCs, advanced mechanics, and individuals. The consistent coordination of programming, gear, and individuals that speeds up, unwavering quality, and stream of data between all frameworks of a maker. Industry 4.0 has made the savvy production line reality, thanks to some degree to the far and wide utilization of computerized innovations in previously manual cycles. Network, mechanization, and enhancement are driving the Industry 4.0 computerized change. The below given diagram explains the fourth industrial revolution related with many industries and the value chain process of industry 4.0 which are as follows:

1. **Industrial Internet of Things (IIoT)**: IIoT is when interconnectivity and cooperation of information, machines, and individuals in the realm of assembling. Basically, it takes IoT—sensors, machines, and information all associated and interfacing flawlessly—and applies it to assembling. Each part of the assembling activity can be associated in the IIoT, and the information it makes can be utilized into enhancing efficiencies across the assembling activity [56]. A definitive objective of an associated processing plant is to expand effectiveness, in this manner boosting benefits. To accomplish that, mechanization should be received into a few or the entirety of the assembling measures. Computerization, through mechanical technology or AI, is made conceivable by the interconnectivity and correspondence that happens across an Industry 4.0 advanced office.

2. **Artificial Intelligence**: Man-made consciousness and its subset AI are essentially a prerequisite for an Industry 4.0-empowered keen industrial facility. The entire reason around this new modern upheaval is to take out manual preparing, and AI is the essential apparatus to use in its place. Man-made intelligence can utilize the information created from an associated production line to streamline apparatus, reinvent work processes, and distinguish generally enhancements that can be made to drive efficiencies and eventually income. Since each usefulness of the assembling activity is being observed and producing information, there are huge loads of information to filter through. Nonetheless, enormous information investigation frameworks can use AI and AI advancements to rapidly handle information and give chiefs the data they need to make upgrades across a whole assembling activity.

3. **Cloud Computing**: Producers would have or prefer not to utilize the huge measure of room needed to genuinely store tremendous volumes of information made in an Industry 4.0 activity. This is the thing that makes distributed storage and registering an outright need and key gear-tooth in an associated manufacturing plant. Cloud utilization additionally takes into account a solitary wellspring of truth and information sharing across the organization, at lightning speed. Ultimately, distributed storage additionally considers far off access and observing of all information and machine working frameworks, giving incredible deceivability into tasks and efficiencies.

4. **Cybersecurity**: Since each touchpoint in the assembling activity is associated and digitized in Industry 4.0, there is an additional requirement for hearty online protection. Assembling hardware, PC frameworks, information examination, the cloud, and some other framework associated by means of IoT ought to be secured.
5. **Simulations**: Being able to estimate results is one of the greatest distinct advantages in the period of Industry 4.0 and fabricating. Prior to the digitization of the plant, changing over a product offering and upgrading its speed and creation was fairly mystery and consistently blemished. With the present progressed reproduction models fueled by the IoT information and AI, fabricating tasks can improve hardware for their next item run, consequently setting aside time and cash (Fig. 2).

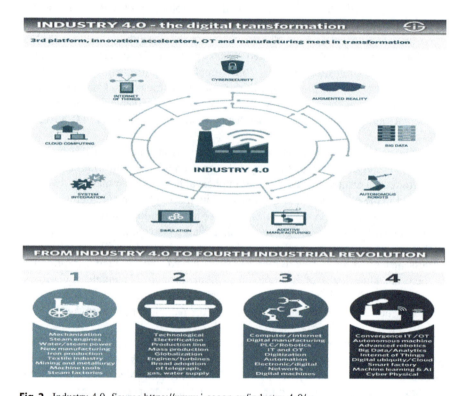

Fig. 2 Industry 4.0. *Source* https://www.i-scoop.eu/industry-4-0/

3.1 Importance of Industry 4.0 in Asset Management Industry

Industry 4.0 has become an ordinary word in the cutting-edge mechanical world these days. This term was authored without precedent for a cutting-edge fabricating project began by the German government in 2011. This term implies the Fourth Industrial Revolution that carefully changes the assembling business by utilizing the force of advanced mechanization in the assembling measures. Industry 4.0 is a piece of the fourth mechanical unrest that utilizes the cutting-edge types of data innovation to upgrade and further develop the assembling offices. The effect of industry 4.0 on the cutting edge worldwide economy has effectively arisen essentially. Additionally, the future effect of I-4 will get tremendous in the coming years on the grounds that practically all enterprises and areas of organizations are investing their full amounts of energy to use the force of industry 4.0. Every significant country and huge organizations are putting tremendously in the R&D focus exercises and the experts of the product advancement group to emerge the true goals of this cutting-edge idea [5, 33, 39, 51–53].

McKinsey Global Institute has as of late anticipated that Finance 4.0, a subset of Industry would assist with developing the arising economies by more than 6%, which will represent about $3.7 trillion by 2025. There are various different areas fueled by the business 4.0 idea that are contributing enormously to the worldwide economy. In the nutshell, the business 4.0 biological system in all spaces of worldwide organizations and ventures would assume a critical part later on and end up being a vital driver in catapulting the future development of the worldwide economy. The fourth mechanical transformation has spread out its wings over all ventures to affect them emphatically. The monetary administrations area is likewise one of the areas that have utilized the force of industry 4.0 widely. Banking, protection, contract, forex, stocks, and numerous other monetary areas are flourishing because of the positive push of advanced development and monetary interaction mechanization in all nations of the word [13, 20, 21, 40].

As per Nielsen Research, cell phones have become another standard for banking exercises. The creating locales like the Asia Pacific and Africa are turning into the significant driver of versatile banking on the planet. In the US, over 70% of the stock exchanging choices are chosen dependent on PC calculations, while just underneath 10% of the stock choices are made by singular specialists. This saves a colossal measure of monetary counseling administrations. Innovation controlled miniature advances in Bangladesh opened up another time of monetary strengthening of the distraught individuals in country regions. The new crypto banking framework dependent on blockchain innovation will radically change the monetary area very soon [59].

We should sum up the effect of industry 4.0 on monetary administrations.

- The quickest monetary exchanges all around the world
- The least expensive monetary assistance charges
- Immense lessening in operational expense of monetary organizations
- The simplest admittance to monetary administrations nonstop

- Expulsion of broker and focal guideline
- The decreased necessity of human tech-ability
- Expanded utilization of mechanical technology and man-made consciousness
- Successful, effective, and designated showcasing
- Compelling client assistance through bots for minimal price
- An expanded space of activities
- Diminished danger factors through information driven appraisals
- Better business knowledge
- More noteworthy client experience
- Diminished spillages in assets and undertakings
- Expanded straightforwardness and trust
- No human blunders in monetary exercises
- The development of new business spaces like FinTech, installment passages, and others
- Monetary incorporation of poor people and hindered individuals
- An expanded volume of liquidity
- Powerful utilization of accessible resources and cash
- Proficient exchanging of stocks and forex
- Diminished number of cheats and robbery
- Expanded consumer loyalty
- Keen agreements fueled by blockchain innovation
- Simpler responsibility and duty fixing.

L&T Technology Services Limited recognized as 'Leader' for its Industry 4.0 offerings

L&T Technology Services Limited, India's driving unadulterated play designing administrations organization, has been perceived as a 'Pioneer' for Development and Verification and Validation administrations in Industry 4.0 by the counseling and examination firm, Everest Group. 'Pinnacle Matrix for Industry 4.0 Service Provider 2020,' Everest Group has distinguished LTTS among the six driving organizations in the market dependent on different boundaries, for example, market appropriation, portfolio blend, esteem conveyed, advancement and speculations, conveyance impression, and vision and methodology. As indicated by Everest Group, LTTS has a solid capacity to drive an undeniable degree of advancement around Industry 4.0 that clients perceive, floated by powerful interests in building up committed framework, fostering a solid accomplice biological system and improving IP resources for Industry 4.0 administrations. The target of the PEAK Matrix is to give an information driven appraisal of administration and innovation supplier's dependent on their general capacity and market sway across various worldwide administrations markets, characterizing them into "Pioneers", "Significant Contenders", and "Applicants".

"LTTS' center around advancement and Verification and Validation (V&V) in the Industry 4.0 section is obvious from the way that this help work represents more than half of the association's income from Industry 4.0 administrations,

"The company's committed interests in V&V labs foundation and CoEs taking into account next generation subjects like IIoT, AI/ML, cloud designing, advanced mechanics, and network safety have fundamentally raised its situating among endeavours as a dependable Industry 4.0 specialist co-op. "Also, LTTS' center around IP advancement is reflected in a set-up of arrangements, including measured structures and stages like Aikno, nB-oN, and DFX, which assists ventures with improving efficiency and speed-to-affect in Industry 4.0 drives. LTTS' vision for this space, alongside centered ventures have assisted them with accomplishing a solid development force in this market,"

4 Role and Impact of Blockchain Technology on Mutual Fund Industry

A blockchain depends on a computerized and disseminated record, which acts in a straightforward climate without the requirement for a confided in power to approve exchanges. Maybe, there are PC hubs that follow some agreements and conventions to work the record in a mechanized manner. A blockchain is additionally ready to execute purported Smart Contracts application, self-executable PC programs that perform yet basic rationale yet can be amassed to deliver refined applications. Blockchain advancement can incite new opportunities and benefit associations through more essential straightforwardness, improved security, and easier perceptibility [22, 61, 62, 66, 81].

Fund Issuers: Blockchain gives huge advantages to guarantors by empowering simpler, less expensive, and quicker admittance to capital through programmable computerized resources and protections. New protections can be given in minutes, with their comparing rights and commitments encoded and mechanized. This permits guarantors and facilitators of new issues to expand the speed of subsidizing occasions. The capacity to program or encode agreements into resources (on account of protections issuance, for instance) gives more prominent adaptability and customization than at any other time. Blockchain innovation can smooth out KYC/AML measures and furnish continuous updates and examination with a solitary interface for financial backers, expanding straightforwardness and productivity. One of the critical benefits of computerized resources is the capacity to fractionalize every resource. Computerized resources can be broken into more reasonable and adaptable units that set out a freedom for more prominent liquidity and financial backer variety in specific business sectors. Besides, the boundaries to give a resource or security are essentially brought down opening up more prominent freedom for more modest backers while existing guarantors profit with new business sectors or types of protections. Finally, the whole lifecycle of a resource can possibly be computerized from financial backer adjusting to occasion taking care of on account of profits.

Asset Fund Managers: Essentially, blockchain empowers the distributed exchanging of any resource on an obvious record. Assets profit with quicker and more straightforward settlement and clearing which diminishes default hazard or fundamental danger in more murky business sectors. Quicker preparing implies that assets and directors have less tied-up capital and can all the more productively use and designate their current capital. Assets will lessen costs from expanded operational efficiencies like the improvement of asset overhauling, bookkeeping, designations, and organization. Expenses paid to outsiders for administrations, for example, reserve bookkeeping and organization, move office, and even guardianship can be decreased or dispensed with through computerized store administrations. Without a doubt there will be various new kinds of monetary items and instruments made utilizing blockchain innovation which will make novel resource classes for capital portion. Despite the fact that there will be a blast of monetary items, the majority of these resources will share explicit modified guidelines, subsequently working on the organizing of new monetary items or instruments. The capacity to give computerized resources and fractionalize existing resources will make a more extensive Investors pool, particularly as more current financial backers are more alright with possessing an arrangement of advanced resources [82–84].

Investors: Blockchain innovation altogether diminishes the boundary to give new resources or monetary items. As the expense of issuance of new protections drops and the speed of issuance builds, guarantors will actually want to tailor new instruments to the bespoke requirements of every financial backer. The improved capacity to all the more precisely match financial backer longing for return, time skyline, and hunger for hazard with custom advanced instruments may significantly affect the connection among financial backer and guarantor, making an immediate connection between capital searchers and financial backers. Financial backers plan to relieve hazard while expanding their possible returns. One of the vital drivers of hazard is an absence of liquidity. This is tended to by the programmable idea of advanced resources and monetary instruments which takes into consideration lower exchange costs, expanding the expected liquidity of a resource and empowering more extensive danger the board. Joined with the expanded availability and efficiencies across capital business sectors, financial backers will see more prominent liquidity and a diminished expense of capital. Furthermore, the straightforward and dispersed blockchain record will empower more strong experiences into resource quality with the possibility to upgrade the due perseverance measure [85–87].

Regulators and Controllers: Controllers are frequently reprimanded for getting too associated with capital business sectors or not getting included quick enough, as on account of the 2008 monetary emergency. Government offices and administrative associations can profit with a blockchain's appropriated record, which is straightforward and irrefutable at all snapshots of the day. The permanent idea of blockchain—which means exchange information can't be modified—empowers controllers to mechanize capacities, for example, examining and consistence. As numerous foundations utilize the equivalent blockchain organization to follow their property and resource lifecycle occasions, controllers will actually want to dedicate more opportunity to examination and hazard forecast, as opposed to on learning the mannerisms of each company's framework climate and bespoke exchange portrayals. The capacity to diminish contact across different work and time-concentrated cycles will smooth out the legitimate and administrative interaction. The improved nature of information and revelations empowered by block chain's record will decrease overhead expenses and possibly forestall explicit sorts of fundamental danger.

4.1 Need of Blockchain on Mutual Fund Industry

Issuance alludes to the way toward offering protections or other speculation resources for financial backers to raise capital. Blockchain empowers the formation of both advanced portrayals of existing customary protections and that of entirely new computerized resources, brought to advertise as tokens. The below given figure state the need of securitization of monetary instruments and protections that will get both more tweaked and smoothed out using blockchain issuance stages. Issuance can be improved across the lifecycle of resources including the digitization of value at consolidation or for the different resources under administration. Regular security-supported resources can be digitized to make tokens addressing singular protections with the improvement of extra programmable usefullness [23–25]. Blockchain empowers new plans of action, for example, decentralized crowdfunding which all the more effectively raises capital and makes a superior circulation of value and administration rights. Another advantage of blockchain all through the securitization lifecycle is the expanded straightforwardness and simplicity of cap table administration which is strategically placed on a solitary circulated record (Fig. 3).

Deals and exchanging are among the principle elements of venture banking. It alludes to the purchasing and selling of protections and other monetary instruments. Blockchain empowers advanced protections to flawlessly go-to-advertise through an assortment of instruments including two-sided dealings, concentrated trades, decentralized trades, coordinating with calculations, and closeouts. Blockchain brings about different additional opportunities including new and bespoke advanced instruments made to coordinate with financial backer requests [57, 58]. These new resources are made conceivable by the quick and adjustable nature of computerized security issuance which can be customized to flawlessly perform various types of business capacities. For instance, computerized and mechanized solicitations or other transient commitments can be empowered using a blockchain network and advanced token or resource. The below given figure explains the importance of blockchain in trade matching as it helps to attract more investors that explores new market with certain new assets and products at a minimum risk.

Current insurance the board measures are moderate and wasteful as a result of manual compromises and actual conveyance of protections which give restricted capacity to react to economic situations. Data is additionally extraordinarily,

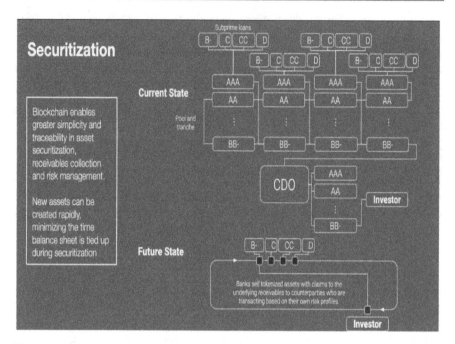

Fig. 3 Need of blockchain in securitization. *Source* https://cdn.consensys.net

making it hard to acquire a brought together image of cross-warehouse, cross-element security property. This structure further restricts an element's capacity to upgrade across guarantee stores or to net adjusts across elements and topographies. Blockchain empowers more effective guarantee the board through the digitization of the security possessions into a solitary, streamlined vault [26–28]. Furthermore, keen agreements can empower the exactness of insurance the board via naturally giving edge calls and conjuring foreordained standards for each two-sided or go-between relationship. The creation and digitization of insurance tokens or resources work with new business sectors and conceivable outcomes. For example, carefully addressed insurances on blockchain can be utilized to redeploy and get comfortable continuous, disposing of deferrals among valuation and call (Fig. 4).

Trades are regularly liable for a heap of undertakings including market administrations (exchanging and the executives of values, fixed pay, subsidiaries, and so on), corporate administrations (IPO, OTC updates, financial backer relations), and authorizing (information or record permitting). Blockchain can possibly improve the business tasks of trades across some of their capacities. Diminished exchanging expenses combined with quicker settlement and clearing can possibly diminish overhead expenses and improve existing cycles. A common, appropriated record empowered by a blockchain organization can upgrade KYC and AML consistence just as give exchange coordinating or affirmation. The blockchain's straightforward record can help trades with information confirmation, access rights, and in the best case give more vigorous admonition frameworks to exchanging movement.

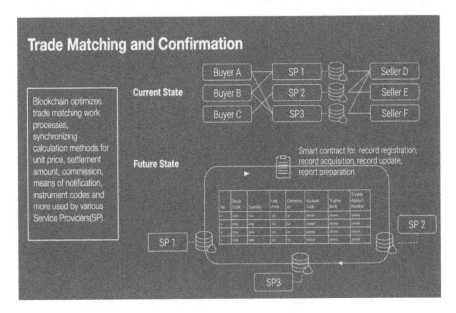

Fig. 4 Need of blockchain in trade matching. *Source* https://cdn.consensys.net

The below given figure explains the blockchain issuance of digitization resources takes into consideration new monetary items and instruments for subsidiaries with upgraded resource adjusting capacities (geo-fencing, whitelists, time-locks, and so forth) Besides, the mix of blockchain and new computerized resources and protections opens the potential for new essential or optional business sectors upgrading liquidity for specific resources. Clearing is the way toward refreshing records and coordinating the exchange of cash and protections. Settlement is the genuine trade of resources and monetary instruments. Shrewd agreements can be modified to coordinate with installments to moves through off-chain cash installments, cryptographic forms of money, or stable coins. For settlement, they can coordinate with an assortment of models that consider hazard resilience and liquidity needs of the market that incorporate nuclear settlement, conceded settlement, and conceded net settlement (Fig. 5) [29–32].

Shared asset organization is involved different cycles including store the board, element enlistment, exchange the executives, and announcing. Asset the executives at present depends on manual handling of asset information and other regulatory assignments that are mistake inclined. The below given figure explains that blockchain upgrades the asset interaction via robotizing and getting store reference information among key partners in close to constant. This extraordinarily builds the straightforwardness and security of asset information and some other reference data. Substance enlistment can be expensive and requires concentrated KYC/AML consistence techniques. Naturally, a blockchain gives a bound together regular record to the substance where records can be consequently put away,

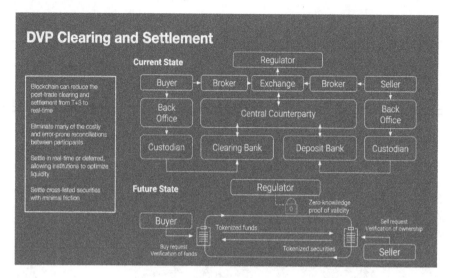

Fig. 5 Need of blockchain in clearing and settlement. *Source* https://cdn.consensys.net

checked, kept up, and dispersed. Furthermore, more cycles in store activity can be smoothed out, for example, the asset unit proprietorship vault, upkeep of financial backer and asset cash adjusts, cash designation, and that's just the beginning. New business sectors empowered by computerized resources and protections give finances a chance to separate their item contributions by making new items and advanced monetary instruments. Asset consistence data can be imparted to controllers or other organization members dependent upon the situation. Controllers and examiners would have the option to check all current data and trust the legitimacy of existing asset information and data. Collateral to the guardianship or holding of protections for care to limit the danger of robbery or misfortune. The high-level security ascribes of blockchain innovation, including its decentralized design and its cryptographically-secure code, guarantee resources are kept incredibly protected (Fig. 6).

4.2 Boom of Blockchain Technology in Mutual Fund Industry

Blockchain assists with all significant strides in the worth chain of the resource the executive's business like client on-boarding, portfolio the board, exchanging and settlement, and announcing among others [40-46]. It can store customer data in an anonymized, changeless way to fulfill administrative guidelines. This aide in KYC as any gathering to the exchange can review client information. A review trail for all exchanges identified with the client improves on the consistence interaction, making hacking for all intents and purposes inconceivable and illegal tax avoidance troublesome. The common asset industry is about abundance age. Common

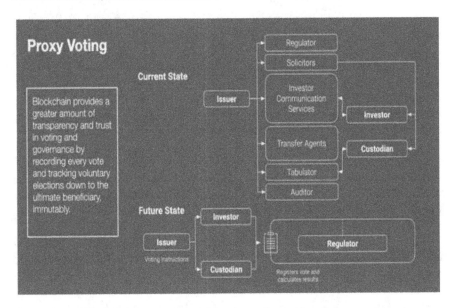

Fig. 6 Need of blockchain in controller and regulators. *Source* https://cdn.consensys.net

asset supervisors settle on ideal venture choices considering hazard. Blockchain interfaces customer needs with resource the board systems vital to common assets with ongoing client information, put away at a solitary, secure area. It assists shared asset supervisors with the speculation dynamic interaction, essential for abundance age, and shields portfolios from hazard.

Blockchain innovation can demonstrate helpful for the shared asset area as far as client on boarding, detailing, portfolio the board, and exchanging and settlement. Having a blockchain-empowered shared asset will help wipe out the halfway advances engaged with store membership. Handling times are decreased. A financial backer will have an advanced wallet which will hold their computerized character and the computerized money to be put resources into the blockchain common asset. The Know Your Customer (KYC) and Anti-Money Laundering (AML) checks should be possible by the blockchain-based advanced personality and not an outsider. Keen agreements can lead these checks electronically and can then consequently buy in to financial backers who have passed the check. A mixture blockchain will be utilized and in the permission private fragment just gives data to approved gatherings. Shrewd agreements will be utilized for data and asset stream between partners [72–76].

The keen agreement is coded in an undeniable level language and is then added to the blockchain, given a one-of-a-kind location, and summoned by approved gatherings. The exchange data is added to the blockchain hub. It is approved in close to constant and consequently will be accessible on the blockchain network for all time, on account of the idea of circulated record innovation. Advanced trade of resource is additionally conceivable. On account of nations like India where managing crypto and virtual monetary forms are not permitted, the RTA would give an exchange solicitation to the bank for an asset move and this occurs through a keen agreement. Blockchain lessens value-based danger and settlement time. As the organization is automatic, there is no requirement for any broad guideline. It wipes out the requirement for a focal trust to give authority as keen agreement executes shrinks without help from anyone else. Data on the blockchain is for all time accessible. Along these lines, permission gatherings can create their assertion of records whenever. Lessens operational weight as financial backer objections can be tended to over the blockchain. Utilizing blockchain innovation saves a ton of time and energy which can be utilized to improve the client experience [37, 38].

The Net Asset Value of the speculation can be figured. Preparing of move of advanced cash in return for reserve venture is conceivable with the blockchain. Client account adjusts can be put away on the blockchain. Financial backers can screen their asset offsets with a UI on a cell phone application.

During the hour of delivering out for profits, the installment is naturally executed by the keen agreement and the returns are stored in the clients' computerized wallet. The whole administration cycle could ultimately be mechanized utilizing blockchain innovation. Blockchain innovation will without a doubt have a positive effect and will unquestionably end up being a distinct advantage in the shared asset industry. The highlights of blockchain like responsibility, straightforwardness, decentralization, protection, and alter obstruction will help partners save both expense and time. Exchange preparing can be robotized utilizing brilliant agreements consequently guaranteeing that refreshed data is accessible on the blockchain consistently. Blockchain is a help for the resource the executive's business as it aids the venture dynamic cycle, shields portfolios from hazard, and is crucial for abundance age.

5 Blockchain Exchange Traded Fund

Blockchain trade exchanged assets (ETFs) own stocks in organizations that have business tasks in blockchain innovation or here and there benefit from it [34, 35, 50]. Blockchain is comprised of complex squares of advanced data, and progressively is utilized in banking, contributing, digital currency, and different areas. While blockchain is a generally new innovation, a large number of the organizations that work in the space are grounded. A few models incorporate International Business Machines Corp. (IBM), Oracle Corp. (ORCL), and Visa Inc. (V). Numerous financial backers might be careful about gambling an interest in blockchain because of the innovation's relationship with the unpredictable digital currency market. Blockchain ETFs put distinctly in supplies of controlled organizations, a large number of which are enormous blue-chip innovation firms. A blockchain ETF is like standard area or subject based corporate securities through trade exchanged assets (ETFs), working by solely putting resources into a container of blockchain-based organizations [54–56, 69, 70]. The organizations claimed in a blockchain

ETF have business activities in blockchain innovation or are those that contribute or benefit from blockchain.

Blockchain is a genuinely new innovation that creates a record, which stores all data with respect to an exchange. This record is decentralized, which means it isn't kept in one area but instead disseminated across an organization that can be seen by general society. The data in the record is additionally morally sound. Blockchain ETFs offer double advantages—pooled interests in bins of stocks like that of a common asset, and ongoing exchanging with tick-by-tick value changes like that of a stock. Blockchain ETFs are a generally ongoing wonder. Thusly, it is hard to decide drifts or get definitive outcomes from their presentation. Notwithstanding, numerous blockchain ETFs have seen positive returns over the most recent few years. That being said, financial backers are as yet worried about the drawn-out possibilities of blockchain ETFs, as some case there is a curiosity to blockchain as an innovation, which may not last. Blockchain ETFs additionally accompany the characteristic danger of putting resources into innovation based new businesses while the blockchain idea is as yet developing, and, consequently, routinely hitting administrative barricades across the globe.

Like other thematic investing like self-sufficient vehicles, distributed computing, and man-made brainpower, retail financial backers can claim blockchain ETFs. Basically, a blockchain ETF holds a container of traded on an open market organization with openness to blockchain innovation. These organizations can either straightforwardly utilize the innovation or benefit from their administrations to help the development. There are two sorts of blockchain ETFs: Passively oversaw and effectively oversaw. In a latently overseen ETF, an asset director purchases a container of blockchain-related stocks that makes up a wide file. Through one of these ventures, you acquire openness to the whole file. This cycle dispenses with the requirement for reserve directors to choose singular organizations at their circumspection. In spite of list reserves, dynamic contributing relies upon an asset administrator's capacity to pick protections and give better than expected returns. Thus, these speculations frequently accompany higher expenses and more noteworthy unpredictability than inactively oversaw ETFs.

Top blockchain ETFs
This specialty space of the ETF market stays uncrowded, with just a small bunch of major parts in the space. Since there are no unadulterated play blockchain organizations, the majority of the property in these assets will in general cover with other expansive based ETFs. Beneath we feature the ones with the most resources under administration.

Enhance Transformational Data Sharing ETF (BLOK): BLOK is the most noticeable blockchain ETF with $1.2 billion in resources under administration. This effectively overseen reserve chooses organizations engaged with the turn of events and utilization of blockchain advances. Among its top property, the asset possesses protections of Microsoft (MSFT), PayPal (PYPL) and Square (SQ).

About portion of its portfolio puts resources into stocks outside of the United States. It has a cost proportion of 0.70%.

Alarm Nasdaq NexGen Economy ETF (BLCN): BLCN dispatched in January 2020 and has $316 million in resources under administration. The asset tracks a record of worldwide organizations committing material assets to creating, investigating, supporting, or progressing blockchain innovation. A portion of its top property incorporate portions of (IBM), Accenture (ACN) and MasterCard (MA). The ETF has a cost proportion of 0.68%.

First Trust Index Innovative Transaction and Process ETF (LEGR): LEGR offers openness to a worldwide arrangement of organizations with changing levels of contribution in the blockchain. The asset dispatched in January 2018. It has about $90 million in resources under administration. Almost 70% of its speculations are outside of the United States. Among its top property, the asset possesses portions of NVIDIA (NVDA), Texas Instruments (TXN) and Amazon (AMZN). Its cost proportion remains at 0.65%.

6   Challenges of Blockchain in Mutual Fund Industry

Blockchain has uncommon potential yet ought to face different troubles, which possibly stop the wide use of Blockchain. The Blockchain is a scattered conveyed structure that everyone in the association can examine the trade records and add new data to the informational index. The openness and the setback of central coordination are the foundation of the system, which has unfavourable results and restricts the usage of Blockchain [47, 48, 71, 88]. A couple of issues can be raised, similar to adaptability, security, assurance, dormancy and money related business areas really fight to find good plans.

Cost of Initiation, Implementation, and Maintenance: The underlying expense of carrying out a blockchain framework is high. It requires huge information with respect to the product and equipment fundamental for its underlying dispatching. Little venture and banking organizations may not meet such expenses thinking about their monetary status. Subsequently, making and recording exchanges utilizing this framework turns into a weight for such organizations all along. Upkeep cost is additionally high. Account organizations that consider such a framework as risk may not contemplate its high upkeep cost, as it diminishes the general returns for the business.

Changes of Data: It's anything but's an issue with the change of data. The banking and cash region make ordinary changes to the data they store, especially data including a trade. Henceforth, by far most of these cash associations achieve declining its usage for trade recording. The method of data area is furthermore long. Considering that the trades in the cash region may be different reliably, a long procedure may concede the narrative of any such trade, thus, conveying the system inefficient.

Protection: The Blockchain can create numerous addresses rather than genuine character for clients to stay away from data spillage, which is accepted to be very safe for clients. Be that as it may, the Blockchain can't forestall conditional data spillage since all data on exchanges and equilibriums are appeared to general society. Blockchain innovation can make lasting and changeless records for members, however it additionally expands the protection dangers of certain substances. In the interim, privacy is trying to work in open Blockchain-based frameworks, as data is apparent to all members in the organization as a matter of course. Straightforwardness is required for explaining possession and forestalling twofold spending, while clients require protection. Blockchain trades contain individuals' areas, trade regards, timestamps, and sender marks, which makes it possible to follow trade streams to remove customer information through data mining.

Guidelines and law: With the creating use of Blockchain, Australia, US, South Korea, Switzerland, China, the UK, Japan, Singapore, Hong Kong, and Canada give significantly more thought to oversee Blockchain to avoid distortion and other criminal tasks that hurt the interests of customers and the market. Managerial weakness will have various results. The specific trial of Blockchain is that paying little heed to how stunning the Blockchain development is, it can't guarantee the validity of separated data. The data being alluded to will be forever recorded on the Blockchain if there is an issue with the data source. SinceBlockchain is decentralized, without the administration of laws and staff, and it is difficult to change records on the chain, these will wreck a few. A few lawmaking bodies acknowledge advanced monetary standards as an illegal coin in their countries. The support this marvel is that the asset class is new to the point that organizations and banks have not gotten the relating methodology for them. In occasions of blackmail, and various frustrations, the association doesn't have even the remotest clue about the laws and rules. This is particularly perilous for associations working in different wards. Therefore, a couple of risks exist as the assessment assortment status and trading rules of bitcoin could change for the present.

Cybercrime: Public Blockchain advance contest, development and efficiency, however they likewise present difficulties to guideline of tax evasion, psychological militant financing and duty aversion since they don't expect members to verify. Cybercriminals, similarly called PC masterminded bad behaviors, direct criminal activities with the association, causing terrible consequences for losses. Computerized money is the portion methodology for law breakers. Blockchain is applied to Anti-Money Laundering (AML) and Know Your Customer (KYC) necessities for financial applications, for trades on a public; Blockchain is open and pseudonymous to all, while private systems have limitations to individuals.

7   Conclusion and Further Scope

Blockchain innovation can assist all partners in the common asset industry with its straightforwardness, decentralization; alter opposition, responsibility and security. There will be enormous saving of time and cost with all partners. Brilliant agreements computerize exchange handling and guarantee that refreshed data is accessible on the blockchain consistently. NAV calculation will be speedier and explanation age will be dynamic and on the fly. Reclamation installments will get prepared progressively. Realtors and reviewers will be important for the framework and can take out information as and when they need from the blockchain. Generally speaking, we feel that blockchain will have a massive good effect and will to be sure be a distinct advantage in the shared asset industry. Asset directors look for most extreme insurance on settling on speculation decisions and Blockchain could be the appropriate response.

Blockchain could work as an early notice framework, distinguishing blemishes in the dynamic cycle. It could foresee a monetary emergency, or enormous tricks and fakes, saving the common asset industry a few crores and the asset administrator, his work and reward. Blockchain can make stock trades considerably more ideal through computerization and decentralization. It can help lessen gigantic costs imposed on clients as far as commission while accelerating the cycle for quick exchange settlements. The advancement can have reasonable use in clearing and settlement, while securely robotizing the post-trade measure, working with authoritative work of trade and authentic ownership move of the security.

Blockchain can dispose of the need of outsider controller generally, since the principles and guidelines would be in-worked inside savvy contracts and authorized with each exchange request to enlist exchanges with the blockchain network going about as a controller for all exchanges. While the market screens possible administrative turns of events, powerful administration is the way in to the effective execution of blockchain to secure members, financial backers and partners while guaranteeing that the framework is strong notwithstanding foundational hazard, protection concerns and network safety dangers. Blockchain can possibly upset the monetary administrations, especially in mechanizing market reconnaissance occasions preparing and in robotizing post-exchange occasions handling.

Blockchain innovation has become well known because of its effective reception for digital currencies like Bitcoin. This circulated computerized record enjoys many benefits as it can keep the records, all things considered, or cash exchange made between any two gatherings in a solid, unchanging, and straightforward way.

In spite of blockchain is on the highest point of its fame, the work market encounters an absence of blockchain specialists. Upwork, an internet outsourcing data set, has as of late announced a quick expanding request in individuals with "blockchain" abilities. While the innovation is new, there are a set number of blockchain engineers (Fabian 2021).

In blockchain-based crowdfunding, trust is rather made through savvy contracts and online standing frameworks, which eliminates the requirement for a center man. New undertakings can raise assets by delivering their own tokens that address esteem and can later be traded for items, administrations, or money. Numerous blockchain new companies have now raised huge number of dollars through such symbolic deals. Despite the fact that it's still early days and the administrative future or blockchain-based crowdfunding is unsure, it's a region that holds a ton of guarantee.

References

1. AlevtinaDubovitskaya, P. N. (2019). Applications of Blockchain Technology for Data-Sharing in Oncology: Results from a Systematic Literature Review. *Oncology and Informatics – Review*, 403–411. https://doi.org/10.1159/000504325.
2. Alliance, C. (n.d.). Blockchain Opportunities for the Wealth and Asset Management Industry. *Innovation for Financial Services*.
3. Amir Mehdiabadi, M. T. (2020). Are We Ready for the Challenge of Banks 4.0? Designing a Roadmap for Banking Systems in Industry 4.0. *International Journal of Financial Studies, 8*(32), 2–28.
4. Andy Daecher, B. S. (2018). *The Industry 4.0 paradox Overcoming disconnects on the path to digital transformation.* Deloitte Consulting LLP's.
5. Arnab Kumar, T. M. (2020). Blockchain: The India Strategy.
6. Ashish Sharma, D. B. (2019). Literature Review of Blockchain Technology. *International Journal of Research and Analytical Reviews, 6*(1), 430–437.
7. O., Ally, M., & Dwivedi, Y. (2020). The state of play of blockchain technology in the financial services sector: A systematic literature review. *International Journal of Information Management, 54,* 102199.
8. Ayan Bhattacharya, M. O. (2020). ETFs AND SYSTEMIC RISKS.
9. B., T. E. (2015). Exchange Traded Funds (ETF): history, mechanism, academic literature review and research perspectives. 1–22. Retrieved from https://ssrn.com/abstract=2819646.
10. BadrMachkour, A. A. (2020). Industry 4.0 and its Implications for the Financial Sector. In P. C. Science (Ed.), *The 7th International Symposium on Emerging Information, Communication and Networks. 177*, pp. 496–502. Elsevier.
11. Bajpai, P. (2021). 6 ETFs for Investing in the Blockchain. Retrieved from https://www.nasdaq.com/articles/6-etfs-for-investing-in-the-blockchain-2021-05-14.
12. Bidnur, C. V. (2020). A Study on Industry 4.0 Concept. *International Journal of Engineering Research & Technology, 9*(4), 613–618.
13. Bikramaditya Ghosh, D. P. (2012). Exchange Traded Fund-Are They Popular Enough Among Indian Investors. *International Journal of Research in Management, Economics and Commerce, 2*(11), 388–394.
14. Bilan Yu., R. P. (2019). The Influence Of Industry 4.0 On Financial Services: Determinants Of Alternative Finance Development. *Polish Journal Of Management Studies, 19*(1), 70–93.
15. Blockchain ETF List. (n.d.). Retrieved from https://etfdb.com/themes/blockchain-etfs/.
16. Blockchain in asset management. (2016).
17. Blockchain in Institutional Capital Markets. (n.d.). Retrieved from https://consensys.net/blockchain-use-cases/capital-markets.
18. Blockchain was developed as a solution for an efficient, cost-effective, reliable, and secure system for conducting and recording transactions. (2021). *Blockchain Technologies ETF*, pp. 2–6.
19. Blockchain-Enabled Trade Finance Innovation: A Potential Paradigm Shift on Using Letter of Credit. (2019). *Sustainability, 12,* 2–16.
20. Bo˙zenaGajdzik, S. G. (2021). A Theoretical Framework for Industry 4.0 and Its Implementation with Selected Practical Schedules. *Energies, 14,* 1–24.

21. C. Todd Gibson, T. K. (2016). Blockchain 101 for Asset Managers. *The Investment Lawyer Covering Legal and Regulatory Issues of Asset Management, 23*(10), pp. 7–14.
22. Cherecwich, P. (2016). What does the integration of blockchain mean for mutual funds?
23. ChiaraFranciosi, B. S. (2018). Maintenance for Sustainability in the Industry 4.0 context: a Scoping Literature Review. *IFAC-PapersOnLine, 51*(11), 903–908.
24. Chinmaya Goyal, J. F. (2021). Discussion paper on blockchain technology and competition. pp. 2–48.
25. Chiu, J., &Koeppl, T. V. (2018). *Blockchain-based settlement for asset trading.* Bank of Canada Staff Working Paper, No. 45, Bank of Canada, Ottawa.
26. Danda B. Rawat, V. C. (2020). Blockchain Technology: Emerging Applications and Use Cases for Secure and Trustworthy Smart Systems. *Journal of Cybersecurity and Privacy, 1*, 4–18.
27. Dave Dowsett, H. W. (2020). Blockchain and the reshaping of investment management.
28. DOUGLAS MILLER, P. M. (2019). *BLOCKCHAIN Opportunities for Private Enterprises in Emerging Markets.* Washington, U.S.A.: International Finance Corporation World Bank Group.
29. Dr. ArunaPolisetty, V. K. (2019). Growth And Development Of Exchange Traded Funds (Etfs) In India. *International Journal of Advance and Innovative Research, 6*(1), 90–94.
30. Dr. PrashantaAthma, B. M. (2017). Performance of Equity Exchange Traded Funds in India: An Analysis. *International Journal of Business and Management Invention, 6*(11), 59–66.
31. Elisa, M. (2020). How DLT & Blockchain is shaping the future of wealth & asset management. *wealth tech views report.*
32. Fabian, S. (2021). Decentralized Finance: On Blockchain- and Smart Contract-Based Financial Markets. *103*(2), 153–174.
33. Fengwei Yang, S. G. (2021). Industry 4.0, a revolution that requires technology and national strategies. *Complex & Intelligent Systems, 7*, 1311–1325.
34. Fiergbor, D. D. (2018). Blockchain Technology in Fund Management. *Chapter in Communications in Computer and Information Science* (pp. 310–319). Springer. https://doi.org/10.1007/978-981-13-2035-4_27.
35. Fiergbor, D. D. (2018, March). Blockchain technology in fund management. In International Conference on Application of Computing and Communication Technologies (pp. 310–319). Springer, Singapore.
36. Park, J. K., & Kim, I. (2017). A Study on Adoption and Policy Direction of Blockchain Technology in Financial Industry. *Journal of Information Technology Services, 16*(2), 33–44.
37. Fran Casinoa, T. K. (2019). A systematic literature review of blockchain-based applications: Current status, classification and open issues. *Telematics and Informatics, 36*, 55–81.
38. Huasheng Zhu, Z. Z. (2016). Analysis and outlook of applications of blockchain technology to equity crowdfunding in China. *Financial Innovation, 2*(29), 2–11. https://doi.org/10.1186/s40854-016-0044-7.
39. Industry 4.0: Building the digital enterprise. (2016). *2016 Global Industry 4.0 Survey*, pp. 2–34. Retrieved from www.pwc.com/industry40.
40. Investing in Companies Involved in Blockchain Technology. (2021). *AMPLIFY TRANSFORMATIONAL DATA SHARING ETF.*
41. *Investment management firms management firms.* (2017). The Deloitte Center for Financial Services.
42. Jesse Yli-Huumo, D. K. (2016). Where Is Current Research on Blockchain Technology?—A Systematic Review. *PLoS ONE, 11*(10), 1–27. https://doi.org/10.1371/journal.pone.0163477.
43. Jo Van de Velde, A. S. (2016). Blockchain in Capital Markets.
44. JörgWeking, M. M. (2019). The impact of blockchain technology on business modelsa taxonomy and archetypal patterns. 285–305. https://doi.org/10.1007/s12525-019-00386-3.
45. Kaur, A., Nayyar, A., & Singh, P. (2020). Blockchain: A path to the future. Cryptocurrencies and Blockchain Technology Applications, 25–42.
46. Khadka, R. (2020). *The impact of blockchain technology in banking: How can blockchain revolutionize the banking industry?* Centria University Of Applied Sciences.

47. Kimberlyn George, P. N. (2020). The Blockchain Evolution and Revolution of Accounting. Retrieved from https://ssrn.com/abstract=3681654.
48. Kishore Kumar Das, S. A. (2020). The role of digital technologies on growth of mutual funds industry: An impact study. *International Journal Of Research In Business And Social Science, 9*(2), 171–176.
49. Krishnamurthi, R., Kumar, A., Gopinathan, D., Nayyar, A., & Qureshi, B. (2020). An overview of IoT sensor data processing, fusion, and analysis techniques. Sensors, 20(21), 6076.
50. Kumar, D. B. (2016). Exchange Traded Fund In India Performance Analysis With Mutual Fund And Global Perspectives. *International Journal of Marketing & Financial Management, 4*(7), 22–35.
51. Kumar, A., & Nayyar, A. (2020). si 3-Industry: A Sustainable, Intelligent, Innovative, Internet-of-Things Industry. In A Roadmap to Industry 4.0: Smart Production, Sharp Business and Sustainable Development (pp. 1–21). Springer, Cham.
52. Lisa Bosman, N. H. (2019). How manufacturing firm characteristics can influence decision making for investing in Industry 4.0 technologies. *Journal of Manufacturing Technology Management*.
53. Marco Bettiol, M. C. (2019). *Impacts of industry 4.0 investments on firm performance Evidence From Italy*. DSEA.
54. Martin Lettau, A. M. (2018). Exchange Traded Funds 101 For Economists. *Nber Working Paper Series*.
55. Mats Isaksson, S. Ç. (2018). The Potential for Blockchain Technology in Public Equity Markets in Asia. *OECD Capital Market Series*, pp. 3–42. Retrieved from www.oecd.org/corporate/capital-markets.
56. Mori, T. (2016). Financial technology: Blockchain and securities settlement. *Journal of Securities Operations & Custody, 8*(3), 208–227.
57. Mhlanga, D. (2020). Industry 4.0 in Finance: The Impact of Artificial Intelligence (AI) on Digital Financial Inclusion. *International Journal of Financial Studies, 8*(45), 2–14.
58. Michael Crosby, N. P. (2015). *BlockChain Technology Beyond Bitcoin*. Sutardja Center for Entrepreneurship & Technology Technical Report.
59. Michela Piccarozzi, B. A. (2018). Industry 4.0 in Management Studies: A Systematic Literature Review. *Sustainability*, 2–24. https://doi.org/10.3390/su10103821.
60. Min Xu, X. C. (2019). A systematic review of blockchain. *Financial Innovation, 5*(27), 2–14. https://doi.org/10.1186/s40854-019-0147-z.
61. Mugdha Kulkarni, K. P. (2020). Block Chain Technology Adoption for Banking Services-Model based on Technology-Organization-Environment theory. *Proceedings of the International Conference on Innovative Computing & Communications (ICICC) 2020*. Retrieved from https://ssrn.com/abstract=3563101.
62. Natalia Dashkevich, S. C. (2020). Blockchain Application for Central Banks: A Systematic Mapping Study. *IEEE Access, 8*. https://doi.org/10.1109/ACCESS.2020.3012295.
63. Nayyar, A., & Kumar, A. (Eds.). (2020). A roadmap to industry 4.0: Smart production, sharp business and sustainable development. Springer.
64. Nayyar, A., Rameshwar, R. U. D. R. A., & Solanki, A. (2020). Internet of Things (IoT) and the Digital Business Environment: A Standpoint Inclusive Cyber Space, Cyber Crimes, and Cybersecurity. The Evolution of Business in the Cyber Age, 10, 9780429276484-6.
65. Nikita Karandikar, R. A. (2021). Blockchain-based prosumer incentivization for peak mitigation through temporal aggregation and contextual clustering. *Blockchain: Research and Applications*, pp. 1–35. https://doi.org/10.1016/j.bcra.2021.100016.
66. Olaniyi, E. (2019). Blockchain Technology and the Financial Market: An Empirical Analysis. *Actual Problems of the Economy No. 211*, 82–101.
67. Peck, M. (2017). Reinforcing The Links Of The Blockchain. *Future Directions Blockchain Initiative White Paper*. Retrieved From Blockchainincubator.Ieee.Org.
68. Petra Maresova, I. S. (2018). Consequences of Industry 4.0 in Business and Economics. *Economies, 46*(6), pp. 2–14.

69. Prasanna, P. K. (2012). Performance of Exchange-Traded Funds in India. *International Journal of Business and Management, 7*(23), 122–143.
70. Harper, J. T., Madura, J., & Schnusenberg, O. (2006). Performance comparison between exchange-traded funds and closed-end country funds. *Journal of International Financial Markets, Institutions and Money, 16*(2), 104-122.
71. Rebecca Lewis, J. M. (2017). Blockchain and financial market innovation. *Economic Perspectives.*
72. RIO, R. (2019). Asset Management KPIs with Industry 4.0.
73. Schär, F. (n.d.). Decentralized Finance: On Blockchain- and Smart Contract-based Financial Markets. 1–24. Retrieved from https://ssrn.com/abstract=3571335.
74. Segran, G. (2020). Blockchain applications: Hype or reality?.
75. Fernandez-Vazquez, S., Rosillo, R., De La Fuente, D., & Priore, P. (2019). Blockchain in FinTech: A mapping study. *Sustainability, 11*(22), 63–66.
76. Fosso Wamba, S., Kala Kamdjoug, J. R., Epie Bawack, R., & Keogh, J. G. (2020). Bitcoin, Blockchain and Fintech: a systematic review and case studies in the supply chain. *Production Planning & Control, 31*(2-3), 115–142.
77. Singh, P., Nayyar, A., Kaur, A., & Ghosh, U. (2020). Blockchain and fog based architecture for internet of everything in smart cities. Future Internet, 12(4), 61.
78. Stefanuk, A. (n.d.). Industry 4.0 And Its Impact On The Financial Services. *Fintech Weekly.* Retrieved from https://fintechweekly.com/magazine/articles/industry-4-0-and-its-impact-on-the-financial-services.
79. Tam, P. T. (2020). Impacting Industry 4.0 On The Banking Service: A Case Study Of The Commercial Banks In Dong Nai Province. *Journal of Entrepreneurship Education, 23*(6), 1–8.
80. Tayal, A., Solanki, A., Kondal, R., Nayyar, A., Tanwar, S., & Kumar, N. (2021). Blockchain-based efficient communication for food supply chain industry: Transparency and traceability analysis for sustainable business. International Journal of Communication Systems, 34(4), e4696.
81. Varma, J. R. (2019). Blockchain in Finance. *VIKALPA The Journal for Decision Makers, 44*(1), 1–19.
82. Vijaya Killu Manda, V. K. (2020). Blockchain-based Net Asset Value (NAV) calculation for Mutual Funds. *International Journal of Advanced Trends in Computer Science and Engineering, 9*(2), 1668–1673.
83. Vijaya Kittu Manda, P. R. (2018). Blockchain Technology for the Mutual Fund Industry. *National Seminar on Paradigm Shifts in Commerce and Management 2018 in Congruence with Block Chain Accounting,* (pp. 12–17). Retrieved from https://ssrn.com/abstract=3276492.
84. Xiaoying Wanga, R. S. (2021). Industry 4.0 and intellectual capital in the age of FinTech. *Technological Forecasting and Social Change, 166.*
85. Yermack, D., & Fingerhut, A. (2019, May). Blockchain technology's potential in the financial system. In Proceedings of the 2019 Financial Market's Conference.
86. Federal Reserve Bank of Atlanta. (https://www.atlantafed.org/)
87. Yoo, S. (2017). Blockchain based financial case analysis and its implications. *Asia Pacific Journal of Innovation and Entrepreneurship, 11*(3), 312–321.
88. Zhou, R., & Le Cardinal, J. (2019). Exploring The Impacts Of Industry 4.0 From A Macroscopic Perspective. *International Conference On Engineering Design, Iced19,* (Pp. 2111–2120).

Industry 4.0 Internet of Medical Things Enabled Cost Effective Secure Smart Patient Care Medicine Pouch

Sourav Singh, Sachin Sharma, Shuchi Bhadula, and Seshadri Mohan

1 Introduction

Distinct patients have different medical needs in today's globe. In order to improve medical services, proper bespoke solutions for unique needs are urgently required. On the other hand, these solutions must be developed in such a way that they are accessible to the average person [1]. One method is to make innovative solutions more cost-effective in general. It is relatively required to construct new medical solutions for patients using cost-effective and readily available components [2]. The usage of new technologies such as IoT, embedded systems, Wi-Fi, Bluetooth, cloud computing, and others also aids developers in developing unique IoMT solutions [3]. The standard under which this innovation and manufacture of automated cost-effective solutions is carried out is known as Industry 4.0. The proposed work is also based on this standard, as as a novel IoMT solution for patients who have specific needs for an automated, cost-effective medical pouch that notifies the scheduled intake of medicines is proposed and and the system also predict the specific lifesaving medicine from the medicine pouch in the event of an emergency. The system also makes use of new wearable IoT technologies, with which with which is designed as a wearable-band to track patient vitals including heart rate and temperature. The sensors used for collecting data using the wearable band are heartbeat sensor and temperature sensor.

S. Singh · S. Sharma (✉) · S. Bhadula
Department of Computer Science and Engineering, Graphic Era Deemed to Be University, Dehradun, Uttarakhand, India
e-mail: sachin.cse@geu.ac.in

S. Mohan
Systems Engineering Department, University of Arkansas at Little Rock, Little Rock, AR, USA
e-mail: sxmohan@ualr.edu

In this chapter, a literature review is conducted wide range of research papers based on the architecture and design of modern advance pouch-based pill box solutions. The work identifies and attempts to improve the capabilities of modern pill boxes. New technologies are intergrated into cutting-edge pill boxes, such as wearable IoT, real-time monitoring, cloud, Android application, GPS, and so on. In the healthcare domain, a new IoMT system architecture is proposed. The work is dedicated to support new innovations in IoMT devices that adhere to Industry 4.0 standards.

Problem Definition

The proposed system is intended to address the issue of developing a cost-effective custom solution in the healthcare domain in accordance with Industry 4.0 standard. The system solves the problem with cutting-edge pill boxes and dispensers that only serve as an alarm system for medication intake. These systems do not incorporate new methods such as wearable technologies, real-time monitoring, anti-left-behind Bluetooth features, GPS medication location, and so on. Wearable technology allows for real-time monitoring of a patient's or user's vitals. By design, the proposed system also addresses the issue of "availability." It is designed in such a way that it can be developed by researchers and developers from anywhere in the world, as its architecture and design rely solely on readily available, low-cost off-the-shelf hardware components and software platforms.

Contributions

The contributions for this chapter can be summarized as below:

- To compare recent related works in the healthcare domain that concentrate on pill dispensers and new innovations in IoMT devices,
- To present a detailed overview of modern IoMT devices and Industry 4.0,
- To propose a novel architecture for a low-cost IoMT-based medical pouch system,
- To discuss the preliminary design requirements of the proposed system's electronic parts and development board modules, as well as their software platforms,
- To utilize Proteus software tool to simulate the wearable band module electronic circuit of the proposed system, and
- To compare existing techniques with proposed system.

Organization of Chapter

The chapter is organized as: Sect. 2 highlights some recent works of different authors related to the IoMT. Section 3 discusses the overview of Industry 4.0 and IoMT devices. Section 4 highlights some impacts, challenges and applications of proposed system. Section 5 explains the architecture and implementation of the proposed system. Section 6 give a performance and comparative analysis of the proposed system. Finally, Sect. 7 concludes the chapter with future scope.

2 Literature Review

Rumi et al. [4] focused on aging problems related to senior citizens. The authors proposed a smart medical box which is equipped with a camera. The camera reads the medical prescription. After that some processing techniques were applied such as "Maximally-Stable-External-Regions" and "String-Manipulation" (applied on extracted data). Then the extracted information is stored in the database. The medicine-box uses this database information to notify the patient using a buzzer-system. It also displays the relevant information on LCD-display. The patient has to undergo a fingerprint verification process, so as to collect the medicines. The system also periodically updates the database with medicine timings, and notify the user when low on medicines.

Al-Mahmud et al. [5] proposed a similar medical box concept as discussed above. This medical box sends an email to the patient, which is the indication of taking the timely medicine. Detailed information about prescription, doctor details, patient-details and appointment-details were stored in a laptop, which acts like a server. The patient temperature is also stored in server-laptop, this temperature acts like a reference to doctor. The doctor can take necessary action in case of emergency, by using the data collected by the system.

Srinivas et al. [6] implemented an IoT medical box, for health diagnostics and monitoring purposes. This medical box is also equipped with sensors. The box remains wirelessly connected to an android application. The doctor and any preregistered guardian (of patient) has access to this android application. The application is connected to the internet and provides timely updates for proper medicine consumption.

Kumar and Kumari [7] developed a "Smart-medicine-dispenser" system. Basically, it keeps the track of daily medicine dose timings and notify the user on appropriate schedule of medicine intake. For achieving this, the authors used a medicine box with small pockets in it, which was used for storing different medicines. The system used real-time clock module and an alarm system to notify the daily medicine timing. The system also utilized Wi-Fi connectivity technology.

Karagiannis and Nikita [8] proposed a smart 3D-printed pill-box. It can hold 48 different types of pills in small compartments, with built-in LED indications. The platform architecture consists of a server, which stores patient information in "MariaDB". The doctor can change and view patient's medicine-schedule over internet. The system consists of a microcontroller, "ESP32-CAM", a clock module, a buzzer (for signaling), addressable LEDs, temperature sensor and humidity sensor. Sensors were used for the "patient-monitoring".

Ayshwarya and Velmurugan [9] proposed a medical box with 6-compartments. These compartments can organize 6-pills of different varieties. The reminders were supposed to be displayed as notification from android application in a handheld phone device. For patient monitoring a heartbeat-sensor and temperature sensor is used. Along with 6-motor driver modules were used for pill compartments.

Sinnapolu and Alawneh [10] demonstrated a prototype which monitors the car driver's heart-rate and use this data in detecting emergency situations. Authors used used wearable technology by using "Apple-watch" for medical system. In case of some emergency, the wearable device initiates the vehicle controls, so as to escort the patient-driver to the hospital. The driver had to download "healthdetect" and "healthlocate" apps from "Apple–app-store". The vehicle also is equipped with a microcontroller ("electron-board" is used here), which is connected to BCM and high-speed CAN bus, of the vehicle. System's proper working requires that vehicle-driver should always wear an Apple-watch and keep the vehicle ignition in ON-state, while driving the vehicle. If the heart-rate crosses a predefined threshold, then the "heartdetect" app initiates the emergency scenario and waits for a button press. If button press is missed (no activity sign), then the system guides the driver for an optimized route towards the hospital. Or in case of serious conditions the system took control of driver's car also, by activating autopilot system.

A large number of researchers worked on an intelligent patient health monitoring system. It monitors patients' vitals in real time using bio-sensors such as an oxygen level sensor, a pulse rate sensor, and a body temperature sensor. The data is then sent to the cloud via the ThingSpeak cloud service. The device prototype has been tested on more than 50 patients. When compared to current healthcare monitoring systems, the result is up to 90% accurate. Researchers have also investigated the domain of Green-IoT, where they analyzed and conducted an in-depth study on the principle of Green-IoT over a sustainable environment. The work provided an overview of various ICT technologies. In addition, an investigation is conducted into the potential of real-time development and the challenges associated with Green-IoT implementations [11]. One of the challenges in various versions of healthcare is the implementation of fog computing with industrial IoT in terms of Industry 4.0. Various case studies were discussed in order to demonstrate the significance of fog computing in Industry 4.0 [12]. With today's IoT, there are challenges in IoT sensor network analytics and various technical concepts of touch enabled handheld devices. Real-world IoT applications with touch enabled devices are outfitted with advanced technologies such as Wi-Fi, Bluetooth, ZigBee, LoRa, and sensor technology for medical monitoring.

3 Overview of Industry 4.0 and IoMT Devices

Industry 4.0 medical IoT devices demonstrate the ability to design new specialized tools and equipment for the medical IoT domain. It provides a low-cost electronic or digital hospital as well as a monitoring system that meets the specific needs of the medical industry [14]. Industry 4.0 is an intriguing approach to generating new concepts and undertaking development in the medical industry by combining technologies, smart devices, and other applications [15]. Industry 4.0 produces controlled, high-quality medical equipment that is highly customized to the consumer's needs. This transition is the result of automation, and it opens up new opportunities for the growth of medical IoT. It creates a new digital habitat with the help of IoMT. It promotes connectivity and the exchange of Big Data by utilizing new development technologies, software, sensors, and other advanced AI and embedded technologies [16]. Industry 4.0 has the potential to expand opportunities for health-care innovation.

3.1 Internet of Medical Things in Industry 4.0

Industry 4.0 refers to the automation process as it relates to manufacturing and smart devices. IoT, connected devices, human-resources for monitoring, Data Acquisition, robots for the automated world, new sensors, new actuators, communication protocols, and cyber-threat-control are all part of it. It employs AI techniques to disseminate useful information and provides high-quality data on inventory, materials, by products, output management, and customer fulfilment. It fosters positive relationships between distributors and producers. Industry 4.0 can meet the needs of healthcare and its associated IoMT domains [17]. Industry 4.0 refers to a self-managed manufacturing and development process in which devices and end-products are aware of their limits and capabilities, as well as options for optimizing production. It generates a versatile development system [18]. In the field of IoMT, Industry 4.0 aids in the production and development of any custom device for an individual doctor/patient through the use of software and specific new technologies [19]. It satisfies the requirements of a personalised IoMT device solution. Figure 1 depicts the overall features of Industry 4.0. Internet of Medical Things is a complex combination of medical equipment and software APIs. These sums up with wireless healthcare systems connectivity. There is a wave of vast number of IoT devices which are benefitting multiple industrial domains. Among these IoT devices, the standalone and wearable technology devices are making their way towards the brighter future. The IoMT devices are generally can be classified under following categories:

- **Fitness tracking Wearable devices:** These are the consumer wearables which consists of some sensors and some central processing microcontroller with some embedded software. These are mostly paired with some mobile App and all the data can be visualized and processed through the app interface only.

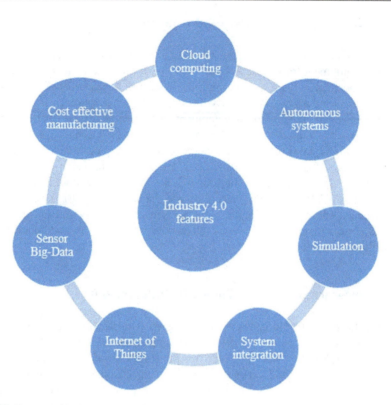

Fig. 1 Features of industry 4.0

- **Remote Patient Monitoring:** So as to keep an eye on the fitness of discharged patients, these types of remote sensing, monitoring and tracking IoMT devices were utilized. These in a way allow a doctor to visit a patient statistic virtually and check the patient health status.
- **Smart IoMT consumable Pills:** These are based on the concept of having ingestible pill devices. These are very small in size and can be taken as capsules. These type of IoMT get activated after being swallowed by the patient.
- **IoMT installed on Hospitals:** Today's hospital infrastructure is full of connected devices ranges from Various monitoring equipment to sophisticated MRI machines. Also, almost all major healthcare providers keep track of their patients and keeps a record of patient healthcare data.
- **Clinical Devices:** These devices used to store patient's vitals. These were also capable of transmitting patient real-time data to some mobile application or to the cloud. An example of such IoMT device is digital stethoscope.

4 Impact, Challenges and Applications of IoMT Enabled Cost Effective Secure Smart Patient Care Medicine Pouch

4.1 Impacts of Proposed Smart Patient Care Medicine Pouch

Economic impact: In the IoMT domain, the proposed system, which is based on Industry 4.0 standards, provides a low-cost option for patient care. It is made up of off-the-shelf electronic components that are readily available all around the world [20]. In addition, with medical diagnostics and monitoring, the system assists the patient in taking medications on time. This functionality could save patients money on typical human "nurse-support" medical expenditures.

- **Recognize the patient's medical needs:** Our technology gives timely assistance in remembering and suggesting relevant life-saving medicines from the medical bag. These types of qualities were critical for today's patients, especially those who were living alone in their homes.
- **Technology integration in healthcare:** The proposed system brought together a variety of technologies to offer the essential medical service. Our efforts will inspire new developers and researchers to combine new technologies in order to deliver a variety of healthcare services [21].
- **Improve patient quality of life:** This low-cost patient monitoring and medicine schedule reminder device aids patients in remembering when to take their medications. Continuous monitoring can also aid in the monitoring of the patient's condition, which can be beneficial to the patient's guardians. The users' quality of life is improved as a result of these characteristics [22].
- **Low-cost IoMT development:** The proposed system is built with off-the-shelf components such as low-cost sensors and Arduino boards for microcontroller and wireless communication. This reduces the system's cost and makes it more accessible to the general public. Such behaviors encourage the development of low-cost IoMT methods [23].

4.2 Challenges in Designing the Proposed Smart Patient Care Medicine Pouch

- **Avoid false-positive alarms:** The incorrect sensor readings can cause the system to issue a false alarm. Environmental conditions, such as high temperatures, can also cause false-positive readings in the system.
- **Energy:** The majority of IoMT systems were battery-powered. These batteries have a tendency to discharge quickly. As a result, this could be fatal to the patient's life, as the patient's life is entirely dependent on the proper operation of monitoring services provided by these IoMT devices.

- **Efficient medicine prediction**: Medicine prescribing IoMT devices should adhere to certain guidelines in order to provide proper service. AI approaches that are more accurate should be used.
- **Connectivity**: For proper server connectivity, new and tested technologies such as LoRa, Wi-Fi, BLE, and others should be employed so that the system does not disconnect during emergency situations [24].
- **Affordability**: The new generation of IoMT devices can only be successful if they can address high-cost challenges in end products.
- **Energy requirements**: Appropriate battery solutions must be included, allowing the IoMT to run indefinitely.

4.3 Applications of Proposed Smart Patient Care Medicine Pouch

- **Memory-disorder support**: The proposed medical-bag can be utilised by patients who have memory problems and are unable to recall their doctor's advised dosage schedule.
- **Support for the elderly**: The elderly is prone to become frail and helpless when it comes to remembering medication dosages. These patients also required continual medical supervision in order to avert emergency scenarios.
- **Activity tracking**: The proposed technology can track patient's heart rate and body temperature via a wearing band. This will aid in determining the patient's current condition.
- **Emergency assistance**: The suggested system uses sensor data to make pharmaceutical recommendations. This function can help specific patients in an emergency.
- **For the patient**: The patient will receive a timely notification about the medicine intake schedule. Also, the feature of an alarm when the pouch of a Bluetooth-connected wearable band is left behind will ensure that the medicines are available 24 h a day, seven days a week.
- **For Doctors**: The proposed method will offer doctors with certainty that the patient will not miss any scheduled drug doses. Doctors can also seek rapid assistance in an emergency.
- **Companies in the Insurance Industry**: The real-time condition of the patient will be recorded if the patient is properly tracked. This will make it easier for insurance companies to keep track of their clients' health. This will also help to prevent fraud claims by capturing real-time data from systems.
- **For hospitals**: The real-time monitoring feature of the patient can be used by hospitals. In the event of an emergency, the GPS location of the medical pouch can also assist in locating the patient's medications.

5 Industry 4.0 IoT Enabled Cost Effective Secure Smart Patient Care Medicine Pouch: Architecture and Implementation

The goal is to create a system that allows a patient to get a quick medicine recommendation based on his or her medicine pouch during an emergency at home. The medical-pouch also reminds you to take your medicine on time. There are four basic modules that make up the entire system:

- IoT sensor-enabled wearable band.
- IoT Medicine Pouch with Local Database and Connectivity.
- Mobile App to manage inputs and outputs.
- Cloud to store and show data and outcomes.

5.1 System Architecture

The system architecture is made up of all essential parts that are linked together via various wireless protocols. Simple networking solutions were provided using IEEE 802.15 and IEEE 802.11 protocols. Wearable band, mobile device, medicinal pouch, and cloud used as various modules [25]. The proposed system design is shown in Fig. 2.

The hardware modules requirements for the given system are described below [13].

- *Arduino Uno*: It is used as an extension to increase GPIO pins and as a central controller in the project modules. It has a microcontroller ATmega328p, operating voltage is 5, 7 to 12 V input voltage, total GPIOs is 14, 6 analog input pins, 16 MHz clock speed.
- *ESP8266*: Consists of 10 GPIOs, Pulse Width Modulation, I2C, USB-TTL logic etc. It is Wi-Fi enabled easy to integrate development board based on Arduino. Power ratings 4.5–9 V max.
- *LM35 Temperature Sensor*: Temperature range from −55 to 150 °C.
- *LDC display*: It is used for displaying test results and data. It has 16 characters × 2 lines, blue backlight, 4-bit or 8-bit interface, 6-GPIOs needed for interfacing, operates on 5 V.
- *PCF8574 LCD I2C board*: Arduino I2C interface adapter was developed to reduce the Input/Output port usage. It can also use to control backlight through Pulse Width Modulation or with a resister.
- *Heart Rate Sensor*: It is combined with supporting electronics and works on 3 or 5 V power ratings. It is a low cost and easy way of getting data regarding live-heart-rate in projects.

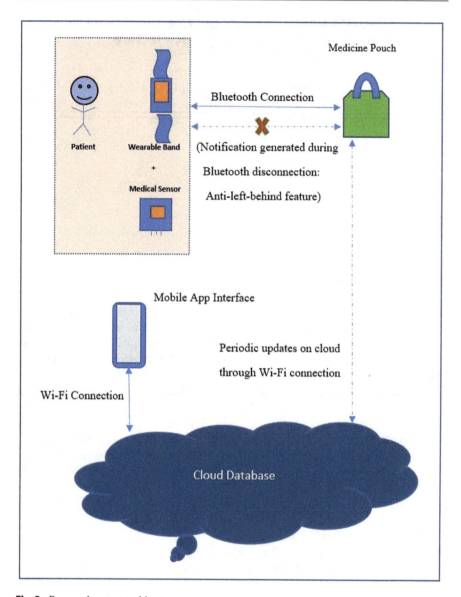

Fig. 2 Proposed system architecture

- *GPS Module*: It is latest technology which provides acceptable readings for positional coordinates. Its position update rate is 5 Hz. 3.3 V supply voltage. Default baud rate is 9600. Has 18 × 18 mm GPS antenna separately attached.
- *Piezo Buzzer*: It rings out whenever it gets electric signals. The sound is created by resonators and diaphragms. It operates on 3.3–5 V power supply.

The software requirements for the proposed system are given below:

- *Arduino IDE*: Arduino IDE is an open-source software which is used to program the microcontrollers in an easy to write codding system. Its program structure consists of generally two methods, named as "setup()" and "loop()". In setup part we initialize various variables, pin modes etc. of the electronic circuitry supporting microcontroller. On the other hand, the loop part manages what the system has to do continuously.
- *Proteus Simulator*: Proteus-Design-Suite is a Microsoft-Windows application used to simulate embedded systems prototypes. It uses "hex file" to burn the Arduino Uno's microcontroller inside the Proteus environment as shown in Fig. 5. The circuit is virtually designed in Proteus simulator and the cloud connection is done through Arduino Libraries in the Arduino board code. Simulating IoMT virtually is presently one of the efficient ways of rapid prototyping and development of such devices and systems.
- *ThingSpeak cloud service*: It is an open-source software which provides API to connect to cloud and facilitates the user to log sensor data, display sensor data in graphical representation etc. We can create various channels for displaying data on a dashboard.
- *Android Studio*: It is an official IDE for Android app development which is developed by Google. Any application compiled with Android studio can be published in Google Play Store.

The proposes system is observed as a layered model where each layer specifies a task. Different layers of the system are given below:

- *Physical Layer*: The system's physical layer consists of fundamental modules such as sensors, medicine pouches, wearable bands, and mobile devices.
- *Network layer*: Different protocols are used to maintain connectivity between different components. This layer includes IEEE 802.15 (Bluetooth connection) and IEEE 802.11 (Wi-Fi connection) technologies.
- *Technology Integration Layer*: In the development of the system, variety of technologies are used such as cloud computing, artificial intelligence, the internet of things and data analytics.
- *Applications Layer*: The end-user interface is introduced in the application layer to control the system's inputs and outputs via an Android app.
- *Security*: We used a proper authentication system that works on each system layer. To keep the system safe, user-id and password are used for authentication. Figure 3 depicts a block schematic of the system' various tiers. Security is applied at every layer of the system.

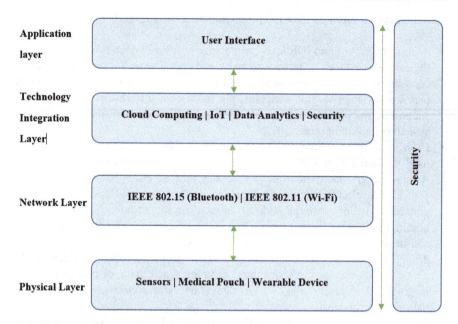

Fig. 3 Block-diagram for different layers of our proposed system

5.2 Implementation of IoMT Enabled Smart Medical Patient Medicine Pouch

The implementation of proposed IoMT enables cost effective and secure smart patient medicine pouch need integration of various sensors and modules. Different sensors and modules used for implementation of the proposed system are given below.

A. Wearable Band

NodeMCU/ESP module: The wearable band's wireless connectivity is enabled by this component. The ESP module was chosen since it is small and requires little power.

Arduino Uno: The Arduino UNO is exclusively utilized in the wearable band for prototyping. Its primary purpose is to facilitate circuit integration by providing enough General Purpose Input Output (GPIO) pins for sensor interfaces.

Bluetooth module: It is employed to make the connection between the wearing band and the medical pouch possible.

Pulse Sensor: It is essential in order to obtain real-time heart rate readings.

Temperature Sensor: LM35 Temperature Sensor is used for monitoring temperature of the patient.

Battery Module: Li-ion battery module is used to power up the system.

B. Medicine Pouch

NodeMCU/ESP module: The ESP module is utilized in the pouch for wireless communication.

Arduino Uno: It is utilized here for rapid prototyping and obtaining a suitable number of GPIO pins. We can also use the Arduino Mega development board in this case.

GPS Module: The medical pouch is equipped with a GPS module that provides real-time latitude and longitude data. This information may be fed into an Android app to build a global map of pouch locations.

Piezo Buzzer: A buzzer has been added to provide sound beep alarms for certain events.

Battery Module: A lithium-ion battery module is also included to power the medical pouch's embedded electronics.

C. Mobile Device

Android application: The android application is designed for the system to manage the system's inputs and outputs as well as to provide an effective interface. This app shows the patient's scheduled prescription consumption times, as well as the pouch's GPS location, heart rate, and temperature data. It also sends out timely notifications for various occurrences.

Internet Access: In order to use an Android application, you must have access to the internet. However, the system is built in such a way that it is not reliant on internet connectivity to work.

Cloud Service Enabled: The android application includes a cloud service that uploads data to a server on a regular basis.

D. Cloud

Medication Prediction Service: In addition to the system's local offline lightweight prediction algorithm, the cloud can access various online AI based algorithms for medicine prediction and patient status monitoring on an as-needed basis.

Medicine Pouch Database: To provide dependable service, a database of medicine pouch data is kept on the server.

Patient Health Monitoring Database: Patient sensor data and patient information are also stored on the server, allowing for remote monitoring if necessary.

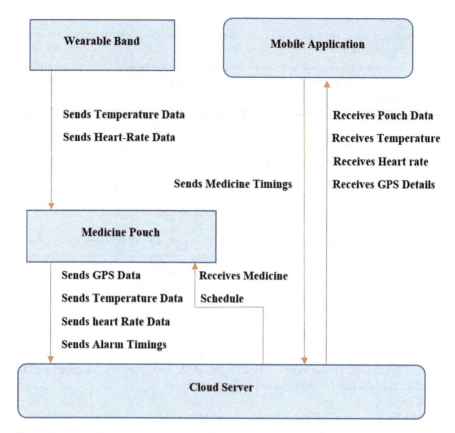

Fig. 4 Data-flow diagram of proposed system

The flow of the implementation of given system is explained with a Data Flow Diagram (DFD). The proposed system connects several modules using technologies such as Bluetooth Low Energy (BLE) and Wi-Fi connectivity. Data is shared often and fully across different units. The mobile device, medication pouch, wearable band, and cloud are all part of our system architecture. To begin, data generated by sensors in the wearable band. It is then replaced with a medical-pouch. After then, the data from the wearable band is analyzed and transferred to the cloud. The mobile device, on the other hand, communicates with the cloud as well. Figure 4 depicts the dataflow diagram for the system.

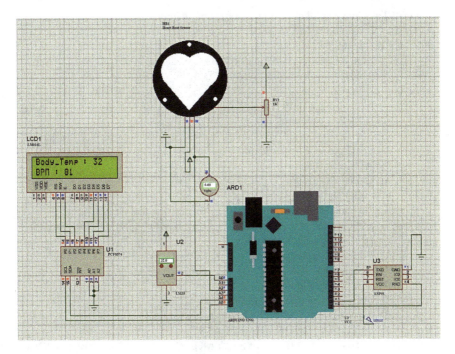

Fig. 5 Simulation setup of wearable-band module on Proteus software

The following algorithms and Pseudocodes developed for execution of different features of proposed system:

A. *Bluetooth based Anti-Left-Behind feature algorithm*:

Goal: To achieve anti left behind feature using Arduino modules.
1. **SETUP** Pinmode and Baud-rate
2. **WHILE** in loop()
3.   Set value incoming = Serial.read()
4.   Establish Bluetooth Connection
5.   If incoming == 1
6.   Update print "BT connection stable : pouch inside   range"
7.   ELSE IF incoming == 0
8.   Update print "BT connection Disconnected : pouch-left-behind"
9.   Seek Bluetooth Reconnection
10.   READ GPS value
11.   UPDATE TO CLOUD : GPS value
12. End **WHILE**

B. Wearable band monitoring and medicine suggestion feature algorithm:

Goal: To achieve medicine suggestion through patient monitoring using Arduino modules.
1. **SETUP** Pinmode and Baud-rate
2. **WHILE** in loop()
3.   **READ** "Heart-Rate-Value"
4.   **READ** "Temperature-Value"
5.   **IF** Heart-Rate-Value >= "100" beats-per-minutes
6.   **NOTIFY:** "HIGH HEART RATE"
7.   **SUGGEST MEDICINE**: "TAKE BP MEDICINE FROM POUCH – ONLY   IF PRESCRIBED EARLIER BY DOCTOR"
8.   **ELSE IF** Temperature-value >= "101" degree Fahrenheit
9.   **NOTIFY:** "SYMPTIOMS OF FEVER DETECTED"
10.   **SUGGEST MEDICINE**: "TAKE FEVER MEDICINE FROM POUCH – ONLY IF PRESCRIBED EARLIER BY DOCTOR"
11.   **ELSE IF** Temperature-value <= "94" degree Fahrenheit
12.   **NOTIFY:** "LOW BODY TEMPERATURE DETECTED"
13.   **NOTIFY:** " LOW BODY TEMP EMERGENCY"
14.   **ELSE IF** Heart-Rate-Value <= "60" beats-per-minutes
15.   **NOTIFY:** "EMERGENCY: LOW HEART RATE DETECTED"
16.   **UPDATE TO CLOUD :** "Heart-Rate-Value"
17.   **UPDATE TO CLOUD :** "Temperature-Value"
18. **End WHILE**

To effectively create the proposed system, we use the Proteus software to mimic the system's key parts. In Proteus, we first draw the electrical connections of "wearable-band." The heart-beat sensor, LM35 temperature sensor, LDC display, PCF8574 LCD I2C board, ESP-module, and Arduino UNO are all included in the wearable-band. The wearable-band module simulation in Proteus is shown in Fig. 5. The heart-rate and temperature measurements were analysed and transferred to the cloud for cloud-analytics and monitoring. The values of "Temperature" and "Beats-per-minute" were also displayed on the LCD display for testing purposes.

6 Performance Analysis of Proposed System

The proposed system comprises of a wearable band that the patient is expected to wear. Patient health monitoring sensors are also included in this band. Patient heart rate and temperature are monitored using sensors such as a pulse sensor and a temperature sensor. This sensor data is subsequently sent to an Android mobile app through a low-energy Bluetooth connection. The sensor data from the patient band is uploaded to the cloud using this mobile's internet connection. The system, on the other hand, has a unit called Medicine Pouch. Specific medicines are organised in the marked pockets of this medicine pouch. This pouch is also connected to the cloud, which sends information about medicine availability. The medicine prediction service predicts the appropriate medicines that were present in the pouch during the patient's emergency medical condition, which is sensed through the sensors interfaced with the wearable band. Pouch also contains a GPS module that displays the pouch's GPS coordinates on the cloud dashboard.

The medical-pouch module of the system, on the other hand, is also replicated in the Proteus programme. LCD display, I2C display module, ESP-module, and Arduino UNO are also included. We built two distinct sorts of pockets in our medical pouch to retain the patient's prescribed medicine. To distinguish between two types of drugs, these pockets were designated POCKET A and POCKET B. When the sensor values surpass a predetermined threshold, the LCD also shows the pocket name. The system was then put through its paces with various sensor values. The final results were examined and determined to be satisfactory. The medical-pouch module simulation in Proteus is shown in Figs. 6, 7 and 8 depict snapshots of heart-rate and temperature test results, respectively.

6.1 Alzheimer Patients as Potential Users of Proposed System: A Case Study

Alzheimer's disease is caused by a neurological illness that causes the brain to shrink. It is a terrible condition that causes dementia, in which a person's social skills, mental ability, and behavior skills deteriorate. This has an impact on a person's life to the point where he or she is unable to live independently without assistance. Around 44 million Alzheimer's sufferers are estimated to exist worldwide. In such a dire circumstance, new technologies such as the Internet of Things (IoT) should be developed to assist and improve the quality of life of such people. Anyone over the age of 65 is at risk of developing such symptoms. As a result, if these persons were on medicine, remembering medical dosage schedules became extremely difficult. Continuous monitoring is also essential for such vulnerable folks. Approximately 1–4 family members act as caretakers for such patients in the families, according to estimates. This process and setup can be difficult to handle and upsetting at times. Our suggested solution overcomes these concerns by including wearable monitoring bands that can track a patient's heart

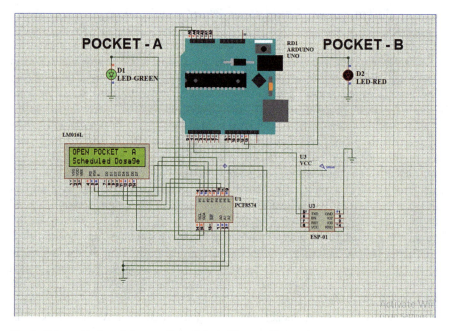

Fig. 6 Schematic of medicine-pouch module on Proteus software

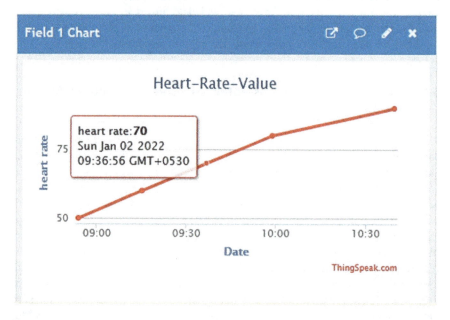

Fig. 7 Heart-rate channel test values on ThingSpeak dashboard

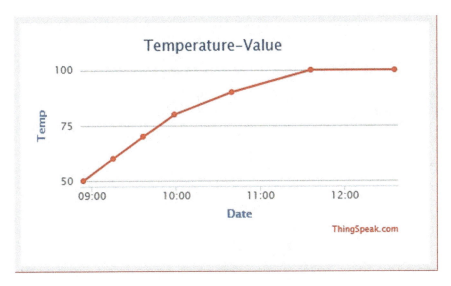

Fig. 8 Temperature test values channel on ThingSpeak dashboard

rate and other vitals. In addition, an IoT-based medical pouch system that is coupled with these wearable monitoring bands notifies for scheduled drug ingestion. A Bluetooth-based anti-left-behind feature with GPS support is also included in the medical pouch [26]. The proposed medical pouch provides 24×7 connectivity to the patient, tracking his conditions and alerting him to take medicines on time. This continuous monitoring and control process will prevent the patient from further and sudden health issues by alerting the patient if he forgot to take his medicine on time. It also reduces the dependency of Alzheimer patients on others to give medicines. As a result, more cost-effective solutions should be created and delivered. Our proposed system is built on a cost-effective end-product development process, such that the entire globe can profit from and have access to such IoMT device systems for unique healthcare needs and cost-effective patient monitoring.

During literature research, lots of recent works by various researchers was studied to gain a sense of the state-of-the-art existing devices. A few parameters to compare the proposed work with current state-of-the-art techniques is done and comparison of proposed system is summarized in Table 1.

7 Conclusion and Future Scope

The prototype of the system was demonstrated in the simulation. The IoT-based prototype can be built and used in any case when a patient requires assistance ingesting life-saving pills from his or her medicine pouch. The sensor-equipped IoT band should be worn at all times by the patient. In an emergency, the Android

Table 1 Comparative analysis

References	Real-time operation	Analysis of data	Decision making	Cost effective	Automated process	IoT	Cloud computing	Security	Big data address
[4]	✗	✗	✗	✓	✓	✓	✗	✓	✗
[5]	✓	✓	✗	✗	✓	✓	✓	✗	✗
[6]	✓	✗	✗	✗	✓	✓	✓	✗	✗
[7]	✓	✗	✗	✗	✓	✓	✓	✗	✗
[8]	✓	✗	✗	✗	✓	✓	✓	✗	✗
[9]	✗	✗	✗	✗	✓	✓	✓	✗	✗
Proposed system	✓	✓	✓	✓	✓	✓	✓	✓	✓

app connects to the band to receive patient sensor data and aid in the prescription of exact treatment. When designing the system, which adhered to the industry 4.0 standard, new technologies were integrated, smart production with cost-effective measures, and unique solutions from innovations were all taken into account. As a result of our research, a variety of academics and developers can get new ideas for building such systems. Furthermore, our work can be used by a variety of companies as a model for developing new IoMT devices.

In the future, the system can be improved by adding more functionality to the wearable band. To stay watchful, other sorts of sensors can be used. Sensors such as gyroscopic sensors should be included into future versions of the system to track patient physical activities. A more effective medicine indication system can also be added to medicine pouches. More features, as well as the addition of a doctor as a user, could be added to the Android app. For the time being, however, this is just a prototype for an IoT system that can save lives before and during an emergency.

References

1. Aman, A. H. M., Hassan, W. H., Sameen, S., Attarbashi, Z. S., Alizadeh, M., & Latiff, L. A. (2021). IoMT amid COVID-19 pandemic: Application, architecture, technology, and security. Journal of Network and Computer Applications, 174, 102886.
2. Chetan Pandey, Sachin Sharma, and Priva Matta. "Body Sensor Network Architectures in Healthcare Internet-of-Things (HIoT): A Survey." In 2021 6th International Conference on Communication and Electronics Systems (ICCES), pp. 494–499. IEEE, 2021.
3. Chetan Pandey, Sachin Sharma, and Priya Matta. "DME: An Efficient Encryption Technique for Body Sensor Network in Healthcare Internet of Things (HIoT)." In 2021 2nd International Conference on Intelligent Engineering and Management (ICIEM), pp. 80–85. IEEE, 2021.
4. Rumi, R. I., Pavel, M. I., Islam, E., Shakir, M. B., & Hossain, M. A. (2019, December). IoT Enabled Prescription Reading Smart Medicine Dispenser Implementing Maximally Stable Extremal Regions and OCR. In 2019 Third International conference on I-SMAC (IoT in Social, Mobile, Analytics and Cloud)(I-SMAC) (pp. 134–138).
5. Al-Mahmud, O., Khan, K., Roy, R., & Alamgir, F. M. (2020, June). Internet of Things (IoT) based smart health care medical box for elderly people. In 2020 International Conference for Emerging Technology (INCET) (pp. 16). IEEE.
6. Srinivas, M., Durgaprasadarao, P., & Raj, V. N. P. (2018, January). Intelligent medicine box for medication management using IoT. In 2018 2nd International Conference on Inventive Systems and Control (ICISC) (pp. 3234). IEEE.
7. Kumar, A., & Kumari, M. (2021). Non-living Caretaker as a Medicine Box Based on Medical Equipment Using Internet of Things. In Soft Computing: Theories and Applications (pp. 155–161). Springer, Singapore.
8. Karagiannis, D., & Nikita, K. S. (2020, August). Design and development of a 3D Printed IoT portable Pillbox for continuous medication adherence. In 2020 IEEE International Conference on Smart Internet of Things (SmartIoT) (pp. 352–353). IEEE.
9. Ayshwarya, B., & Velmurugan, R. (2021, March). Intelligent and Safe Medication Box In Health IoT Platform for Medication Monitoring System with Timely Remainders. In 2021 7th International Conference on Advanced Computing and Communication Systems (ICACCS) (Vol. 1, pp. 1828–1831). IEEE.
10. Sinnapolu, G., & Alawneh, S. (2018). Integrating wearables with cloud-based communication for health monitoring and emergency assistance. Internet of Things, 1, 40–54.

11. Nayyar, A., Puri, V., & Nguyen, N. G. (2019). BioSenHealth 1.0: a novel internet of medical things (IoMT)-based patient health monitoring system. In International Conference on Innovative Computing and Communications (pp. 155–164). Springer, Singapore.
12. Krishnamurthi, R., Kumar, A., Gopinathan, D., Nayyar, A., & Qureshi, B. (2020). An overview of IoT sensor data processing, fusion, and analysis techniques. Sensors, 20(21), 6076.
13. Nayyar, A. (2016). An Encyclopedia Coverage of Compiler's, Programmer's & Simulator's for 8051, PIC, AVR, ARM, Arduino Embedded Technologies. International Journal of Reconfigurable and Embedded Systems, 5(1).
14. Saksham Gera, Mridul Mridul, and Sachin Sharma. "IoT based Automated Health Care Monitoring System for Smart City." In 2021 5th International Conference on Computing Methodologies and Communication (ICCMC), pp. 364–368. IEEE, 2021.
15. Sakshi Painuly, Sachin Sharma, and Priya Matta. "Future Trends and Challenges in Next Generation Smart Application of 5G-IoT." In 2021 5th International Conference on Computing Methodologies and Communication (ICCMC), pp. 354–357. IEEE, 2021.
16. Elbasani, E., Siriporn, P., & Choi, J. S. (2020). A survey on RFID in industry 4.0. In Internet of Things for Industry 4.0 (pp. 1–16). Springer, Cham.
17. Lu, Y. (2017). Industry 4.0: A survey on technologies, applications and open research issues. Journal of industrial information integration, 6, 1–10.
18. Kumar, A., & Nayyar, A. (2020). si 3-Industry: A Sustainable, Intelligent, Innovative, Internet-of-Things Industry. In A Roadmap to Industry 4.0: Smart Production, Sharp Business and Sustainable Development (pp. 1–21). Springer, Cham.
19. Kumar, A., Krishnamurthi, R., Nayyar, A., Sharma, K., Grover, V., & Hossain, E. (2020). A novel smart healthcare design, simulation, and implementation using healthcare 4.0 processes. IEEE Access, 8, 118433–118471.
20. Shreya Aggarwal, and Sachin Sharma. "Voice Based Deep Learning Enabled User Interface Design For Smart Home Application System." In 2021 2nd International Conference on Communication, Computing and Industry 4.0 (C2I4), pp. 1–6. IEEE, 2021.
21. Amar Johri, Shuchi Bhadula, Sachin Sharma, and Amal Shankar Shukla. "Assessment of factors affecting implementation of IoT based smart skin monitoring systems." Technology in Society (2022): 101908.
22. Ranu Tyagi, Sachin Sharma, and Seshadri Mohan. "Blockchain Enabled Intelligent Digital Forensics System for Autonomous Connected Vehicles." In 2022 International Conference on Communication, Computing and Internet of Things (IC3IoT), pp. 1–6. IEEE, 2022.
23. Ishita Gupta, Sarishma Dangi, and Sachin Sharma. "Augmented Reality Based Human-Machine Interfaces in Healthcare Environment: Benefits, Challenges, and Future Trends." In 2022 International Conference on Wireless Communications Signal Processing and Networking (WiSPNET), pp. 251–257. IEEE, 2022.
24. Sourav Singh, Sachin Sharma, and Shuchi Bhadula. "Automated Deep Learning based Disease Prediction Using Skin Health Records: Issues, Challenges and Future Directions." In 2022 International Conference on Electronics and Renewable Systems (ICEARS), pp. 638–643. IEEE, 2022.
25. Shreya Aggarwal, and Sachin Sharma. "Voice Based Secured Smart Lock Design for Internet of Medical Things: An Artificial Intelligence Approach." In 2022 International Conference on Wireless Communications Signal Processing and Networking (WiSPNET), pp. 1–7. IEEE, 2022.
26. Sharma, S., Dudeja, R. K., Aujla, G. S., Bali, R. S., & Kumar, N. (2020). DeTrAs: deep learning-based healthcare framework for IoT-based assistance of Alzheimer patients. Neural Computing and Applications, 1–13.

Design and Automation of Hybrid Quadruped Mobile Robot for Industry 4.0 Implementation

Sivathanu Anitha Kumari, Abdul Basit Dost, and Saksham Bhadani

1 Introduction

The global manufacturing industry enters its next generation widely known as Industry 4.0 in which innovative technologies such as robotics, automation and artificial intelligence play an increasingly important role. Active industrial robots are increasing worldwide by approximately 14% every year and automation continues to create new types of robots with improved utility and function.

Migration of robots towards Industry 4.0 components. Industrial robot must fulfil certain requirements in order to be considered as Industry 4.0 component. In industries, mobile robots are used for performing tasks like carrying things repeatedly from one location to another specific location, overseeing the manufacturing line. One of the difficulties for mobile robot is that it might fail to cross the obstacle. Rovers and Unmanned ground vehicles break out of these bounds by performing tasks even in outdoor environments with complex terrain to traverse. However, there are limitations to the capabilities of such systems that some terrain might be hard for them to tackle such as climbing stairs, maneuvering over non continuous terrain etc. [1–7].

Challenges in Existing System. The existing mobile robots are either task specific or quite costly. The industrial robots used in warehouses are made to meet specific criteria and can operate only in indoor conditions effectively while the quadruped robots such as Boston Dynamic's Spot costs 70,000 USD, which makes it unaffordable for a majority of the people. Although there are other alternatives being developed that are quite cost effective, this comes at the cost of losing modularity and lesser payloads. Hybrid quadruped mobile robot combines wheel-based

---

S. Anitha Kumari (✉) · A. B. Dost · S. Bhadani
SRM Institute of Science and Technology, Chennai, Tamil Nadu, India
e-mail: anithaks@srmist.edu.in

© The Author(s), under exclusive license to Springer Nature Switzerland AG 2023
A. Nayyar et al. (eds.), *New Horizons for Industry 4.0 in Modern Business*, Contributions to Environmental Sciences & Innovative Business Technology,
https://doi.org/10.1007/978-3-031-20443-2_8

locomotion to the quadruped walking mechanism. The idea of combining wheels with quadruped walking mechanisms has been tested and validated in multiple works beforehand, one of the oldest sources being "Walk and Roll Robot" by NASA [8]. The system proposed is quite unique but at the same time mechanically complex. There are no actuators either on joints or on wheels, but all are driven using actuators in the main chassis frame using complex mechanical transmission. There are other hybrid quadruped robots such ANYmal by ETH Zurich [9], which uses three actuators for the articulated limb, and also hub motor in the wheel, making the control system quiet complex and more power extensive.

Real Time Applications. In the base configuration, the proposed robot can be used for surveillance, mapping, reconnaissance and remotely operated inspections. The modularity allows for a variety of additional systems to be integrated into the primary system with little effort such as articulated robotic manipulators, turret systems, imaging and mapping systems. The system also finds its application in construction, rescue operations and with suitable attachments can also be used in amphibious mode considering the fact that the drive motors being used in the system have an industrial grade hermetic seal. Another application can be the exploration of the unknown terrain. The robot is autonomous and can save the map of an environment. The map can later be used for further studies and research.

Considering the opportunities of Industry 4.0 in the stream of robotics, the objectives of the chapter are:

1. To design hybrid quadruped robot with both rolling and walking mechanism. The system is aimed to be robust with minimal complexity and high modularity. The fabrication cost has to be minimal and at the same time a good payload for the form factor was also aimed.
2. To test the automation of the developed robot while traversing over a long distance in a simulation environment.

The hybrid quadruped mobile robot satisfies design criteria including minimal number of actuators that reduces control complexity, cost effective and high payload carrying capacity by considering various various existing works. The robot has been named as "Gator bot" due to the gait sequence being followed resembles that of reptiles. The walking mechanism specifically used is novel and utilizes a parallel four-bar based mechanism operating using a single rotary actuator for pitch and another actuator for yaw.

Organization of Chapter

The chapter is organized as: Sect. 2 illustrates the structural design of the proposed robot. Section 3 highlights the concept of Robot Automation. Section 4 elaborates experimental Results. Section 5 concludes the chapter with future scope.

2 Structural Design

The idea finds its roots in the basic challenges faced by the wheel based mobile robots. One of the difficulties for mobile robots is passing through the obstacles as it cannot turn any other way but to pass over it [10]. This sometimes causes trouble in wheeled robots as they might fail to cross the obstacle. This is where the robot has to switch from rolling to walking mechanism and then cross the object by walking over it. In the ideation phase of the work, the initial idea was to develop a system with the capability of dynamically maintaining the orientation and level of the chassis, which then can be used as a mobile platform for a variety of tasks with the help of mountable attachments. It was observed that the actuators required for the same if utilized properly, can also allow complete independent motion of each limb in 3D space, opening the possibilities of quadruped based locomotion.

Design challenges and solutions
The challenge in the design of the robot is shown in Fig. 1.

The first challenge in the design is the wheel. If the motor driving the wheel is not powered, the wheel can easily rotate about its axis on application of external

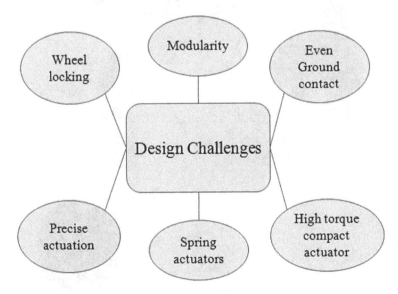

Fig. 1 Design challenges

force. To eliminate the same, a mechanism that could couple the wheel to the motor was required and at the same time allows the transmission of mechanical power only from driver (motor) to the driven component (wheel) i.e., unidirectional and interlocks or behaves as a rigid link, if the force application takes place from the driven end. The simplest of all the mechanisms that fulfilled the requirements were worm and worm gear configuration. This allows the wheel to act as a rigid member of the limb when not in operation.

The next challenge is to maintain a proper even contact of the limb ends i.e., wheels to the ground. If a single rotary actuator was to be connected to a single linkage, then the contact angle of the wheel will also vary with the pitch motion of the limb. To eliminate this, a four-bar mechanism as shown in Fig. 2, that ensure the orientation of the wheel is perfect even if only a single actuator is being used. The whole limb was connected to the chassis of the robot with another rotary joint, which accounts for the yaw motion of the limb. The frame of the chassis was first assembled using a 3-way corner block that meshed using bolts going through the central hollow structure of T-slot extrusions. This was able to hold the chassis frame, but due to joints being held by concentric cylindrical elements, the assembly was susceptible to deformation on application of large enough force. To avoid this, a second iteration with slotted 3-way connector was used, which had extrusions locking into the T-slot of the extrusion, and hence removing any play in the chassis.

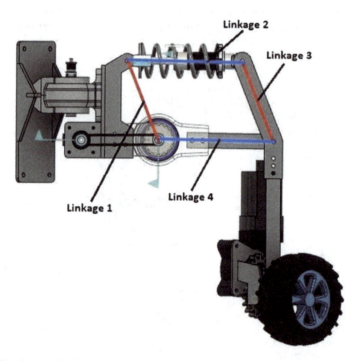

Fig. 2 Four bar mechanism of the robot limb

For the ease of manufacturing and keeping the system modular, T-slot aluminum extrusion channels were used for the chassis frame, which not only allows easy assembly but also the capability to disassemble, reconfigure and flexibility to add new or modular systems over different parts of the frame. At the same time the shape of these slotted extrusions imparts excellent rigidity to the structure, while being inherently lightweight.

Due to the size and weight of the robot, the actuators actuating the links require significant torque. Industrial actuators which have the required torque capabilities were relatively quiet, big, heavy and costly. To keep the whole system cost effective, small and simple, custom cycloidal gearboxes were developed referencing open-source gearboxes. Cycloidal gearboxes are compact and can provide a significant gear reduction. It is also sometimes called wobble drive due to the motion of the gears. The gear is eccentric and the reduction is achieved due to the relative motion between internal and external gears with different teeth or cycloidal lobes. Similar to strain wave gear systems, the cycloidal gearboxes are quite common in industrial applications. The eccentric cam when rotated moves the cycloidal disc along with it. The lobes on the disk mesh with the outer cycloid creating a relative motion. The pin rollers placed in the disk translate this relative motion into rotational output via follower shaft. Since the cycloidal disc is eccentric, this can lead to unwanted vibration and noise in the system. To avoid this, and balance out the eccentric mass, two discs are used with the eccentricity at a difference of 180°. The gearboxes used in the robot were 3D printed using stereolithography technique, which allowed a very high resolution of 0.01 mm for each layer. Resin printing ensures isotropic strength in the parts, unlike Fused Deposition Modeling, in which the strength varies in different orientations [11]. ABS like photopolymer resin from ELEGOO was used for the parts, which has excellent material properties. All the individual components inside the gearbox have deep groove ball bearings or angular contact ball bearings to ensure smooth motion and even load distribution amongst all parts. The gearboxes being used have two stage reductions, of which the first stage has a reduction of 1:15, while the second gearbox has a reduction ratio of 1:16. This provides a total reduction of 1:240 for the entire gearbox. The whole assembly is held together by the attachments at the end of the output. The center piece is an eccentric cam, which drives all the discs. For attaching the gearboxes to the chassis, parts were designed that could easily interface with both elements. For this, mild steel plates were used. Since multiple parts with same dimensions were to be manufactured, simple jigs and fixtures were made to ensure the accuracy in manufacturing. Tig welding was used due to excellent weld strength and ease of processing.

To compensate for spring actuators generally used in the traditional quadruped mechanism, which allows for impact absorption as well as feedback, mechanical damper has been deployed as one of the links in the four-bar mechanism, allowing a similar flexibility with a wider range of motion to the leg. The springs were manufactured using oil tempered spring steel. The compression spring has 6 turns, a wire diameter of 4 mm, height of 120 mm and closed ground edges. The

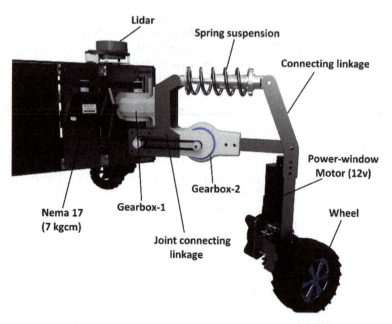

Fig. 3 Limb of Gator bot

springs are assembled over generic bike suspension struts which have further been integrated onto the system as a structural element.

The base motors being used are Nema 17 DC stepper motor, which has a base torque of 7 kg cm. The motor is coupled with the gearbox using timer pulley and timer belt with no reduction. The complete design is shown in Fig. 3.

CAD model. The CAD model of the robot is shown in Fig. 4. The robot chassis has a dimension of 510 × 290 × 190, while the whole robot has a physical footprint of 900 × 1000 × 470, where the dimensions are in the order of length (mm), width (mm) and height (mm) respectively. The whole system was designed with an additional payload of 20 kg, so that a variety of other systems could be integrated without affecting the functionality of the system. The robot achieves the gait by actuating diagonally opposite limbs in synchronous motion. At the initial condition, all the four legs are in contact with the ground and thus the robot is at rest. For the quadruped walking mechanism, the maximum torque required at the joints can be assumed at the point where only two of the limbs are in contact with the ground. In such a case all the weight of the robot will distribute evenly on the two points of support. Table 1 shows the actuator torque calculation, in which it is observed that the maximum torque for yaw actuator and pitch actuator satisfies the required condition of being less than the output torque of the gear box.

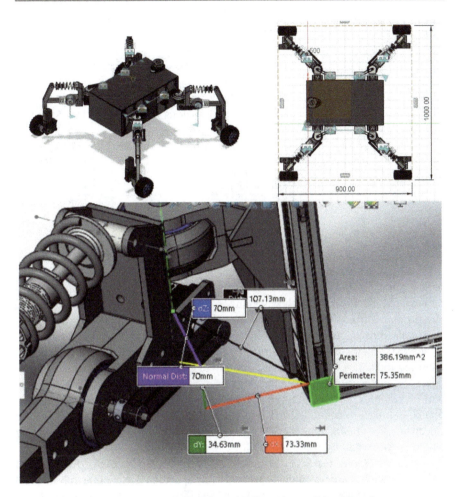

Fig. 4 Gator bot. **a** Isotropic view. **b** Top view with dimensions. **c** Joint dimensions

3 Robot Automation

Sensors for Localization. To take the map of the surroundings, there are many devices that can be utilized. Light Detection and Ranging (LIDAR) is preferred because of its high accuracy [12]. The specific LIDAR that is used for this work is RPLIDAR A1 which has a range of 6 m radius. The motor provides the pulse width modulation to the LIDAR. So, the frequency of rotation can be controlled up to a maximum of 10 Hz. The data is taken by RPLIDAR and it is passed through the scan topic. This data is used to convert to a 2D grid map using gmapping.

Autonomous Navigation. The addition of autonomous mobile robots changes the future manufacturing industries to smart industries. However, the collaboration

Table 1 Torque calculations for actuator

Gear ratio for cycloidal gear box	1:240
Output torque of the gear box	1680 kgcm
Distance between the actuator 1 and actuator 2 of the same limb	121.5 mm
Distance between the actuator 1 and the body	73 mm
Distance between the actuator 2 and the body	194.5 mm
Actuator 1(yaw actuator for the limb) Distance between the diagonally opposite limbs Total weight of one limb Maximum torque	575.41 mm 17.5 kg 1006.96 kgcm
Actuator 2 (pitch actuator of limb) Distance between diagonally opposite limbs Total weight of one limb Maximum torque	820.551 mm 17.5 kg 1436 kg cm

of the robots in Industry 4.0 set-up poses few challenges because of the complex navigation system and the dynamic evolution of the industries. An accurate navigation control is required so that the robot can move in the designated space.

Autonomous Navigation for Industrial Inspection. One of the evident problems in deploying autonomous navigation for inspection in energy and process manufacturing industries is path planning for an environment with frequent technology upgrades which requires frequent updation of maps. In such cases path planning will be complex as there may be infinite ways to achieve the trajectory. A theoretical approach to use ROS navigation stack for path planning and Simultaneous Localization and Mapping (SLAM) is discussed in [13]. The optimal trajectory would be generated by using probabilistic approach.

Adaptive Monte Carlo Localization. Adaptive Monte Carlo Localization (AMCL) adopts particle filter, a probabilistic approach for mobile robot localization [14]. The approach preserves the probability distribution of different poses of the robot and it is updated using the information from laser scan. The 2D map is obtained from the map node and random poses (particles) are generated throughout the map and saved as filter estimate. The estimated pose of the robot is compared against the real LIDAR data. Each time the robot moves, the particles are resampled according to the new probability distribution. If the estimated pose is highly likely then more particles are generated in that pose else the particles are decreased. The particles are generated after each generation and as the probability distribution of the particles become narrower fewer particles would be generated.

Path Planning. A* algorithm is used to find the shortest routes between the initial point and goal point with low computational time [15, 16]. It provides a heuristic estimate of the distance from a node to the goal node. The algorithm requires two lists namely open list and closed list to store information about the nodes. Open list stores nodes for expansions and closed list stores the explored nodes. The node expansion depends on the cost parameters accumulated along a path. The flow chart of the algorithm is given in Fig. 5.

Actuator Control Using Nvidia Jetson Nano. Nvidia Jetson Nano acts as the master device to perform the navigation on the actuators. It computes navigation by processing the information from the sensor array that includes LIDAR, Inertial Measurement Unit and camera. ROS operating system is boarded on the Jetson Nano microprocessor. ROS communicate to Arduino microcontroller using rosserial protocol [17]. The rosserial makes use of topics and services. Topics and services are a part of ROS OS for publishing a message or subscribing to a message. rosserial_python is the recommended package used for connection between host device and the slave device. The physical connection between Nvidia and Arduino can be through many ways including UART, SPI or I2C. However, a simple USB connection is being utilized from Jetson Nano to Arduino, that being the easiest way to communicate.

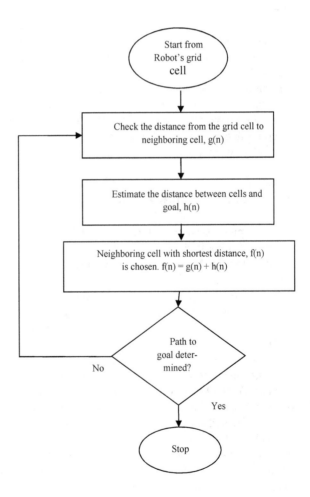

Fig. 5 Flow chart of A* algorithm

Driver Control using Arduino. DRV8825 microstepping driver is used to supply sufficient power to the stepper motors. The smaller size and lighter weight make it a better option for controlling the weight of the robot. It has an operating voltage of 8.2–45 V and can operate at a current up to 2.2 A, provided there is sufficient cooling system attached to the driver. Arduino Mega acts as a node device and also controls all the drivers of the stepper motors used in the joints of the robot. It controls the rotation and direction of the motors. Codes were written using Arduino IDE to communicate to the drivers to drive the motors. Pins are defined for the direction and steps in the software coding. Steps per revolution define the resolution of the steps taken and in turn the control over the joints. The pins that are declared as output pins are activated in a loop to continually rotate the motor. The speed of the motor can be changed based on the delay in the activation of the pin.

Cytron MDDS30 drivers are used to supply sufficient power to the power window motors. The power window motors are used for the wheel rotations. These motors are also controlled by the Arduino microcontroller. The power window motors are operated at 12 V. The Arduino is connected to the driver through PWM input mode. It is faster than the other options. There are different pins of the MDDS30 driver that are to be connected to the microcontroller. There are 4 pins for controlling the motor. Similarly, there are 2 motor power terminal blocks. This is so that we can connect two motors to each driver. One right and one left motor is supplied power through one MDDS 30 driver. The speed of the motors can also be controlled by giving different PWM signals to the driver. The direction is also controlled by coding a positive or negative value in the Arduino IDE to the driver.

4 Experimental Results

The complete system architecture involving the automation of the designed quadruped robot is shown in Fig. 6. Nvidia Jetson Nano acts as the master device. LIDAR, Inertial Measurement Unit and camera are the part of the sensor array interacting directly with the Jetson. The Jetson controls the drive motors via smart motor drivers using I2C communication protocol. The stepper motors are controlled by the Arduino mega, which acts as a node device and in turn controls all the drivers for joint stepper motors. ROS in integration with Gazebo environment is used to control the robot. ROS is used for developing an algorithm to easily communicate to a robot or to call an already present algorithm. The Gazebo environment, is used to visualize the behavior of the robot.

Scan Tests Through RPLIDAR A1. The experimental analysis was performed by choosing a residential apartment as the environment. Scan tests were performed through the RPLIDAR A1 by providing permission to the USB port to connect the RPLIDAR to the master operating system [18]. As RPLIDAR is connected to the port it starts to rotate upon providing permissions for the same. The required drivers of the LIDAR were installed. RPLIDAR moves slowly to get the proper mapping and when the mapping area is completed, the data is allowed to overlap

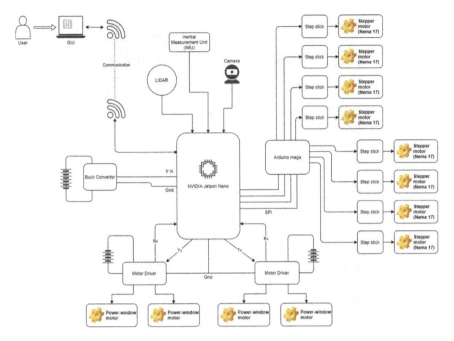

Fig. 6 System architecture

by allowing a few more seconds for it to work before the mapping process is halted.

As a first step the driver files for RPLIDAR, rplidar node is launched and the code is initiated. It reads RPLIDAR raw scan result using RPLIDAR's software development kit and convert to ROS Laser Scan message. The RViz tool of the ROS is launched to view the scan result. This tool gives a visual of what the map would look like. So, there wouldn't be a necessity to wait for LIDAR to scan the whole surrounding and save the map and then see the output. The RViz tool shows the output simultaneously as the LIDAR is moved. A clear map was obtained successfully as shown in Fig. 7 after a few failures. When the data was echoed from the rplidar in the terminal, it was realized that the data had minimum errors and that the position and orientation of the RPLIDAR from the starting position were accurate as shown in Fig. 8.

Gmapping. ROS Gmapping is used to implement SLAM that creates a 2D grid map using the data from RPLIDAR and the odometry data. The reason for using Gmapping is that the data received from odometry is more accurate. Figure 9 shows a map taken through the LIDAR sensor. A static map is not needed for the gmapping to operate properly.

ROS navigation stack uses costmaps to give information about each point on the map, locally and globally [19]. Through this information, the robot would be able to configure the optimal path and the obstacle. Global costmap is created from the map that is available at the start of the localization. Local costmap is created

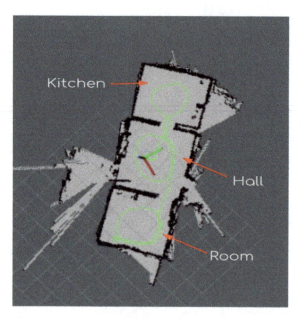

Fig. 7 An apartment scanned through RPLIDAR A1

Fig. 8 The position and orientation of rplidar

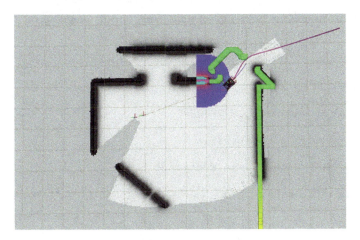

Fig. 9 The robot mapping the surrounding as it is moving forward

and updated from the robot's sensor data. Both the costmaps are configured to set thresholds on the information about the environment, in which certain parameters are common to both the maps. The configured parameters are shown in Fig. 10.

The obstacle_range defines that the robot detects any obstacle within 2.5 m is put into the costmap and given the sensor information, raytrace_range determines that the costmap is cleared and the environment is updated every 3 m. The geometry of the robot in the ROS navigation stack is defined by footprint. The inflation radius at 2.0 m indicates that all path greater than or equal to 0.55 m from obstacles have same obstacle cost.

In global costmap, the static_map parameter is set True for the costmap to be initialized based on the map served to it. The robot already has the map of the surroundings and when a target location is given to the robot, it will create its global path as shown in Fig. 11.

Localization Using Particle Filter Estimation. AMCL node performs localization on the developed robot model. The system works such that it localizes the mobile robot, given the map of the surrounding. Therefore, it is combined with the gmapping or hector_mapping to overcome this limitation. AMCL works on the basis of Monte Carlo Localization, where particles are used to indicate each point on a map [20]. Figure 12 illustrates how particles are generated by the particle filter in the 2D map of the surrounding simulation environment the robot is moving in. The particles are red in color.

The particle filter compares each point on the map with the sensor data. It estimates the likelihood of the robot to be in that pose in the map. The closer a particle is related to the sensor's data; the more is the probability that the robot is in that location. These particles have specific position and direction in the map. This way the particles that in comparison don't have that specific position and direction as the robot are eliminated. More particles are generated at the position where there is higher probability of the robot. After a while, all the particles gather

```
Common parameters between global and local costmaps:
  obstacle_range: 2.5
  raytrace_range: 3.0
  footprint: [[-0.25, -0.145], [-0.25, 0.145], [0.25, 0.145],
              [0.25, -0.145]]
  transform_tolerance: 0.5
  resolution: 0.05
  inflation_radius: 2.0

Global costmap parameters:
  global_frame: map
  robot_base_frame: base_footprint
  update_frequency: 1.0
  publish_frequency: 0.5
  static_map: true
  cost_scaling_factor: 10.0

Local costmap parameters:
  global_frame: odom
  robot_base_frame: base_footprint
  update_frequency: 1.0
  publish_frequency: 2.0
  static_map: false
  rolling_window: true
  width: 2.5
  height: 2.5
  cost_scaling_factor: 5
```

Fig. 10 Costmap parameters

Fig. 11 Initial global path

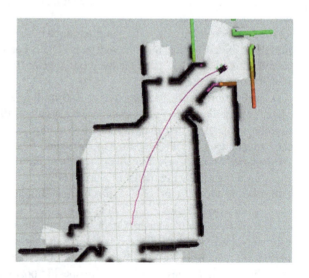

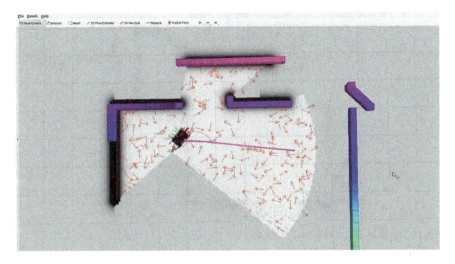

Fig. 12 Particle filter technique used by AMCL

at the exact location where the robot is located. Figure 13 shows all the particles gathered to the exact same pose as the robot. Their direction and orientation match to that of the robot in the simulation environment.

Local Path Planning. The global path doesn't consider any obstacle that might come later in its area. Then the help of local planner is taken to update and correct the global path in the dynamic environment. The local path is created by collecting the data continually from the sensors attached to the robot. The robot faces the obstacle and updates its global path and modifies it as shown in Fig. 14. The local path is created over the local cost map.

Gait Analysis. The gait pattern of the robot is characterized by configuration of specific gait parameters. The maximum velocity of the robot while moving in the

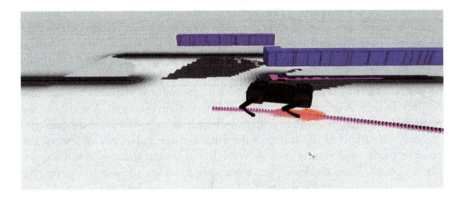

Fig. 13 Red particles gathered at the exact pose of the robot in the map

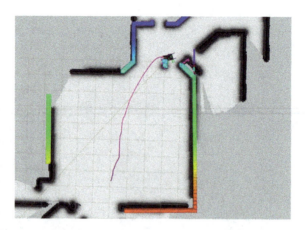

Fig. 14 Local path updation over obstacles

forward and reverse direction is 0.5 m/s and while moving sideways the maximum velocity is 0.25 m/s. The velocity of the robot is scaled to a factor of 0.9 so as to compensate for odometry errors while determining the position of the robot. The maximum rotational speed of the robot is 1 rad/s. Each leg of the robot spends 0.25 s on the ground during the stance phase of the gait cycle. The trajectory height during the swing phase of the gait is 0.04 m. In order to ensure the stability of the robot the walking height is limited to 0.2 m. The mass of the robot is uniform throughout the body hence the reference point in the forward direction is zero.

5   Conclusion and Future Scope

The work presents the design and automation of hybrid quadruped mobile robots with the primary goal being the development of a robust system with minimal complexity, high modularity. The actuators being used were at most efficiency and at the same time a good payload for the form factor was obtained. The robot has been tested in the simulation environment and it traversed over a long distance autonomously. The map created by the LIDAR has higher efficiency and minimum errors. The whole navigation simulation was applied on the real robot to test its autonomy. This was tested with the dynamic environment to check the update of the path as new obstacles are introduced into the environment. The robot takes the decision to change the path until the obstacle is in a very short distance gap. The only constraint to the robot in the simulation world is that it cannot take the optimum decision autonomously when it's in a tight space where it required manual control. Since the design was made to be modular, the future work on robot has the possibility to attach a variety of modules and systems, such as articulated robotic manipulators, turret systems, various imaging, mapping systems etc.

References

1. Batth, R. S., Nayyar, A., & Nagpal, A.: Internet of robotic things: driving intelligent robotics of future-concept, architecture, applications and technologies. In 4th IEEE International Conference on Computing Sciences (ICCS), pp. 151–160 (2018).
2. Nayyar, A., Puri, V., Nguyen, N. G., & Le, D. N.: Smart surveillance robot for real-time monitoring and control system in environment and industrial applications. In Information Systems Design and Intelligent Applications. pp. 229–243. Springer, Singapore (2018).
3. Nayyar, A., Nguyen, N. G., Kumari, R., & Kumar, S.: Robot path planning using modified artificial bee colony algorithm. In Frontiers in Intelligent Computing: Theory and Applications. pp. 25–36. Springer, Singapore (2020).
4. Alzubi, J., Nayyar, A., & Kumar, A.: Machine learning from theory to algorithms: an overview. In Journal of physics: conference series vol. 1142, No. 1, p. 012012). IOP Publishing (2018).
5. Nayyar, A., Kumar, A. (Eds.).: A roadmap to industry 4.0: Smart production, sharp business and sustainable development. Springer (2020).
6. Singh, K. K., Nayyar, A., Tanwar, S., Abouhawwash, M.: Emergence of Cyber Physical System and IoT in Smart Automation and Robotics: Computer Engineering in Automation. Springer Nature (2021).
7. Kumar, A., Nayyar, A.: si 3-Industry: A Sustainable, Intelligent, Innovative, Internet-of-Things Industry. In A Roadmap to Industry 4.0: Smart Production, Sharp Business and Sustainable Development (pp. 1–21). Springer, Cham (2020).
8. Cham.Wilson, A., Punnoose, A., Strausser, K., Parikh, N.: Walk and roll robot. *U.S. Patent No. 8,030,873*. Washington, DC: U.S. Patent and Trademark Office (2011).
9. Hutter, M., Gehring, C., Jud, D., Lauber, A., Bellicoso, C. D., Tsounis, V., Hoepflinger, M.: Anymal-a highly mobile and dynamic quadrupedal robot. In 2016 IEEE/RSJ international conference on intelligent robots and systems (IROS), pp. 38–44 (2016).
10. Köseoğlu, M., Çelik, O. M., Pektaş, O.: Design of an autonomous mobile robot based on ROS. In. International Artificial Intelligence and Data Processing Symposium (IDAP), pp. 1–5 (2017).
11. Covaciu, F., Bec, P., Băldean, D. L.: Developing and Researching a Robotic Arm for Public Service and Industry to Highlight and Mitigate Its Inherent Technical Vulnerabilities. In. Multidisciplinary Digital Publishing Institute Proceedings, vol. 63, No. 1, pp. 25 (2020).
12. Iqbal, J., Xu, R., Sun, S., Li, C.: Simulation of an autonomous mobile robot for LiDAR-based in-field phenotyping and Navigation. Robotics, 9(2), 46 (2020).
13. Guimarães, R. L., de Oliveira, A. S., Fabro, J. A., Becker, T., Brenner, V. A.: ROS navigation: Concepts and tutorial. In Robot Operating System (ROS) pp. 121–160. Springer, (2016).
14. Muñoz–Bañón, M. Á., del Pino, I., Candelas, F. A., Torres, F.: Framework for fast experimental testing of autonomous navigation algorithms. Applied Sciences, 9(10), (2019).
15. Megalingam, R. K., Teja, C. R., Sreekanth, S., Raj, A.: ROS based autonomous indoor navigation simulation using SLAM algorithm. International Journal of Pure and Applied Mathematics, 118(7), 199–205 (2018).
16. Afanasyev, I., Sagitov, A., Magid, E.: ROS-based SLAM for a Gazebo-simulated mobile robot in image-based 3D model of indoor environment. In International Conference on Advanced Concepts for Intelligent Vision Systems. pp. 273–283. Springer, Cham (2015).
17. Candelas-Herías, F. A., Garcia, G. J., Puente Méndez, S. T., Pomares, J., Jara, C. A., Pérez Alepuz, J., Torres, F.: Experiences on using Arduino for laboratory experiments of automatic control and robotics. IFAC-PapersOnLine 48, no. 29, pp. 105–110 (2015).
18. Grzechca, D., Ziębiński, A., Paszek, K., Hanzel, K., Giel, A., Czerny, M., & Becker, A.: How accurate can UWB and dead reckoning positioning systems be? Comparison to SLAM using the RPLidar system. Sensors, 20(13), 3761 (2020).

19. Marin-Plaza, P., Hussein, A., Martin, D., Escalera, A. D. L.: Global and local path planning study in a ROS-based research platform for autonomous vehicles. Journal of Advanced Transportation, (2018).
20. An, Z., Hao, L., Liu, Y., Dai, L.: Development of mobile robot SLAM based on ROS. International Journal of Mechanical Engineering and Robotics Research, 5(1), 47–51 (2016).

To Trust or Not to Trust Cybots: Ethical Dilemmas in the Posthuman Organization

Arindam Das and Debojoy Chanda

1 Introduction

Industry 4.0 has entered the picture and is positively impacting manufacturing, service system, production innovation [1, 2], internationalization of business [3] and global networking [4]. Within a manufacturing environment, Industry 4.0 [5] captures disruptive high-end automation and data-driven mechanization. These innovations conduced to the expanding productivity both in service systems and manufacturing ecosystem. These progressions caused the expansion of the improvements in assembling and data innovation, and these planned and informative advancements are comprised to the term, Industry 4.0 which was first reported from German government as one of the vital drives and features a new modern insurgency. Such a highly networked system combines the digital-cyber and physical-human space. This sort of a Cyber-Physical-System (CPS) which combines mechanized, automated robotics on the one hand and human employees on the other, compels a rethinking of the issues about organizational trust beyond the human-to-human communication model, whether across horizontal or vertical-hierarchical structures. Hence, beyond the conventional concept of human-centric trust, what evolves amidst Industry 4.0 is the question of a technology-centric digital trust. However, with the inevitability of digital trust and its emphasis on innovative organizational performance [6–11], stakeholder relationships (which comprise both human and technological agents) within an organization, are hugely

A. Das (✉)
Faculty of Humanities, Liberal Arts and Social Sciences, Administrative Building, Alliance University Chandapura-Anekal Road, Bangalore, Karnataka 562106, India
e-mail: arindam.das@alliance.edu.in

D. Chanda
P. B. College, RS Kanakpur, Panskura, West Bengal 721171, India

impacted. The mutual knowledge, value sharing, regard, and respect among various agents within an organization, affecting the firm's performance, tend to alter considerably [12].

In this chapter, we make a qualitative theoretical critique of the concept of digital trust as a contradiction in terms within Industry 4.0 as a posthuman organization [13, 14]. Our perception of this contradiction stems from the fact that the digital is driven by technology, while trust is founded on subjective and instinctive human emotion. In this sense, the digital and the human occupy polarities within the dynamics of trust in the posthuman condition of an organization [15]. Indeed, we can choose to view Industry 4.0 as an epitome of digital distrust within the history of organization development if distrust is generally thought to imply betrayal. The betrayal that we hint at is the betrayal of human autonomy and decision-making control by digitized bodies in the form of controlled algorithms [16], job substitution, and re-shaping of employment dynamics [17]. After all, Industry 4.0 makes for workplace impotence for its human employees through practices like biometric examination and the conversion of the employees' identification to data, with this data serving as a means of digital surveillance to categorize supposedly underperforming employees as 'lesser' beings [18, 19]. Such practices of surveillance ominously promise the replacement of the greater number of human employees in an Industry 4.0-based organization with AI-empowered automated cybots [20], what with the greater number of human employees having been found/declared to be underperforming when under digital scrutiny [21]. This narrative of laying off 'deficient' and therefore unwanted employees is in keeping with the characterization of Industry 4.0 as the fourth industrial revolution [22], with the Industrial Revolution (nineteenth century) having been historically perceived as utilitarian in nature [23]. It is this utilitarian character of Industry 4.0 that brings in its wake the distrust surrounding the digital ethics it generates.

Having said that, posthuman organizations are not inherently diabolic. Industry 4.0, as an institution which straddles the synthetic and the human and performs within the liminal technobiological zone of a digital/physical ecosystem, has the potency to critique machinery triumphalism. This is in keeping with the expectation that the posthuman condition is supposed to be emancipatory and democratic for its agents (both man and machine), each respecting the other's position [24]. "Posthumanism questions the banal profitability and commodity fetishism of transhuman global capitalism engendered through computational networks" [25]. For example, the ART (Artificial Reproductive Technology) critiqued by feminists on principles of patriarchal coerciveness [26] may in fact be liberatory for same-sex or unmarried heterosexual couples who can reproduce non-coitally through the intermediacy of donors and surrogates [27, 28]. These simultaneous potentialities of posthuman Industry 4.0 to become "capitalocene" [29] (usurped by the commodity fetish and utilitarian motives of advanced capitalism) on the one hand and paradoxically liberatory and democratic on the other, creates a dilemma around the ethics it generates. To trust or not to trust such ethics in Industry 4.0 becomes a moot issue, and our chapter intends to find a resolution to this dilemma. To delve deep into this inherent dilemma of the ethics of Industry 4.0, we take recourse to French philosopher and thinker Michel Foucault's "ethics of curiosity" which helps us understand the socio-economic impact of Industry 4.0 on its employees. It also allows us to grasp how the employees may engage with the uncertain futures that may impact their livelihoods.

Our intervention is timely because much has already been written about the transition to Industry 4.0 Researchers have examined data analytics, simulation, embedded systems, additive manufacturing, virtualization technologies, the modernization of logistical systems that it brings in its wake, the development of self-management systems conducted with the assistance of the IoT, and the constant development and self-updating of production systems [30, 31]. However, while there is a common consensus that the large-scale automation that Industry 4.0 will bring with it will also lead to layoffs across the board, this point is, at the end of the day, something of a determinist projection which is yet to be fulfilled—a matter that researchers need to probe [32]. We daresay that though this projection of extensive unemployment brings with it a sensation of collective alarm, not much—if at all—has been written about precisely how employees will react to this projection when Industry 4.0 fulfills its goal of complete automation. It is this matter that we examine as a pure projection, if only because complete automation is itself, ultimately, a pure projection.

Thus, the objectives of this qualitative research paper are the following:

a. to theoretically probe the ethical dilemmas around digital trust,
b. to understand the dilemmas of digital trust in the posthuman condition of Industry 4.0,
c. to employ Foucauldian socio-cultural frame of ethics of care in trying to understand how the human employees of Industry 4.0 can move beyond the dilemmas of digital trust, and
d. to probe how within an organization an anthropocentric trust may be generated to counter digicentricity.

Organization of Chapter

The chapter is organized as: Section 2 conceptualized Industry 4.0 as Posthuman Organization. Section 3 stresses on Digital Trust and Posthuman Industry 4.0. Section 4 elaborates Digital Distrust and the Posthuman. Section 5 focusses on the need for Curiosity: The Determinism in Industry 4.0. Section 6 enlightens Michel Foucault's Ethics of Curiosity. Section 7 highlights Two Foucauldian Scenarios: Resolving Dilemmas of Trust in Industry 4.0. Section 8 concludes the chapter with future scope.

2   Industry 4.0 as Posthuman Organization

To understand our characterization of Industry 4.0 as a 'posthuman' organization, we look back to Hayles's [33] coinage of the term 'posthuman.' Hayles's definition of the term bears cautionary hues. Hayles views 'posthuman' as a term implying the privileging of a dispensation which can reduce human consciousness and knowledge to what we can best describe as an automated administration of things. This automated administration is, based on our appropriation of Hayles's definition, a regime which views human consciousness as having been given undue importance. After all, a posthuman dispensation, according to Hayles [33], equates computer simulation, robot technology, and cybernetic mechanisms with the virtues of human consciousness and human knowledge. This is because the first three can, as per the logic of a posthuman dispensation, be made to reduce human consciousness and human knowledge to machinic data, and to consequently displace human presence within a posthuman regime [34]. However, Gareth Morgan, in his *Images of Organization* [34], had already cautioned about any over-digitization of an organization and its left-brain centricity which emphasizes mechanized articulation of AI: "The information processing perspective has created a fresh way of thinking about organization. But there are two major criticisms...The first is that most decision-making and information processing views have had a 'left-brain bias' and an overcentralized view of the nature of the organizational intelligence...The emphasis was placed on rational, analytical, reductive approaches to information processing and problem solving. More intuitive nonlinear approaches, characteristic of a more 'right-brain' orientation, were underemphasized" [35]. This preponderance of machine in an organization undercuts the more democratic level-playing approach of the posthuman organization as promising equal potential for both man and machine.

Our understanding of a posthuman organization owes much to Matthew E. Gladden's *Posthuman Management: Creating Effective Organizations in an Age of Social Robotics, Ubiquitous AI, Human Augmentation, and Virtual Worlds* [13]. Invoking critical posthumanism, Gladden asserts.

Organizational posthumanism does not naïvely embrace all forms of posthumanization; unlike some strains of transhumanist thought, it does not presume that all emerging technologies for genetic engineering or nanorobotics are inherently beneficial and free from grave dangers. But at the same time, organizational posthumanism does not directly join bioconservatism in attempting to block the development of particular technologies deemed to be hazardous or destructive. Instead, organizational posthumanism focuses on analyzing posthumanizing technologies that are already available or whose development is expected in order to assess their (potential) impact on organizations and develop strategies for utilizing such technologies in ways that are ethical, impactful, and efficient. Organizational posthumanism recognizes that emerging technologies are likely to possess both benign and harmful applications, and the role of a manager as such is to identify and creatively exploit the beneficial aspects of a technology within a particular organizational context while simultaneously avoiding or ameliorating the technology's more detrimental effects [36].

In short, we borrow from Hayles's [33] definition that describes posthuman organization as an automated administration of things, and from Gladden's [13] explication which goads us to creatively exploit its positive attributes and warns us against its detrimental effects. Hence, our understanding of Industry 4.0 as an automated posthuman industry is neither a dismissive one which overlooks its benefits nor one that a critically agrees to the absence of impending threats.

3 Digital Trust and Posthuman Industry 4.0

For Gladden, in a digital-physical posthuman organizational ecosystem, five different agents shall co-exist as shown in Fig. 1: anthropic-humans (natural human beings), computronic-humans (cyborgs), anthropic-artificials (bioroids), computronic-artificials (computers), and hybrids (a mix of the previous four) [37].

The biggest challenges for this new techno-biological workspace are the concern of power dynamics among newly formed hierarchies, and the accompanying need for mutual trust [38, 39]. Presuming the leader of such an organization to be a "social-robot", with the much needed "social and emotional intelligence, wisdom, ethical insight, moral courage, and selfless personal commitment" [39], trust, for Gladden, is an upward communication from human subordinates towards an e-CEO. However, as Das [15] reminds us, "trust within the organization can flow in any communication channel—horizontal and/or vertical (both upward and downward)—and among a combination of all performing agents. Hence, trust is not only a strategic vision promulgated by an e-leader but can also be a commitment of a machine towards a man working at the same hierarchy, or of a hybrid employee towards a human CEO.

However, when we talk about any posthuman organizational space, the trust that is thereby generated is a combination of human-socio-emotional-ethical values on one hand and digital-technological on the other hand" [40]. It is this that makes Das claim, "Digital trust is posthuman. And, like any hybrid posthuman element, digital trust, has the potency to subvert the anthropocentricity of human trust" [41]. So much for digital trust that has the postmodernist potency to subvert the anthropocentricity of traditional organization and exalt the digicentricity of a transhuman organization.

Yet, hybrid posthuman digital trust which now theoretically gains a centrality of communication among new-formed agents, may further be invested with power by capitalist agents for profit motives. Das justifiably warns us, "any sense of unquestionable trust on such anthropomorphic subjects…leads to the overreliance and thereby the overdominance of the anthropomorphic agents at workplace. This generates posthuman anxiety that affects the digital trust whereby the apprehensions of digital data breaches and digital panoptic surveillance in an organization

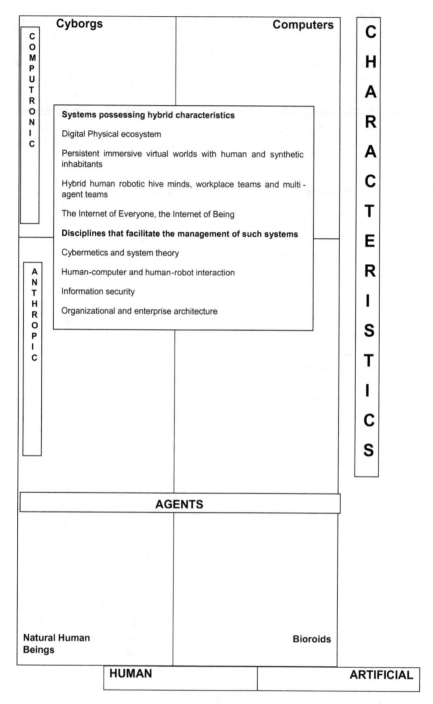

Fig. 1 Gladden's Posthuman Management Matrix for organizational management in the Posthuman Age [13]

loom large" [42]. To put it in a simpler perspective, digital trust tends towards digital distrust.

The overpowering grand narrative of digital trust in Industry 4.0 may thus constitute a regime which deprivileges human employees. This regime may, for example, subject human employees to central surveillance and panoptic mechanisms which in turn threaten to find them 'lacking' in skills and performance. Such surveillance accordingly iterates a call for the replacement of human employees with automation. It promises, through this process of laying off and replacement, that humans, human consciousness, and human knowledge and skills will become the stuff of the past within its automation-based organizational structure [43]. Hence the disturbing aptness of the term 'posthuman' for this regime in Hayles's sense on three counts—1. the equation and reduction of human knowledge and skills to machinic presences, 2. the consequent displacement of human livelihoods, and 3. the concomitant location of human employees in a past moment in the unfurling of organization development. These counts manifest the unsettling appropriateness of the prefix 'post' in 'posthuman' as a descriptor for Industry 4.0, what with human employees apparently becoming figures locatable only in an organizational past, thanks to automation.

4 Digital Distrust and the Posthuman

Industry 4.0's posthuman bearings within the framework we lay out can be traced to its beginnings in the history of organization development. German physicist and businessman, and former Chief Executive Officer of the *Systems, Applications & Products in Data Processing Societas Europaea* (SAP SE), Henning Kagermann, coined the term 'Industry 4.0' with his colleagues. He intended, through his coinage, to describe the German federal government's high-end technological strategy and digitization of manufacturing [44]. This techno-digital regime is, at present, generally accepted as one in which cobots empowered with AI are predicted to replace human employees in the manufacturing scenario, in keeping with what we have stated [45]. Such a chain of replacement is claimed to be continuous, with cobots themselves eventually to be ostensibly substituted with state-of-the-art cybots. These cybots, based on the narrative spun by Industry 4.0 about itself, promise to function like humans and to gradually supplant human employees within the regime as a posthuman organization. The utilitarian justification of this regime for such an act—although it needs none—would be that digital surveillance has found employees wanting in the optimization of their performance [46, 47]. By extension, the narrative clearly indicates that the organization will rob the livelihoods of human employees through its empowerment of cybots with human knowledge and skills.

The projection of a posthuman regime of cybots in Industry 4.0, then, essentially builds an environment of distrust into the organizational framework. After all, human employees who find themselves under constant digital surveillance, get disgruntled with and resistant to the anticipatory conformity to docility expected of them [46]. Having to live in such a spiral of resistance, undermotivation, and inevitable underperformance within the gaze of what we term a 'digital panopticon,' these employees unsurprisingly exist in perpetual uncertainty about the longevity of their employment, apprehensive of their replacement with cybots. Little wonder that distrust of techno-digitization is the stuff of Industry 4.0 as a posthuman organization, with digital trust accordingly proving itself a paradox in the narrative.

Critical posthumanism of the kind that Hayles wields, would view a return to a privileging of the human as impossible, once a posthuman regime has set in. After all, posthuman/cybernetic alterations to anthropocentric socioeconomic infrastructures render restoration to 'human'/"precybernetic" states well-nigh unworkable [48].

Despite this impossibility, allies of critical posthumanism indirectly foreground a longing for a return to anthropocentrism. For example, advocate of biodiversity Donna Haraway, while resisting anthropocentrism and accounts of human exceptionalism, seems to yearn for a return to a human-centric world. Haraway [47] obliquely hints that through such a return, mankind can make up for the sins it has committed against biodiversity [49]. Digital trust within the narrative of Industry 4.0 is, by our understanding, grounded in precisely this kind of a longing for a return to anthropocentrism, if only so that a posthuman organization can be absolved of the wrongs it has perpetuated against its human employees by forcing them into underperformance and threatening to substitute them with automation. After all, a return to anthropocentrism is tantamount to a restoration of pre cybernetic livelihoods and, therefore, to a restitution of human trust in the organization.

5   The Need for Curiosity: The Determinism in Industry 4.0

The inconsistency at the core of restoring digital trust in posthuman organizations or Industry 4.0 is that these organizations' cybernetic futures are founded on narratives bearing fruit in some ambiguous future moment. Furthermore, Özdemir and Hekim [20] correctly point out that there is a determinism associated with futuristic projections of Industry 4.0's apparently inevitable cybernetic automation [50]. Owing to this projected determinism, consumers of the narrative of Industry 4.0 can link two complementary axioms to its future:

1. Human employees in Industry 4.0 will, in the malevolent spiral of underperformance perpetrated by the digital panopticon, lose their jobs to a dystopian AI-emboldened techno-digital administration [50].
2. Industry 4.0 will bestow benevolent capitalist prosperity upon its consumers, thanks to the unflagging efficiency that cybots will apparently bring to the manufacturing ecosystem with their advent and their supplanting of human employees [51].

Both axioms, French philosopher Foucault (1926–1984) would probably have pointed out, are founded on a critical certainty rather than an open curiosity toward what is in fact an uncertain future. Ironically, criticism and critical certainty, Foucault [49] suggests, can only bring themselves to bear on that of which one has complete knowledge [52]. Through a Foucauldian lens, we would consequently indicate that if one must engage with what is in fact a deterministic and uncertain narrative of automation, one must get beyond what Anker and Felski [50] describe as the "chronic negativity of critique" [53]. This would allow us to move towards an ethics of curiosity—an ethics founded on an unconditioned and uncritical openness—when dealing with unknowable futures [52, 54]. The need for such an unconditioned curiosity is grounded in the fact that determinisms are de facto narrative fabrications made from the point of view of the present, refusing to recognize the possibility of future contingencies.

Inversely, a Foucauldian prism can help us suggest that even if the worst befalls human employees who find their livelihoods on the chopping block, they might use their curiosity to build a united front before Industry 4.0's automated dispensation. They can do so by building a network of communication and care with their laid-off fellow-employees—employees with whom they could previously perhaps not communicate despite their possibly genuine curiosity about them. Such communication was, after all, prevented under the omnipresent gaze of the digital panopticon in the confines of the workplace, the idea of a panopticon being to maintain central surveillance upon subordinates who become "object[s] of information, never…subject[s] in communication" [55].

Foucault's ethics of curiosity may allow us to connect with the uncertain materialization of Industry 4.0 and the socioeconomic impacts it may have upon its human employees in the future. Therefore, we will try here to outline Foucault's thoughts on curiosity before using it to rationally participate in a dialogue with the unknown and unknowable futures built into Industry 4.0. It is, after all, necessary to intellectually engage with these futures upon which the sustenance of human employees' livelihoods and the possibility of trust—as opposed to digital distrust—hinge.

6 Toward an Uncritical Curiosity: Michel Foucault's Ethics of Curiosity

Foucault [53] makes it clear that because the known can take a turn for the unknowable, one can never make a definitive claim about the future. Consequently, critics of extant narratives of the future—such as that of Industry 4.0's posthuman automation—may change their beliefs. These critics therefore need to examine the future through an uncritical prism [56]. In keeping with this position, Foucault, in an interview published in 1980, connects the fact of the enigmatic with the need for engagement with the unknowable through the lens of curiosity. In the interview, Foucault [54] indicates that curiosity needs to be open, uncritical, unconditioned, and unconditional because the limits of the knowable often take the most unexpected turns [57]. These turns accordingly demand that one unlearn what one supposedly knows, if only to be able to learn that which is currently unforeseen and unforeseeable [56].

Unfortunately, suggests Foucault [54], unconditional curiosity has been perverted over the ages to produce bodies of knowledge which are apparently fixed [58]. Foucault uses the greater part of his oeuvre to explain how these supposedly authoritative bodies of knowledge—including medicine, psychiatry, and the law—have used this perverted curiosity to make shallow claims about the unchangeability of their findings. These bodies of knowledge superficially negate the necessity for unconditional curiosity because they claim to offer an all-surveillant and therefore holistic knowledge of all there is to know. In the long run, the findings of these bodies of knowledge essentially categorize humans in terms of 'normality' or 'abnormality,' with no grey areas lying in between [59].

The superficial expertise of frameworks of knowledge, Foucault would have argued, has also assisted these frameworks to assert surveillance and power over the bodies and consciousnesses of populations in society who consume them and thus try to 'normalize' their thoughts and actions [60–66]. If one takes a fragmented overview of a cross-section of Foucault's work, one can see this axiom of knowledge-as-power functioning in European contexts. Based on these and more observations, Foucault generally suggests in his work that bodies of knowledge are often founded on truth-claims. These truth-claims have, through their continued use, produced themselves as unfaltering truths. The use of these ostensibly infallible and consequently unalterable truths in society has helped determine the realities that populations presently inhabit. Ergo, these truth-claims now regulate the autonomy of individuals through the establishment of power over their bodies and minds, forcing both into submissiveness and compliance. Unsurprisingly, Foucault terms this regime of truth-claims-as-power a regime of 'knowledge-power'—a regime in which claims of knowledge are used to assert power over all [67].

Knowledge-power, though, proceeds from a perverted and not a sincere and open sense of intellectual curiosity. Consequently, ethical individuals can currently only use their curiosity to critically challenge the bounds of knowledge-power that they live within, caught as they are in the "chronic negativity of critique". They cannot open their curiosity to new bodies of knowledge which may spring from outside the regimes of knowledge-power that they inhabit [68]. Perhaps these burgeoning bodies of knowledge have no connection to power or to the regulation of populations into docility whatsoever. Nevertheless, critical curiosity, having only been exposed to "bad" regimes of knowledge-power, cannot envision the possibility that "good" bodies of knowledge may be forthcoming—bodies of knowledge that can foster uncritical curiosity and even a sense of trust and care among individuals [52, 69]. In other words, unconditional curiosity can allow an individual to build trust in new bodies of knowledge and may even help her see the need to care for others. This caveat about power-knowledge, in a nutshell, constitutes Foucault's ethics of curiosity—at least based on our limited understanding of it.

Foucault, by our interpretation, calls for such an ethics of curiosity to resolve dilemmas of trust in organizations that are founded on burgeoning and unpredictable branches of knowledge. Furthermore, based on our reading of Foucault, such an ethics may also resolve dilemmas of trust created by closed and dispiriting regimes of knowledge-power. The dilemmas of trust we speak of can, we conjecture, be extended to posthuman organizations and the branches of the unknown innate to them, such as that of a seemingly dystopian techno-digital automation.

7 Two Foucauldian Scenarios: Resolving Dilemmas of Trust in Industry 4.0

One component of the narrative of knowledge-power that Industry 4.0 projects about itself, is that of imposing a digital panopticon upon its human employees to normalize them into docility. This panopticon, unfortunately, leads the employees into a spiral of a lack of motivation, a rise of resistance, and consequent performance deficits in the workplace, as we have discussed [70]. The employees' apprehension of thus being laid off, owing to their underperformance when extreme digitized automation enters the picture to supplant them, is part and parcel of Industry 4.0 as their working environment [71]. This narrative of automation forces the employees to live in a perpetual state of distrust of the digital panopticon and of automation—a state of digital distrust not permitting them to overcome worries about job security due to determinist prophesies about the onslaught of cybots. In such a critical state of mind, these employees are unable to grasp the ethical stance of care that Foucault claims an uncritical curiosity can stimulate, though care should be innate to ideal workplaces through the fostering of communities such as employee resource groups [72, 73] which ultimately work to enhance company goodwill [74]. This crippling inability to care in the omnipresence of the digital panopticon is hardly astonishing—an employee uncertain about her livelihood will inevitably be unsure of whether she can care for herself, let alone for others in the workplace as "part of the family" [qtd. in 80].

The future in Industry 4.0, though, is equally uncertain [75]. It lies outside the boundaries of the knowable. Foucault, were he alive today, would therefore probably call for the employees of Industry to practice an ethics of unconditioned and unconditional curiosity and face the prospect of automation with an open mind. Based on our reading of Foucault's thoughts on curiosity, this openness would lead the employees to face the future possibilities of Industry 4.0—futures outside the parameters of the known—through two maxims:

1. The cynicism built into a critical curiosity will give way to an ethical sense of care for others among the employees in an Industry 4.0-based workplace [76].
2. The employees will accordingly face the future with an unmitigated curiosity founded on a sense of trust and accommodation. This sense of trust stems from the fact that the future in question is unknown and may therefore prove unthreatening when it crystallises [69].

These two Foucauldian maxims, we argue, correspond to two scenarios—one 'bad' and one 'good'—either of which human employees may face in Industry 4.0's self-projected future, based on contingencies. Both these scenarios, however, resolve the dilemma of trust involved in the possibility of a techno-digital regime, although in different ways:

1. In the 'bad' scenario, the employees will, owing to the digital panopticon and the concomitant vortex of underperformance induced in them, be laid off to make way for AI-empowered cybots. However, these cybots will not necessarily gain complete possession of the knowledge and skills that human employees enjoy. This, we assert, is because the employees are themselves conditioned to be 'posthuman'—that is, by our reading, beyond the parameters of the 'human'—by the knowledge-power regimes they inhabit. They may consequently, though unknown to themselves, perhaps prove resilient to cybernetic suppression.

    Empowered by the knowledge of their resilience, when finding themselves jobless, these 'posthuman' employees may, we predicate, launch a collective resistance against the dispensation of cybots to reclaim their livelihoods. This resistance will see the employees building a community based on mutual communication, trust, and care that they weave amongst themselves to ensure that their professional and material security remains protected.

2. In the 'good' scenario, the knowledge-power regime of Industry 4.0 may expose its projected truth-claim about automation to be merely a truth-claim and no more. Within this scenario, automation will not come about, or will come about in a fragmented and incomplete fashion. This will help the employees come face to face with digital futures sans distrust and hostility. It will perhaps even lead the employees to eventually trust the digital, thus resolving the dilemma of digital trust.

    We shall discuss the two scenarios at greater length in the two sections that follow.

7.1 The Centrality of Care: The 'Bad' Scenario in the Posthuman Organization

As we have stated in the previous section, the 'bad' scenario is predicated on the proposition that the forthcoming model of Industry 4.0 will inevitably inaugurate complete automation through the assault of AI-empowered cobots and, ultimately, of cybots. Current narratives of Industry 4.0 project these cybots as posthuman figures vested with the strength to supplant all inevitably underperforming human employees. Cybots can ostensibly perform this act of ousting by robbing human employees of their knowledge and skills to replace them in the workplace [77–79]. The rise of the cybots is tantamount to the rise of a posthuman kind—a kind that is 'post'-human' because it has ostensibly displaced human employees by dislocating them into a past moment.

With the advent of cybots, human employees will realize the centrality of care to the realities they inhabit, as we now show. By exercising their critical curiosity to understanding this posthuman kind, the employees might recognize parallels between themselves and the cybots within the fields of knowledge-power surrounding them in Industry 4.0. This recognition might gear up the employees' will and ability to fight for their livelihoods through the three following steps, although not necessarily in the order we have laid out:

A. According to Foucault [58], medicine as a field of knowledge-power has validated itself overtime because it has, despite its limitations, kept the degeneration, decay, and death of the human body and its organs at bay. It has achieved this feat by being administered to the human body at the first appearance of a symptom of what has been classified as a disease within a medical regime of knowledge-power [80]. The field of medicine thus allows human bodies to outlast their 'natural' corporeal mortality, forcing them to live rather than be left to die in the usual course of things [69].

If the act of making live counts as the care of the human body, then medicine, through this seeming act of care, essentially reconfigures the human body by using techniques to stop letting it remain merely mortal/'human.' Through the advent and development of medicine, then, the human body becomes posthuman—more than human—with the naturally mortal vessel of the 'human' having been made a thing of the past.

The act of becoming posthuman in this sense, we accordingly propose, might arm human employees to resist other posthumans—like cybots—in a posthuman organization. This resistance becomes particularly pertinent in a dispensation in which these other posthumans—the cybots—are trying to exercise power over vulnerable employees to substitute and suppress them.

B. Fields of knowledge-power evidently keep on regenerating their efforts to equip themselves with infallible information about human consciousness because they keep failing at this endeavor. For instance, attempts to extend the reach of psychoanalysis over individuals had the founding fathers of the field complaining that they were unable to completely uncover what the individuals thought [72, 73].

If self-proclaimed experts in a field of truth-claims, despite their apparent power over the human consciousness, confess to their failures to enact this power, the human consciousness perhaps has a mechanism of self-protection and resistance—and therefore of self-care—built into it. Ergo, if cybots try to grasp human consciousness, knowledge, and skills soundly, they may end up with a similarly insufficient grasp over these components. This deficiency may render cybots unable to replace human employees in Industry 4.0. Perhaps, then, we should do better than to underestimate the human consciousness' power to resist its cybernetic usurpers.

C. Should cybots nevertheless end up rendering human employees passe in Industry 4.0, all the employees, trapped in the same disempowering regime of posthuman knowledge-power, will realize that they are equally helpless before cybots. Their burgeoning uncritical curiosity about and unconditionally empathetic knowledge of each other's equally powerless lot will, as a corollary, allow them to comprehend that they can trust each other unreservedly. Such unconditional curiosity about, and open trust and care for each other had been absent among them in the digitally panoptic workplace. The burgeoning of such a sense of community among the employees will situate them at long last in a state of reciprocal human trust, as opposed to digital distrust. This trust may help the employees develop an ethical community of mutual care amongst themselves, almost like an ex-employee resource group of sorts. The advent of this community of care, located in an as-yet unknown future, will perhaps see the employees fighting for each other's professional and material interests. In the long run, this community may even reach its goal of helping its members be collectively retained in an Industry 4.0-based posthuman setting. Its arrival at this goal will negate the endangerment of the employees' livelihoods.

Foucault would probably predict that this ethical and unconditional knowledge, trust, and care of each other will help human employees break out of the docile and conforming patterns of behavior in which a posthuman organization has them trapped through surveillance. This act of breaking out will thus help the employees use their ethics of curiosity about each other to "get free of oneself," as Foucault puts it [52].

Whether the provisional steps we have outlined through our use of a Foucauldian lens will indeed unfold in this 'bad' scenario, remains to be seen, if only because an ethics of curiosity rests on the fact that determinations of the future are uncertain.

7.2 Not-Yet-Post Humanized Trust: The 'Good' Scenario in Industry 4.0

The 'good' scenario, as we have stated in a previous section, is based on the proposition that the projected narrative of automation in Industry 4.0 proves itself shallow. At best, within the parameters of this proposition, the organization may inaugurate a fragmented techno-digital automation—an automation which drags in a network of cybots in its wake. These cybots may continue what the AI-empowered cobots are currently doing—helping human employees with mundane tasks without being involved in (and at the most helping indicate) decision making.

The reason we make this unusual proposition—a proposition which runs contrary to the determinist narrative of automation built into Industry 4.0—is because Foucault has himself instructed us to maintain an uncritical openness when using our curiosity to engage with unknowable futures. Such an openness helps us build trust in these futures and in the bodies of knowledge they bring with them, as Foucault's ethics of curiosity would have it.

We will admit that this 'good' proposition seems unduly optimistic, because AI-empowered cobots may prove enough to lay off and thus betray the trust of some employees involved in Industry 4.0-based organizations. For example, even in their present stage, such organizations bear employees down with the anxiety of supplanting their positions as cashiers in retail stores and as drivers of cabs and trucks [81]. However, Industry 4.0 currently only *"appears to be* promising new solutions" of this sort—solutions which are *"envisioned* to impact society and daily life in unprecedented ways" at an indeterminate future moment [74].

In short, Industry 4.0 presently makes for an incomplete automation, discounting the installation of absolute automation in multiple mainstream organizations, surveillance regardless [75]. To put it differently, since cybots that threaten human employees' livelihoods in a partially or otherwise techno-digitised regime, are locatable in "envisioned" futures, digital trust as a paradox of betrayal may ultimately not emerge in the organization. Thus, the incomplete automation we speak of at least partly negates Industry 4.0's portrayal of itself as a posthuman organization. This is because human knowledge and skills are evidently still required for the sustenance and development of the organization.

As far as the partial presence of the digital goes, cobots presently help sustain human employees' trust in the dispensation by rendering the employees' work less laborious and by freeing them from routine tasks [75]. They thus evoke trust among the employees, rather than threatening their livelihoods. As for digital surveillance, it may prove futile in a setting in which human employees are, in the final reckoning, found to be indispensable.

8 Conclusion and Future Scope

Much as Foucault would advise human employees to maintain an openness of curiosity to future bodies of knowledge in a posthuman organization, a threatening body of knowledge that counteractively continues to produce itself against this curiosity is that which celebrates automation [76]. More problematically, this celebratory narrative is spilling over into liberatory projections not only about Industry 4.0 but also about Industry 5.0 [77]. The irony is that this narrative is no more than a prescriptive projection—a fact that its oracular prophets themselves often seem ignorant of.

Let us set these projections aside by identifying them as fields of knowledge-power which are trying to assert their authority and continually failing in the process, much like psychoanalysis. It is, after all, this continued failure which forces the spinning of these yarns about the future ad nauseum. Let us confess that the futures in question are unknowable and therefore unauthoritative. Let us, then, sustain our openness, care, and trust while allowing our uncritical curiosity to patiently wait for automated futures to materialize, if and in whatever form they do. That is, after all, what Foucault would have recommended.

Hence, in the way of conclusion, we are theoretically optimistic about the human employees' capacity to avoid digital (dis)trust either through community-care resistance or adoption of the fragmented process of digitization. We perceive an unknown future where anthropocentric trust will exhibit the full potential to undercut the digi-centricity of anthropomorphic conditions.

Talking about the future scope of research that our paper may generate; we have to initially admit that the socio-cultural-anthropological approach to digital studies is far from adequate. The 4IR, characterized by disruptions such as AI, cybernetics, blockchain, IoT, machine learning, etc., is too absorbed in its digi-centric development to have much strategic vision in the ways it impacts the human stakeholders. This paper of ours just picks up the issue of digital (dis)trust and human employees' adjustment with the same amidst the uncertainties of 4IR. However, a lot of unique issues around the human stakeholders' adoption/negotiation of and resistance to Industry 4.0 ought to be probed. Critics have predicted that the roadmap to Industry 4.0 will create a new business and industrial model through its high-end technological disruptions, experiences for its existing employees, and even new workforce and consumers [78, 79]. All these will not only shape up the future but create several research questions and agendas. The social, ethical, cultural, and psychological perspectives of impacted homo sapiens, by 4IR, may be reconsidered in new research. Future research on 4IR, essentially positivist, structuralist, and quantitative in nature, may feel the necessary compulsion to add a section about the impact on the human-humane side of the research.

References

1. Rajput,S. & Singh, S. P. (2019). Connecting circular economy and industry 4.0. *International Journal of Information Management, 49,* 98–113
2. Koh, L., Orzes, G., & Jia, F. (2019). Guest editorial: The fourth industrial revolution (Industry 4.0): Technologies disruption on operations and supply chain management. *International Journal of Operations and Production Management,* 39 (6/7/8), 817–828
3. Sung,T. K. (2018). Industry 4.0: A Korea perspective. *Technological Forecasting and Social Change,* 132, 40–45
4. Frank,A. G., Dalenogare, L. S., & Ayala, N. F. (2019). Industry 4.0 technologies: Implementation patterns in manufacturing companies. *International Journal of Production Economics,* 210, 15–26
5. Kagermann, H., Wahlster, W., & Helbig, J. (2013). Securing the future of German manufacturing industry. *Acatech.* Retrieved August 7, 2021, from, http://forschungsunion.de/pdf/industrie_4_0_final_report.pdf
6. Kodama, M. (2005). Knowledge creation through networked strategic communities: Case studieson new product development in Japanese companies, *Long Range Planning,* 38(1), 27–49
7. Kodama, M. (2005). How two Japanese high-tech companies achieved rapid innovation viastrategic community networks, *Strategy and Leadership,* 33(6), 39–47
8. West, J., & Gallagher, S. (2006). Challenges of open innovation: the paradox of firm investmentin open-source software, *R and D Management,* 36(3), 319–331
9. Cheng, S., Evans, J.H., & Nagarajan, N.J. (2008). Board size and firm performance: the moderating effects of the market for corporate control. *Review of Quantitative Finance Accounting,* 31, 121–145

10. Hartmann, P.M., Zaki, M., Feldmann, N., & Neely, A. (2014). Big data for big business? A taxonomy of data driven business models used by start-up firms. *Working Paper: Cambridge Service Alliance.* Retrieved August 15, 2021, from, http://www.nsuchaud.fr/wp-content/uploads/2014/08/Big-Data-for-Big-Business-A-Taxonomy-of-Data-driven-Business-Models-used-by-Start-up-Firm.pdf
11. Brennan, L., Ferdows, K., Godsell, J., Golini, R., Keegan, R., Kinkel, & S., Taylor, M. (2015),Manufacturing in the world: where next? *International Journal of Operations and Production Management,* 35(9), 1253–1274
12. Mubarak, M. F., & Petraite, M. (2020). Industry 4.0 technologies, digital trust and technological orientation: What matters in open innovation? *Technological Forecasting and Social Change,* 161: 120332. Retrieved 07 July 2021, from, https://doi.org/10.1016/j.techfore.2020.120332
13. Gladden, M. E. (2016). *Posthuman Management: Creating Effective Organizations in an Age of Social Robotics, Ubiquitous AI, Human Augmentation, and Virtual Worlds.* 2nd edn., Indianapolis: Synthypnion Press LLC
14. Gladden,M. E. (2018). *Sapient Circuits and Digitized Flesh: The Organization as Locus of Technological Posthumanization,* Indianapolis: Synthypnion Press LLC
15. Das, A. (2020). Posthuman Organizations and Digital Trust. *Conference Proceedings of 4th Service Management Congress, Ostafalia University of Applied Sciences.* Suderburg, Germany. 51–58. Retrieved 16 August 2021, from, https://www.researchgate.net/publication/351270478_Posthuman_Organization_and_Digital_Trust_Conference_Proceedings_from_4th_Service_Management_Congress_Ostfalia_University_Suderburg_Germany#fullTextFileContent
16. Dworschak,B., & Zaiser, H. (2014). Competences for cyber-physical systems in manufacturing. *Procedia CIRP,* 25, 345–350
17. Romero, D., Bernus, P., Noran, O., Stahre, J., & Fast-Berglund, Å. (2016). The Operator 4.0: human cyber-physical systems and adaptive automation towards human-automation symbiosis work systems. *IFIP International Conference on Advances in Production Management Systems.* pp. 677–686
18. Goode, A. (2019). Digital identity: Solving the problem of trust. *Biometric Technology Today.* Retrieved 07 July 2021, from, https://doi.org/10.1016/S0969-4765(19)30142-0. pp. 5–6
19. Ball, K. (2010). Workplace surveillance: An overview. *Labor History,* 51(1), 87–106. Retrieved 07 July 2021, from, https://doi.org/10.1080/00236561003654776. pp. 88–92
20. Özdemir, V. & Hekim, N. (2018). Birth of industry 5.0: Making sense of big data with artificial intelligence, 'the internet of things' and next-generation technology policy. *OMICS: A Journal of Integrative Biology,* 22(1). Retrieved 07 July 2021, from, https://www.liebertpub.com/doi/10.1089/omi.2017.0194. p. 5
21. Ball, K. (2010). Workplace surveillance: An overview. *Labor History,* 51(1), 87–106. Retrieved 21 July 2021, from, https://doi.org/10.1080/00236561003654776. p. 93
22. Schwab,K. (2016). *The Fourth Industrial Revolution.* Geneva: World Economic Forum. p. 12
23. Zlotnick, S. (1998). *Women, Writing, and the Industrial Revolution,* Maryland and London: The Johns Hopkins University Press. p. 43
24. Banerji,D. & Paranjape, P.R. (2016). *Critical Posthumanism and Planetary Futures,* E-book: Springer. p. 2
25. Das, A. (2020). Posthuman organizations and digital trust, *Conference Proceedings of 4th Service Management Congress, Ostafalia University of Applied Sciences.* Suderburg, Germany. 51–58. Retrieved 07 July 2021, from, https://www.researchgate.net/publication/351270478_Posthuman_Organization_and_DIgital_Trust_Conference_Proceedings_from_4th_Service_Management_Congress_Ostfalia_University_Suderburg_Germany#fullTextFileContent. p. 53
26. Saravanan,S. (2018). *A Transnational Feminist View of Surrogacy Biomarkets in India,* Singapore: Springer
27. Robertson, J. A. (2004). Gay and lesbian Access to assisted repr technology. *Case Western Reserve Law Review,* 55(2), 323–372

28. Haraway, D. (2015). Anthropocene, capitalocene, plantationocene, chthulucene: Making kin, *Environmental Humanities*, 6 (1), 159–165
29. Salkin, C., Oner, M., Ustundag, A., & Cevikcan, E. (2018). A conceptual framework for Industry 4.0, In Ustundag, A., Cevikcan E., Eds., *Industry 4.0: Managing The Digital Transformation.* Springer. pp. 3–23
30. Sukhodolov, Y. A. (2019). The notion, essence, and peculiarities of Industry 4.0 as a sphere of industry, In Popkova, E.G., Ragulina, Y. V., & Bogoviz, A. V. Eds.. *Industry 4.0: Industrial Revolution the 21st Century*, Springer. pp. 3–10
31. Sumer, B. (2018). Impact of Industry 4.0 on occupations and employment in Turkey, *European Scientific Journal*, 14(10), 1–17
32. Hayles, N. K. (1999). *How We Became Posthuman: Virtual Bodies in Cybernetics, Literature, and Informatics*, Chicago and London: Chicago University Press
33. Hayles, N. K. (1999). *How We Became Posthuman: Virtual Bodies in Cybernetics, Literature, and Informatics*. Chicago and London: Chicago University Press. pp. 2–3
34. Morgan, G. (2006). *Images of Organization*. Beverly Hills: Sage
35. Gladden, M. E. (2018). *Sapient Circuits and Digitized Flesh: The Organization as Locus of Technological Posthumanization.* Indianapolis: Synthypnion Press LLC. pp. 26–27
36. Gladden, M. E. (2018). *Sapient Circuits and Digitized Flesh: The Organization as Locus of Technological Posthumanization.* Indianapolis: Synthypnion Press LLC. p. 216
37. Gladden, M. E. (2018). *Sapient Circuits and Digitized Flesh: The Organization as Locus of Technological Posthumanization.* Indianapolis: Synthypnion Press LLC. p. 251
38. Gladden, M. E. (2018). *Sapient Circuits and Digitized Flesh: The Organization as Locus of Technological Posthumanization.* Indianapolis: Synthypnion Press LLC. p. 305
39. Das, A. (2020). Posthuman organizations and digital trust, *Conference Proceedings of 4th Service Management Congress, Ostafalia University of Applied Sciences.* Suderburg, Germany. 51–58. Retrieved 03 September 2021, from, https://www.researchgate.net/publication/351270478_Posthuman_Organization_and_Digital_Trust_Conference_Proceedings_from_4th_Service_Management_Congress_Ostfalia_University_Suderburg_Germany#fullTextFileContent. p. 55
40. Das, A. (2020). Posthuman organizations and digital trust. *Conference Proceedings of 4th Service Management Congress, Ostafalia University of Applied Sciences.* Suderburg, Germany. 51–58. Retrieved 03 September 2021, from, https://www.researchgate.net/publication/351270478_Posthuman_Organization_and_Digital_Trust_Conference_Proceedings_from_4th_Service_Management_Congress_Ostfalia_University_Suderburg_Germany#fullTextFileContent. p. 56
41. Özdemir, V. & Hekim, N. (2018). Birth of industry 5.0: Making sense of big data with artificial intelligence, 'the internet of things' and next-generation technology policy, *OMICS: A Journal of Integrative Biology*, 22(1), 65–77. Retrieved 03 September 2021 https://www.liebertpub.com/doi/10.1089/omi.2017.0194. p. 2
42. Özdemir, V. & Hekim, N. (2018). Birth of industry 5.0: Making sense of big data with artificial intelligence, 'the internet of things' and next-generation technology policy. *OMICS: A Journal of Integrative Biology*, 22(1), 65–77. Retrieved 03 September 2021, from, https://www.liebertpub.com/doi/10.1089/omi.2017.0194. p. 4
43. Özdemir, V. & Hekim, N. (2018). Birth of industry 5.0: Making sense of big data with artificial intelligence, 'the internet of things' and next-generation technology policy, *OMICS: A Journal of Integrative Biology*, 22(1), 65–77. Retrieved 03 September 2021, from, https://www.liebertpub.com/doi/10.1089/omi.2017.0194. p. 7
44. Ball, K. (2010). Workplace surveillance: An overview. *Labor History*, 51(1): 87–106. Retrieved 03 September 2021, from, https://doi.org/10.1080/00236561003654776. p. 93
45. *AI Trends: The Newsletter of the Artificial Intelligence Industry*, (1985). 2–3. p. 17
46. Hayles, N. K. (1999). *How We Became Posthuman: Virtual Bodies in Cybernetics, Literature, and Informatics.* Chicago and London: Chicago University Press. p. 118
47. Haraway, D. (2016). *Staying with the Trouble: Making Kin in the Chthulucene.* Duke University Press

48. Özdemir, V. & Hekim, N. (2018). Birth of industry 5.0: Making sense of big data with artificial intelligence, 'the internet of things' and next-generation technology policy. *OMICS A Journal of Integrative Biology,* 22(1), 65–77. Retrieved 03 September 2021, from, https://www.liebertpub.com/doi/10.1089/omi.2017.0194. p. 6
49. Foucault, M. (1985). *The History of Sexuality Volume 2: The Use of Pleasure,* (R. Hurley Trans.). London: Random House, Inc. (Original work published 1984). p. 8
50. Anker, E. S. & Felski, R. (2017). Introduction, In E. S. Anker and R. Feslki (Eds.), *Critique and Postcritique.* Durham and London: Duke University Press. 1–28. p. 11
51. Das, A. (2020). Posthuman organizations and digital trust, *Conference Proceedings of 4th Service Management Congress, Ostafalia University of Applied Sciences.* Suderburg, Germany, 51–58. Retrieved 03 September 2021, from, https://www.researchgate.net/publication/351270478_Posthuman_Organization_and_Digital_Trust_Conference_Proceedings_from_4th_Service_Management_Congress_Ostfalia_University_Suderburg_Germany#fullTextFileContent. pp. 56–57
52. Foucault, M. (1977). *Discipline and Punish: The Birth of the Prison,* (A. Sheridan, Trans.). London: Random House, Inc. (Original work published 1975). p. 200
53. Foucault, M. (1989). *The Archaeology of Knowledge,* (A. M. Sheridan Smith., Trans.) London: Routledge. (Original work published 1969). p. 19
54. Foucault, M. (1990). The masked philosopher. (A. Sheridan, Trans.). *Politics Philosophy Culture: Interviews and Other Writings 1977–1984.* London: Routledge. (Original work published 1980). p. 328
55. Foucault, M. (1978). *The History of Sexuality Volume 1: An Introduction,* (R. Hurley, Trans.). London: Random House, Inc. (Original work published 1976). p. 108
56. Foucault, M. (1977). *Discipline and Punish: The Birth of the Prison,* (A. Sheridan, Trans.). London: Random House, Inc. (Original work published 1975). p. 11
57. Foucault, M. (1977). *Discipline and Punish: The Birth of the Prison,* (A. Sheridan, Trans.). London: Random House, Inc. (Original work published 1975). p. 136
58. Foucault, M. (1989). *The Birth of the Clinic: An Archaeology of Medical Perception,* (A. M. Sheridan, Trans.). London: Routledge. (Original work published 1963). p. 217
59. Foucault, M. (1989). *The Birth of the Clinic: An Archaeology of Medical Perception,* (A. M. Sheridan, Trans.). London: Routledge. (Original work published 1963). p. 233
60. Foucault, M. (1989). *The Birth of the Clinic: An Archaeology of Medical Perception,* (A. M. Sheridan, Trans.). London: Routledge. (Original work published 1963). p. 40
61. Foucault, M. (2003). *Abnormal: Lectures at the Collège de France 1974–1975,* (G. Burchell, Trans.) London: Verso. (Original work published 1999). p. 32
62. Foucault, M. (2008). *The Birth of Biopolitics: Lectures at the Collège de France 1978–1979,* (G. Burchell, Trans.) Palgrave Macmillan. (Original work published 2004). pp. 19–20
63. Foucault, M. (1978). *The History of Sexuality Volume 1: An Introduction,* (R. Hurley, Trans.). Random House, Inc. (Original work published 1976). p. 112
64. Özdemir, V. & Hekim, N. (2018). Birth of industry 5.0: Making sense of big data with artificial intelligence, 'the internet of things' and next-generation technology policy. *OMICS: A Journal of Integrative Biology,* 22(1). Retrieved 03 September 2021, from, https://www.liebertpub.com/doi/10.1089/omi.2017.0194. p. 2
65. Goode, Shelton, & Dixon, Isaac. (2016). Are employee resource groups good for business? *SHRM: Better Workplaces Better World.* Retrieved 03 September 2021, from, https://www.shrm.org/hr-today/news/hr-magazine/0916/pages/are-employee-resource-groups-good-for-business.aspx
66. Black, A., Davis-Carden, C., & Momplaisir, J. (2020). 5 ways to incorporate community care in the workplace, *She+Geeks Out.* Retrieved 03 September 2021, from, https://shegeeksout.com/5-ways-to-incorporate-community-care-in-the-workplace/
67. Hunt, R. (2019). 5 Reasons company community involvement builds employee wellbeing, *Buck.* Retrieved 08 August 2019, from, https://buck.com/5-reasons-corporate-community-involvement-builds-employee-wellbeing/

68. Hooker, John. & Kim, T. W. (2019). Ethical implications of the fourth industrial revolution for business and society, *Business Ethics (Business and Society 360, Vol. 3)*, Emerald Publishing Limited, Bingley. Retrieved 08 August, 2019, from, https://doi.org/10.1108/S2514-175920190 000003002. p. 4
69. Magruk, A. (2016). Uncertainty in the sphere of the Industry 4.0 – potential areas to research, *Business, Management and Economics Engineering*, 14(2), 275–291
70. *AI Trends: The Newsletter of the Artificial Intelligence Industry, Vol. 2–3*, (1985). p. 17
71. Ball, K. (2010). Workplace surveillance: An overview, *Labor History*, 51(1), 87–106. Retrieved 08 August, 2019, from, https://doi.org/10.1080/00236561003654776. p. 93
72. Foucault, M. (1989). *The Birth of the Clinic: An Archaeology of Medical Perception*. (A. M. Sheridan, Trans.). London: Routledge. (Original work published 1963). pp. 194–197
73. Foucault, M. (2003). *Society Must Be Defended: Lectures at the Collège de France 1975–76*, (A. I. Davidson, Trans.). New York: Picador. (Original work published 1997). p. 247
74. Foucault, M. (1978). *The History of Sexuality Volume 1: An Introduction*, (R. Hurley, Trans.). Random House, Inc. Original work published 1976). p. 23
75. Özdemir, V. & Hekim, N. (2018). Birth of industry 5.0: Making sense of big data with artificial intelligence, 'the internet of things' and next-generation technology policy, *OMICS: A Journal of Integrative Biology*, 22(1), https://www.liebertpub.com/doi/10.1089/omi.2017.0194. p. 5
76. Özdemir, V. & Hekim, N. (2018). Birth of industry 5.0: Making sense of big data with artificial intelligence, 'the internet of things' and next-generation technology policy. *OMICS: A Journal of Integrative Biology*, 22(1). Retrieved 03 September 2021, from, https://www.liebertpub.com/doi/10.1089/omi.2017.0194. p. 5
77. Özdemir, V. & Hekim, N. (2018). Birth of industry 5.0: Making sense of big data with artificial intelligence, 'the internet of things' and next-generation technology policy. *OMICS: A Journal of Integrative Biology*, 22(1). Retrieved 03 September 2021, from, https://www.liebertpub.com/doi/10.1089/omi.2017.0194. p. 8
78. Özdemir, V. & Hekim, N. (2018). Birth of industry 5.0: Making sense of big data with artificial intelligence, 'the internet of things' and next-generation technology policy. *OMICS: A Journal of Integrative Biology*, 22(1). Retrieved 03 September 2021, from, https://www.liebertpub.com/doi/10.1089/omi.2017.0194. pp. 2–3
79. Nayyar, A. & Kumar, A. (2020). Eds. *A Roadmap to Industry 4.0: Smart Production, Sharp Business and Sustainable Development*. Springer
80. Kumar, A. & Nayyar, A. (2020). Si, In Nayyar, A. and Kumar, A. Eds. *A Roadmap to Industry 4.0: Smart Production, Sharp Business and Sustainable Development*. Springer. pp. 1–2
81. Sjöholm, C. (2010). Foucault, Lacan, and the question of technique, In Jens De Vleminck & Eran Dorfman (Eds.), *Sexuality and Psychoanalysis: Philosophical Criticisms*. Leuven: Leuven University Press.183–196. p. 189

Hydrogel Based on Alginate as an Ink in Additive Manufacturing Technology—Processing Methods and Printability Enhancement

Magdalena B. Łabowska, Ewa I. Borowska, Patrycja Szymczyk-Ziółkowska, Izabela Michalak, and Jerzy Detyna

1 Introduction

Nowadays we are experiencing the unfolding of the next era, termed as the Fourth Industrial Revolution, termed as Industry 4.0, where factory machines meet advanced computing technologies in order to improve technology, make it more efficient, and increase accuracy level of fabricated objects. The data collected

---

M. B. Łabowska (✉) · J. Detyna
Department of Mechanics, Materials and Biomedical Engineering, Faculty of Mechanical Engineering, Wroclaw University of Science and Technology, Wroclaw, Poland
e-mail: magdalena.labowska@pwr.edu.pl

J. Detyna
e-mail: jerzy.detyna@pwr.edu.pl

E. I. Borowska
The College of Inter-Faculty Individual Studies in Mathematics and Natural Sciences (MISMaP), University of Warsaw, Banacha 2C, 02-097 Warsaw, Poland

Faculty of Geology, Geophysics and Environmental Protection, AGH University of Science and Technology, Reymonta 23, 30-059 Cracow, Poland

Hyphae, 415 1St Ave N, Seattle, WA 98109, USA

P. Szymczyk-Ziółkowska
Centre for Advanced Manufacturing Technologies (CAMT/FPC), Faculty of Mechanical Engineering, Wroclaw University of Science and Technology, Wroclaw, Poland
e-mail: patrycja.e.szymczyk@pwr.edu.pl

I. Michalak
Department of Advanced Material Technologies, Faculty of Chemistry, Wroclaw University of Science and Technology, Wroclaw, Poland
e-mail: izabela.michalak@pwr.edu.pl

© The Author(s), under exclusive license to Springer Nature Switzerland AG 2023
A. Nayyar et al. (eds.), *New Horizons for Industry 4.0 in Modern Business*, Contributions to Environmental Sciences & Innovative Business Technology,
https://doi.org/10.1007/978-3-031-20443-2_10

from physical machine sensors, processed by specially designed algorithms in virtual environment, aim to automate technology as much as possible. This provides increased productivity, stability, and high repeatability of manufactured products, thus the role of people in factories is redefined [1–4]. The manufacturing process traditionally focused on the processing side of products has been improved by the introduction of modern devices and increased computing power, and has also improved the entire supply chain, through business development, standards, and logistics. Through the connection of artificial intelligence and machines industry, the production process can be more efficient, cost-effective and even more environmentally friendly. The only limitation seems to be the machine maximum capacity and restrictions related to material properties [3, 5–9].

One of the branches that perfectly matches the Industry 4.0 framework is additive manufacturing technology that uses computer aided design by which it is possible to obtain any complex shapes of manufactured objects by embedding consecutive layers of materials on a substrate. This technology is intended for both mass production and personalized manufacturing, according to the requirements of a particular customer. Therefore, it can be successfully applied in various fields, including aerospace, automotive, robotics, chemical, biomedical field [5, 10, 11]. Compared to traditional production techniques (i.e. subtractive manufacturing), AM technology enables the production of a three-dimensional object without material waste and the need for additional tools (e.g., drill bit in a drill, cutters in milling machines, grinders in the grinding process) [1]. The small size of devices and their relatively low price allow the equipment to be placed anywhere—from large factories to small private businesses and even households. In addition, they are simple to operate; thus creating objects requires only 3D computer models, which are usually open-source available [12–14].

Although AM technology has an extensive range of processing methods, the deciding factor in choosing an incremental technique is the potential application and the type of material used. Techniques based on semifluidic materials are mainly inkjet-based, extrusion-based, laser-assisted, and stereolithography [15]. The ease of material processing, as well as the biocompatibility and nontoxicity of the used material, enable applications in the medical industry. One of the basic requirements for AM technology is the fidelity of shape reproduction, which is necessary in precise structure manufacturing, especially in the biomedical field, where the precision of fabrication to mimic tissues and organs is extremally important [16].

Alginate is one of the most widely used natural materials in biomedicine due to its versatility, similarity to mimic extracellular matrix (ECM), and excellent physicochemical and biological properties. This polymer is capable of gelling in the presence of water. In addition, the ability to form a hydrogel enables its use as an ink in several AM techniques, which are based on semi-liquid materials [17–19]. However, the rheological properties, molecular weight, and low viscosity of alginate solution influence its printability and make it not a perfect candidate for additive manufacturing. During processing on the bioplotter, alginate-based printouts lose their shape and spread on the printing area [20]. Therefore, it is impossible to produce three-dimensional complex structures. Shape fidelity can be achieved by the cross-linking process immediately after the material flows out of the nozzle. On the other hand, printing with alginate in a liquid solution with a cross-linking agent can result in subsequent clogging of the printing nozzle [16, 21].

The alginate structure consists of polymer blocks of (1–4)-linked β-d-mannuronic (M) and α-l-guluronic (G) acids, which are arranged in various sequences. These units form groups of MM, GG or MG blocks, where their arrangement and amount determine the properties of the obtained hydrogel [17, 19, 22]. Furthermore, the mechanical properties of alginate hydrogels can be modified at the cross-linking level [23]. A solution of alginate has to be cross-linked to retain a particular shape of the obtained hydrogel. It can be carried out in several ways: ionically, covalently, or by thermal gelation [24].

Challenges of alginate hydrogel processing for applications in AM technology have led to the development of several methods that facilitate alginate ink modifications, allowing the output to retain its shape until its cross-linked. Therefore, it is possible to create a three-dimensional object without worrying about the accuracy of the model reproduction. These methods can be based on alginate ink additives, special support baths, or materials that become sacrificial outliners [25–28]. The first mentioned technique interferes with the ink composition, but, on the other hand, it can be helpful in creating scaffold structures, which must have adequate mechanical strength and will be destined for cell growth. Alginate alone is bioinert, which means it does not react with cells, and there is a lack of adhesion. Therefore, in some cases, additives in alginate ink can improve the viscosity and cytocompatibility of the created hydrogel [29]. Other methods use thermoreversible materials, which are easily removed after the manufacturing process by changing the temperature. These techniques allow the use of alginate alone as an ink because the suspended printing in a sacrificial material is often mixed with a cross-linking agent. Consequently, it is possible to obtain single layers with high precision, as well as elaborate structures of any shape [30, 31].

In this work, methods based on semifluidic materials for biological applications and techniques used in extrusion-based applications are discussed. These include ink additives that improve printability, as well as sacrificial materials used to create support baths, or as temporary frameworks for complex structures. The most important features, together with their mechanical properties that are necessary for integrity and strength depending on their applications, are also included.

1.1 Overview of Additive Manufacturing Technologies Using Hydrogel Inks

Depending on the material used, its processing capabilities, and potential applications, there are many techniques that support AM technologies. Among them are techniques that use metals, ceramics, polymers, but also semiliquid materials such

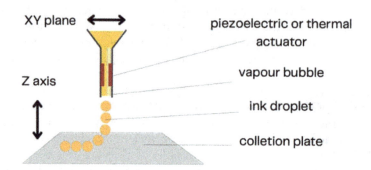

Fig. 1 Schematic representation of inkjet technology (own source)

as hydrogels. The methods differ in the way the material is processed and solidified. Some methods use a high temperature or light source generator, or there are some that utilize vats filled with powder or photosensitive resins. But the goal for all of them is to create a three-dimensional object, layer by layer, with the highest autonomy and maximum accuracy in shape and dimension representation [32, 33].

The use of incremental technology for hydrogel materials includes techniques such as inkjet-based, extrusion-based, laser-assisted, and stereolithography [15, 34]. Inkjet-based (Fig. 1) is a technique for depositing a liquid binder as droplets on a substrate in a continuous (CJ) or drop-on-demand (DOD) manner. The DOD system is preferred for biological applications due to its good controllability and low potential for contamination. The ink can be settled on a powder or cross-linked with ultraviolet (UV) light. The biggest advantage of this technique is the high resolution of the output; unfortunately, the system elements are very expensive. The most commonly used materials in this technology include alginate and collagen [15, 34, 35].

The extrusion-based technique (Fig. 2) is based on the depositing of semi-liquid materials (hydrogel or molten materials), extruded from a nozzle or needle (print head) operating in the X and Y axis. The extrusion of the material from the print head can be performed pneumatically or by piston displacement. Material is cross-linked ionically, thermally, or photoinduced. It is the most commonly used method in bioprinting (it uses cell-containing hydrogel ink), but cell survival is constrained by the pressure of the material extruded from the nozzle. The materials used in this technique include agarose, alginate, collagen gelatin, gelatin-methacryloyl (GelMA), methylcellulose, poly(ethylene glycol) diacrylate (PEGDA), poly(ethylene glycol) dimethacrylate (PEGDMA) [15, 34, 36, 37].

The technique that uses a pulsed laser source to deposit materials on a substrate is called laser assisted (Fig. 3). The major advantage of this technology is its resolution, which covers single layers on a pico to micro scale. The materials commonly used in this method are collagen, Martigel® (solubilized basement membrane preparation) [15, 38].

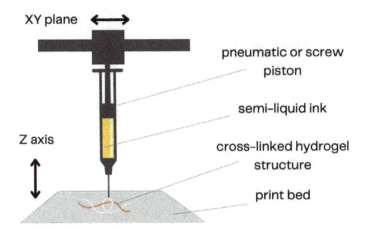

Fig. 2 Schematic representation of the extrusion-based technique (own source)

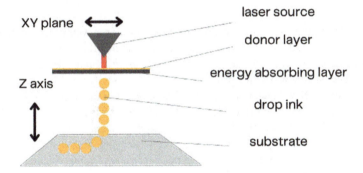

Fig. 3 Schematic representation of the laser-assisted method (own source)

The last one, stereolithography, is the oldest AM technique that consists of polymerization of a photocurable resin by a laser source, layer by layer. The resin is placed in a container on a table moving in the Z direction, while the laser head moves in the XY plane. Typical materials used are hyaluronic acid, alginate, and poly(ethylene glycol) (PEG) based acrylates. The schematic representation of this technique is shown in Fig. 4. [15, 37].

1.2 Application of Extrusion Techniques in the Biomedical Field

Many AM technology methods are successfully used in various fields because of their versatility, lack of material loss compared to other technologies, relative cost-effectiveness, and ease of use. Furthermore, AM technologies offer a wide range

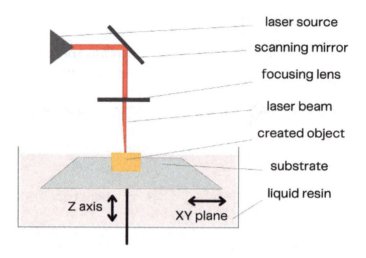

Fig. 4 Schematic representation of the stereolithography technology (own source)

of materials (natural, synthetic, and hybrid) that are non-toxic, biocompatible, and mostly biodegradable [39]. Extrusion-based is the most widely used technique due to its ability to process thermoplastics, hydrogels, and also cell-containing bioinks. The selection of the printing nozzle diameter from 20 μm, enables to obtain relatively thin layer thicknesses. This allows the creation of precise, high-reproducibility porous scaffolds used in tissue engineering [15, 37, 40]. Certain materials, such as thermoplastic polymers, require the use of high temperatures to melt the material in the print head. The utilization of hydrogel as an ink does not require the application of elevated temperatures, which makes it readily available for processing with cells. In the case of hydrogels, the printing is followed by the cross-linking process, where the shape of the obtained structures is solidified, and the print acquires mechanical strength [41]. Ionic cross-linking using calcium chloride is the most commonly used for the synthesis of hydrogels based on sodium alginate. The process, as well as the substrates utilized, are non-invasive and non-toxic to biological structures [42, 43].

Extrusion-based is the most widely used technique among all using semi-fluid materials due to its simple use and low cost of components. However, the most important advantage is the ability to process inks with a high-density cell content (bionics) with relatively low cell damage during fabrication [36].

The use of hydrogels allows for the precise manufacture of porous structures with sufficient mechanical properties for culture of cells for tissue engineering. This enables cells to be suspended across their entire volume. The biological properties of the hydrogel support proper cell growth, and the ability to biodegrade the hydrogel structures provides space for growing cells. Wide-ranging personalization is feasible due to the computer-aided design and resolution capabilities of this method. Therefore, it is possible to select the right amounts of active substances in the production of drug delivery systems or manufacturing size-matched tissue grafts, for example, for burned areas [44, 45].

Tissue Engineering and Regenerative Medicine (Scaffold)
With the progress in tissue engineering, alginate-based hydrogels find new applications. In recent years, new hydrogel-based techniques have become important for tissue biological construction and regenerative medicine. Bioartificial tissues produced from hydrogels can serve as scaffolds for tissue fabrication. Some crucial features of hydrogels, such as biodegradability, bioadhesion, biocompatibility and durability, imply their wide biomedical applications [46]. The quality of hydrogels can be enhanced by additional desirable components, depending on the design and type of gels. Thus, their capability to easily shape and change the surface, simplify the maintaining the integrity and consistency. Hydrogels have been presumed to be one of the most cutting-edge candidates used in tissue engineering, with particular emphasis on natural scaffolds for regenerative medicine. Depending on the application area, hydrogel-based scaffolds with an accurate composition can be designed. The versatile applications in biomedicine make hydrogels one of the most important natural polymers. 3D mimic microenvironments create peculiar characteristics for tissues and probably enable faster regeneration. Together with tools such as bioimaging and previously scanned damaged area, fitting and adjacency can be more effective for tissues repair. Precision in making 3D tissue model construction enables better material selection and parameters for bioprinting. To increase the fidelity of the ambient microenvironment of the built complex tissue, extracellular matrix with physiological properties that can be adapted to the proper structure is used [47, 48]. Thus, the key element to be considered in the synthesis of hydrogels containing naturally occurring polymers, such as alginate, collagen, fibrin, chitosan, gelation, and hyaluronic acid, is the future application/location. Alginate is processed with cells to produce tissue, muscle, aortic valve, pancreatic islets, cartilage [49], proteins carriers, controlled-release drugs [50, 51]. Notwithstanding, the maintenance of stable mechanical properties remains at a low level. Therefore, the majority of polymers used for the hydrogel's composite production is improved by modified and derived chemical compounds from natural polymers, such as methylcellulose, hydroxypropyl methylcellulose (HPMC), or nanoparticles (NPs) for better stabilization of hydrogels, for example in cancer therapy [52–55].

Cell Culture and Other Applications
Hydrogels have found applications in cell-culture techniques due to the similarity and ability to mimic the extracellular matrix. The specific composition of the microenvironment designed for cells should be constructed considering such elements of cell culture as cell adhesion, cell proliferation, and cell differentiation to build functional cells. The alginate hydrogel structure facilitates cell growth, provides oxygen exchange and nourishment supply, and ensures the transfer of metabolic products [49, 56]. Thus, hydrogels create potentially easy to manipulate, cultivate and fabrication of environment under cell culturing. The development of an easy method for cells encapsulation in an extracellular matrix compatible with

them remains a challenge. One of the most problematic aspects is the construction of a nontoxic and compatible environment, which will help cells in faster proliferation and establishment of cells culture assemblages. The complexity and diversification of in vitro cell's culture cause a pivotal arrangement, and gels are widely used to mimic the accurate structure and formation of ECM. The juncture of wide-range techniques to build compatible surroundings for cells is related to the relevant biophysical properties and biological functions and, therefore, is mapping many features of native, nascent ECMs [45, 55]. However, it is challenging to accreate and maintain clinical samples obtained from patients in cell culture based on a typical matrix, which may reflect in vivo models such as three-dimensional (3D) models—spheroids. Another example of the use of cultures and hydrogels are 2D models with bacteria on plates. To increase or decrease desirable effects important to adapt microbes, different types of hydrogel compositions for microorganisms-surface interactions such as adhesion, rate of colonization, or biofilm formation have been studied. The modeling and innovative approaches to bacteria culture are associated with the control strategies, which is important in the case of bacteria's biofilm. The biofilm can affect all 2D culture model and its microenvironment. What is more, the interaction and cooperation of microorganisms and also their secretion of compounds have an eerie impact on the matrix [57, 58]. Specific hydrogels also enhance the microenvironment for other microorganisms, such as photosynthetic microalgae or cyanobacteria. The immobilization of cells causes durable and stable conditions for culture growing (Fig. 5). Thereby, it developed its features for metabolite production, improvement of culture collections handling, nutrients and pollutant removal from aquatic environments, biofuel production or system production for different directions [59, 60].

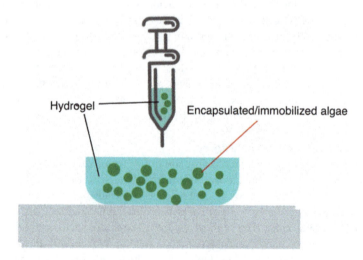

Fig. 5 Cell's culture as an example of improving the cell cultivation (own source)

Drug Delivery

Additive manufacturing technologies offer many opportunities, especially adjustment of the right amount of material during object fabrication. The high precision of material dosing affects the possibility of personalization and adjustment of a particular dose to the patient (e.g., age, weight, sex, state of health), for example, in the drug delivery system [11, 61]. Moreover, this solution can be used for both single-unit and mass production used in industry. The greatest advantage is the ability to modify the model at the stage of computer aided design for individual requirements, to select customized components, and the ability to implement arbitrary shapes and structures that can affect the release profile of the active substance [62].

The drug can be incorporated into the carrier in several ways, depending on the technique used. The drug release profile depends on the distribution of the active substance in the carrier. The active substance can be placed directly into the filament, before the manufacturing process, and also independently, as a separate component. In one carrier, one active substance can be included, as well as several different ingredients, forming together a carrier called polypill or multidrug (Fig. 6) [11, 63, 64]. Bioplotters with several printing nozzles are used to produce such multidrugs. Each nozzle is provided with a separate ink, having, respectively, a hydrogel, a carrier base, and an appropriate amount of active ingredients. Furthermore, at the drug delivery system (DDS) design level, it is possible to achieve a distinct drug release profile, whereby each included active ingredient can be released at a particular time in a controlled manner. Therefore, several ways of placing drugs can be distinguished, especially in the case of multigradient drugs, in which each component may have a distinct release profile [65, 66]. At the same time, the use of AM technology will enable precision in the adjustment of the therapeutic dose, therefore improving drug delivery.

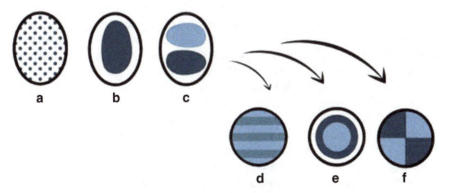

Fig. 6 Methods of the active ingredient encapsulation in the drug carrier, **a** drug merged with the carrier base, **b** inclusion of active ingredient separately, **c** multidrug formulation, **d** placement of active ingredients in a layered manner, **e** centralized placement of multiple drugs, **f** equal radial arrangement of several active ingredients (own source)

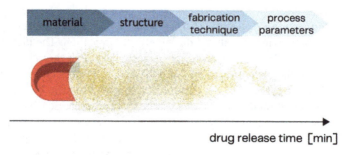

Fig. 7 Factors influencing the time of drug release (own source)

The release of the drug can be influenced by various factors that should be considered during the development of the drug carrier, as well as the selection of the manufacturing process parameters. Therefore, the control of the release of active ingredients (constant, immediate or delayed) depends on the amount of active substances and the sequence of release of the individual components [61]. Decisive elements, in addition to the DDS structure selection, involve also the adjustment of process parameters, such as the speed of the printhead and the pressure in the printing nozzle, as well as the manner and time of hydrogel cross-linking used to create the base of the drug carrier (Fig. 7). Thus, all parameters can influence the obtained mechanical properties of the carrier, which are important to ensure strength to compression and abrasion.

Additive manufacturing technologies offer many techniques that are dedicated to different materials, among which several can be used to fabricate objects using hydrogels. The aim of this chapter is to present one branch of Industry 4.0, which is additive manufacturing technology, in the context of alginate hydrogel processing for biomedical applications. Processing alginate ink using AM techniques is a challenging process. Therefore, this chapter includes ways to improve printability with these inks, as well as methods used to improve the accuracy of shape reproduction. Implementation techniques such as sacrificial materials and support baths enable obtaining a high precision result. Hydrogel additives can further enhance the mechanical properties of this ink. Hence, the ample perspectives and versatile possibilities of the use of these methods. In overall conclusions we would like to stress the importance of hydrogel in biofabrication. It enables formation of an adapted 3D biofabricate what can reduce eventuality of incompatibility between cells/tissues and used material in the long prospect. Bioprinter devices in further study can improve in situ active substances delivery, directly based on organism's cells. The final goal will be the progress of tissue fabrication that can mimic the extracellular matrix in vivo to maximize sequestration of necessarily components into cells.

Low viscosity is a characteristic of alginate-based solutions and using them as ink presents certain challenges for achieving accurate reproduction. Therefore, this chapter's objective is to outline methods for fabricating stable hydrogel structures with suitable mechanical properties for their applications. Combining alginate-based solutions with other polymer additives is one method that can improve the processability of inks used in additive manufacturing (AM), as well as the mechanical strength of completed structures and cell adhesion in cell culture. Alginate-based hydrogels have widespread applications in the medical industry due to their biological properties, nevertheless, their weak mechanical properties appear to be a limitation. The current chapter focuses at the factors that influence the change in characteristics during the fabrication process and discusses approaches to increase the durability of alginate-based hydrogel structures.

Organization of Chapter
The chapter is organized as: Sect. 2 describes the use of alginate as an ink in additive manufacturing technologies and discusses methods of improving its properties and printability. Section 3 discusses the influence of process parameters on mechanical properties of Hydrogel Structure. Section 4 concludes the chapter with future scope.

2   Improvement of Properties of Alginate-Based Hydrogel Ink

Alginate is an anionic polysaccharide typically obtained from brown seaweed, capable of gel formation in the presence of water and crosslinking when exposed to divalent and trivalent cations. After cross-linking, alginate assumes the form of a hydrogel, i.e. three-dimensional networks containing hydrophilic polymers in their structure, which are capable of absorbing large amounts of liquids [19, 24, 67, 68].

Alginate, in spite of its benefits in biomedical applications, as a hydrogel ink is difficult to process in manufacturing. The reason for this is that the alginate solution behaves as a low viscosity thick liquid that is incapable of forming and maintaining repeatable shapes prior to the cross-linking process, where the shape is established by divalent or trivalent cations [16, 27].

There are few methods to support shape fidelity of alginate-based hydrogel prints. Among these can be highlighted additives intended to improve the printability of the alginate solution, as well as to provide better mechanical properties of alginate structures. These additives can also enhance biological characteristics of hydrogels can be highlighted in various biomedical applications, such as tissue engineering. Therefore, alginate inks are used in combination with other materials (e.g., gelatin, collagen, chitosan), as well as fiber-filled (e.g., methylcellulose) or nondegradable particles as additives [26, 27, 69].

Another method used in alginate processing on bioplotters is the application of sacrificial materials that do not interfere with the composition of the ink, but simply allow it to retain its shape until cross-linking. This includes the utilization of thermoreversible support baths, which usually contain a cross-linking agent. The solution provides the possibility to embed complex three-dimensional structures using the alginate ink without additions. After the object has been fabricated, the

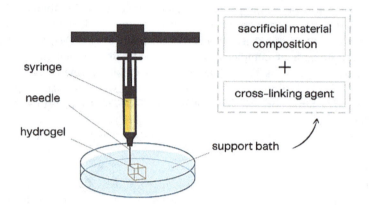

Fig. 8 Scheme for using the support bath as a sacrificial material to improve alginate ink processing using an extrusion-based technique (own source)

support material can be easily removed by raising the temperature. This technique is also called FRESH (free-form reversible embedding of suspended hydrogels) [25, 28]. The method is visualized in Fig. 8.

Another technique is also based on a sacrificial material with the ability to change the state by temperature. The relatively low melting temperature values of the materials considered as sacrificial materials allow them to be easily removed after the printed structure is cross-linked. In this case, not a support bath, but a stiffer outline holds the shape of the printed hydrogel. Although not as much material is used as in the first method, an additional printhead is needed to produce the outline [23]. In this way, it is possible to create a form for printing with low-viscosity materials. The sacrificial material is easy to remove and has no toxic effect on the hydrogel; therefore, it is safe for biological use. A schematic representation of the method is shown in Fig. 9.

2.1 Additives Improving Printability

Extrusion-based techniques often do not use alginate in its initial form. It is usually combined with other polymers or additives, because of the topoor strength of the cross-linked alginate structures. Another reason is also the lack of cell adhesion in biological applications. Additives or material compositions with alginate can ultimately provide a mechanically robust object and also promote the development of cells cultured on such substrates [29].

Depending on the application, various compositions of alginate-based inks are used. The selected material composition determines the cross-linking method. Table 1 shows examples of alginate-based composite mixtures together with their potential applications.

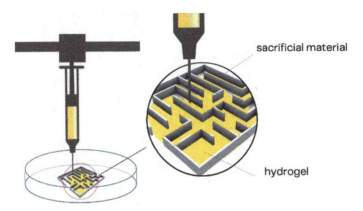

Fig. 9 Scheme for using sacrificial material as an outline to improve alginate ink processing using extrusion-based techniques (own source)

2.2 Methods Using the Support-Bath

Support bath materials are usually easily gelling materials that form suspensions in which 3D structures can be created with a printing nozzle. They are easily removed by temperature changes and usually include a cross-linking agent to reinforce the printed objects. The most commonly used materials are summarized in Table 2.

2.3 Sacrificial Material as an Outline

Among the sacrificial materials are mainly thermoreversible ones. This means that they can be easily removed under temperature changes and can form an outline in which the target material will be placed. Materials for biological applications should be nontoxic and biocompatible to ensure no undesirable reactions with the target ink. In Table 3, the most often used sacrificial materials for bioprinting are presented.

3 Influence of Process Parameters on Mechanical Properties of Hydrogel Structure

The final performance of hydrogel output fabricated with AM technology is affected by several factors, including material properties, manufacturing process parameters, and post-process operations. All factors related to the material manufacturing are managed from the computer-aided level, beginning with the design of the 3D model through the parameters set by the additive manufacturing software. This is a significant advantage of smart manufacturing in Industry 4.0 [89]. The

Table 1 Use of additives for alginate inks in extrusion-based techniques

Ink composition	Cross-linker	Ratio of ink components	Applications	References
Alginate-agarose	CaCl$_2$	3.5:2	Cartilage scaffolds	[70]
		2:0.6 2:1.2 2:1.8	Endothelial scaffolds	[71]
Alginate-agarose-carboxymethyl-chitosan	CaCl$_2$	5:1.5:5	neural scaffolds	[72]
Alginate-chitosan	HCl	10:1 10:10	Scaffolds	[20]
Alginate-gelatin	CaCl$_2$	7.5:5	3D cell culture	[73]
		1:10 2:10 4:10	3D cell culture	[74]
Alginate-gellan gum	CaCl$_2$	2:3	Cartilage and bone scaffolds (nose, human ear, vertebral disk, meniscus)	[75]
Alginate-hyaluronic acid	CaCl$_2$	2.5:0.25	Neural scaffolds	[76]
		2:1	Cartilage scaffolds	[77]
Alginate-methylcellulose	CaCl$_2$	1.5:1.5	Scaffolds intended for 3D cell culture (fibroblasts, mesenchymal stem cells (MSCs), and chondrocytes)	[22]
Alginate-nanocellulose	CaCl$_2$	2.5:2.5	Cartilage scaffold (human ear)	[78]

determination of the mathematical model can be helpful in optimizing the manufacturing process parameters to obtain structural integrity, increase dimensional and shape accuracy assumed porosity, and thus the mechanical properties of the manufactured object. Then the identified parameters are validated by experimental measurements [90, 91].

Table 2 Examples of a support bath for the alginate inks manufacturing process in extrusion-based techniques

Support bath material	Processing temperature	Applications	References
Acrylamide	84.5 °C	3D cell culture structures	[79]
Agarose	90–95 °C (melting temperature) 34–38 °C (gelling temperature)	Support for alginate-collagen bioink	[80]
Agar	85 °C (melting temperature) 32–45 °C (gelling temperature)	cartilaginous structures (nose, human ear)	[81]
Carbopol	116 °C	cell structures based on alginate/GelMA	[82]
Gelatin	35 °C	3D structures based on bioinks	[25]
Gellan	>80 °C (melting temperature) 20–30 °C (gelling temperature, depending on concentration)	Support material for fluid bioink	[83]
Laponite	116 °C	Nanoclay support bath	[84]

The three-dimensional structures for biomedical applications manufactured by AM technology should be characterized by adequate mechanical properties, suitable for their applications. For example, hydrogel scaffolds for use in regenerative medicine, which are intended to replace a host tissue defect to provide support and promote cell divolution and faster regeneration of the graft site, should indicate mechanical properties similar to the healthy tissue area [92–94]. This is required because the defect is protected from external mechanical damage and also due to the importance of ensuring integrity between tissues. These structures should gradually degrade, providing space for proliferating cells. However, during degradation, structures lose their mechanical properties [94, 95]. Another example is a hydrogel drug carrier, which, depending on the destination of active compound release, should have sufficient hardness to avoid premature release. In the case of orally administered carriers, it is important to note that compression forces in the gastrointestinal tract, as well as frictional forces, influence the path of the pharmaceutical drug until it degrades and releases the active ingredient. However, the mechanical properties also affect the release kinetics of the drug. Pharmaceuticals also lose their mechanical properties through degradation in the human body, which should be taken into account during the design process of a specific carrier [96, 97].

The mechanical properties of manufactured objects depend not only on the selected material and, in the case of hydrogel, also on the biopolymer concentration and the cross-linking process, but can also be influenced by the manufacturing

Table 3 Examples of sacrificial material for the alginate inks manufacturing process in extrusion-based techniques

Sacrificial material	Gelling temperature (°C)	Melting temperature	Fabrication method/concentration	Application	References
Agarose fiber	<32	>40 °C	n.a	Hydrogel construction for vascularization	[85]
Carbohydrate glass		>100 °C	Mixture of glucose, dextran, sucrose, and water	Casting of patterned vascular networks	[86]
Gelatin	<35	>37 °C	10% concentration	Fluidic channel for vascularization	[87]
Polivinyl alcohol	<20	>200 °C (glass transition temperature, Tg, of 80 °C)	n.a	Hydrogel construction for vascularization	[71]
Pluronic F127	<37	>57 °C	n.a	Hydrogel construction for vascularization	[88]

n.a.—not available

process parameters [92, 98, 99]. Obtaining robust alginate-based hydrogel structures is achieved by a cross-linking process (e.g., ionically by calcium chloride). Cross-linking time plays an important role for the stiffness of the structures and the possibility of ion diffusion into the hydrogel. The concentration of the cross-linking solution is responsible for the number of ions that bind to the polymer chain. The higher the number of ions in the solution, the shorter the cross-linking time; however, a short period of curing can affect the heterogeneity of the hydrogel. Optimization of cross-linking time and cross-linking agent concentration is sought to obtain a tensile and compressive resistant structure [100–102].

With regard to AM technology, parameters related to the process of material extrusion can be distinguished, as well as the design of a manufactured element. Factors taken into account in the fabrication process in extrusion-based technology include: print head feed speed, pressure of material ejection from nozzle, build orientation, nozzle diameter [103, 104]. In certain circumstances, depending on the material requirements, it is also necessary to consider temperature changes. Increased material processing temperature or reduced print bed temperature may be a necessity. These parameters can also affect the final mechanical properties of manufactured structures [105]. Parameters associated with the morphology of manufactured elements are layer thickness, filling density, porosity, and dimensional integrity [103]. A helpful tool for determining mechanical properties is the use of finite element modeling (FEM), where it is possible to determine the mechanical properties of structures that change over time during the degradation process. However, there are many different factors influencing the degradation process, which may influence the predicted values not to be exact, but only approximate [93].

4   Conclusion and Future Scope

Technology and its development in recent years have led to automation of the manufacturing process with high-efficiency. The methods of raw materials processing and products manufacturing are adapted to the characteristics of raw materials and customers' requirements. Due to the high level of computerization, the obtained products are designed and tested in a virtual environment, which eliminates waste on the production path. Computerization in combination with physical machines allows optimization, making production faster, repeatable, and cost-effective.

AM technology is one of the fastest growing industries that has recently been widely used in the biomedical industry. Among the methods that use semifluidic material processing are inkjet, stereolithography, as well as laser-assisted and extrusion-based technology. However, the last-mentioned technique is the most widely used, and it finds applications in tissue engineering and regenerative medicine, cell culture, drug delivery, and others. It is based on materials characterized by low toxicity and biocompatibility, but in most cases also biodegradability. The ability to produce any shape with high resolution or to dose the appropriate amount of material enables personalized production. The development of methods supporting the production of 3D structures, such as the application of additional materials as temporary supports, is an important step towards the use of demanding materials (e.g., alginate-based hydrogels) for processing in AM technologies.

The current advancement of materials coupled with the development of Industry 4.0 offers enormous opportunities for applications. The future scope of using additive technologies may increase these applications by improving the properties of hydrogel materials manufactured by these technologies as well as the techniques used in bioprinting. The miniaturization of actuators creates the opportunity for increasingly precise manufacturing of printouts. Furthermore, the possibility of biomaterials with cell content utilization is moving towards the fabrication of tissue structures, which in the future may result in the fabrication of organs for human transplantation. Therefore, future applications using improved hydrogel structures may fulfill their purpose in sectors such as tissue engineering, regenerative medicine, 3D cell cultures, and most importantly, contribute to the development of personalized medicine.

References

1. Horst, D.J., Duvoisin, C.A., Vieira, R.D.A.: Additive Manufacturing at Industry 4.0: a Review. Int. J. Eng. Tech. Res. 8 (8), 3–8 (2018).
2. Butt, J.: Exploring the interrelationship between additive manufacturing and industry 4.0. Designs. 4 (13), 1–33 (2020). https://doi.org/10.3390/DESIGNS4020013.
3. Kumar, A., Singh, G., Singh, R.P., Pandey, P.M.: Role of Additive Manufacturing in Industry 4.0 for Maintenance Engineering. In: Martinetti, A., Demichela, M., and Singh, S. (eds.) Applications and Challenges of Maintenance and Safety Engineering in Industry. pp. 235–254 (2020). https://doi.org/10.4018/978-1-7998-3904-0.CH013.
4. Nayyar, A., Kumar, A.: A Roadmap to Industry 4.0: Smart Production, Sharp Business and Sustainable Development. Springer Nature (2020). https://doi.org/10.1007/978-3-030-14544-6_1.
5. Haleem, A., Javaid, M.: Additive manufacturing applications in Industry 4.0: a review. J. Ind. Integr. Manag. 04 (4), 1930001 (2019). https://doi.org/10.1142/S2424862219300011.
6. Korner, M.E.H., Lambán, M.P., Albajez, J.A., Santolaria, J., Corrales, L. del C.N., Royo, J.: Systematic literature review: Integration of additive manufacturing and industry 4.0. Met. 10 (8), 1061 (2020). https://doi.org/10.3390/MET10081061.
7. Kumar, A., Nayyar, A.: Industry: A sustainable, intelligent, innovative, Internet-of-Things industry. In: A Roadmap to Industry 4.0: Smart Production, Sharp Business and Sustainable Development. pp. 1–12. Springer, Cham. (2020).
8. Krishnamurthi, R., Kumar, A., Gopinathan, D., Nayyar, A., Qureshi, B.: An Overview of IoT sensor data processing, fusion, and analysis techniques. Sensors. 20 (21), 6076 (2020). https://doi.org/10.3390/S20216076.
9. Singh, K.K., Nayyar, A., Tanwar, S., Abouhawwash, M.: Emergence of Cyber Physical System and IoT in Smart Automation and Robotics. Springer International Publishing, Cham (2021). https://doi.org/10.1007/978-3-030-66222-6.
10. Reeves, P., Tuck, C., Hague, R.: Additive manufacturing for mass customization. In: Fogliatto, F.S. and da Silveira, G.J.C. (eds.) Mass Customization. pp. 275–289. Springer, London (2011). https://doi.org/10.1007/978-1-84996-489-0_13.
11. Annaji, M., Ramesh, S., Poudel, I., Govindarajulu, M., Arnold, R.D., Dhanasekaran, M., Babu, R.J.: Application of extrusion-based 3D printed dosage forms in the treatment of chronic diseases. J. Pharm. Sci. 109 (12), 3551–3568 (2020). https://doi.org/10.1016/J.XPHS.2020.09.042.
12. Geraedts, J., Doubrovski, Z., Stellingwerff, M.C.: Three views on additive manufacturing: Business, research, and education. In: Horvath, I., Albers, A., Behrendt, M., and Rusak, Z. (eds.) TMCE 2012 (2012).
13. Zanardini, M., Bacchetti, A., Zanoni, S.: Benefits and costs of additive manufacturing applications: An evaluation guideline. In: Proceedings of 20th Summer School "Francesco Turco" (2015).
14. Nayyar, A., Rameshwar, R., Solanki, A.: Internet of things (IoT) and the digital business environment: A standpoint inclusive cyber space, cyber crimes, and cybersecurity. In: The Evolution of Business in the Cyber Age. pp. 111–152. Apple Academic Press (2020). https://doi.org/10.1201/9780429276484-6.
15. Unagolla, J.M., Jayasuriya, A.C.: Hydrogel-based 3D bioprinting: A comprehensive review on cell-laden hydrogels, bioink formulations, and future perspectives. Appl. Mater. Today. 18 100479 (2020). https://doi.org/10.1016/J.APMT.2019.100479.
16. Temirel, M., Hawxhurst, C., Tasoglu, S.: Shape fidelity of 3D-bioprinted biodegradable patches. Micromachines. 12 (2), 195 (2021). https://doi.org/10.3390/MI12020195.
17. Axpe, E., Oyen, M.L.: Applications of alginate-based bioinks in 3D bioprinting. Int. J. Mol. Sci. 2016, Vol. 17, Page 1976. 17 (12), 1976 (2016). https://doi.org/10.3390/IJMS17121976.
18. Datta, S., Barua, R., Das, J.: Importance of Alginate Bioink for 3D Bioprinting in Tissue Engineering and Regenerative Medicine. In: Pereira, L. and Cotas, J. (eds.) Alginates - Recent

Uses of This Natural Polymer. IntechOpen (2019). https://doi.org/10.5772/INTECHOPEN.90426.
19. Zhang, M., Zhao, X.: Alginate hydrogel dressings for advanced wound management. Int. J. Biol. Macromol. 162 1414–1428 (2020). https://doi.org/10.1016/J.IJBIOMAC.2020.07.311.
20. Liu, Q., Li, Q., Xu, S., Zheng, Q., Cao, X.: Preparation and properties of 3D printed alginate–chitosan polyion complex hydrogels for tissue engineering. Polym. 2018, Vol. 10, Page 664. 10 (6), 664 (2018). https://doi.org/10.3390/POLYM10060664.
21. Piras, C.C., Smith, D.K.: Multicomponent polysaccharide alginate-based bioinks. J. Mater. Chem. B. 8 8171–8188 (2020).
22. Zhang, H., Cheng, J., Ao, Q.: Preparation of alginate-based biomaterials and their applications in biomedicine. Mar. Drugs . 19 (5), 264 (2021). https://doi.org/10.3390/MD19050264.
23. Kaczmarek-Pawelska, A.: Alginate-based hydrogels in regenerative medicine. In: Pereira, L. (ed.) Alginates - Recent Uses of This Natural Polymer. IntechOpen (2019). https://doi.org/10.5772/INTECHOPEN.88258.
24. Lee, K.Y., Mooney, D.J.: Alginate: Properties and biomedical applications, (2012). https://doi.org/10.1016/j.progpolymsci.2011.06.003.
25. Hinton, T.J., Jallerat, Q., Palchesko, R.N., Park, J.H., Grodzicki, M.S., Shue, H.-J., Ramadan, M.H., Hudson, A.R., Feinberg, A.W.: Three-dimensional printing of complex biological structures by freeform reversible embedding of suspended hydrogels. Sci. Adv. 1 (9), e1500758 (2015). https://doi.org/10.1126/SCIADV.1500758.
26. Hazur, J., Detsch, R., Karakaya, E., Kaschta, J., Teßmar, J., Schneidereit, D., Friedrich, O., Schubert, D.W., Boccaccini, A.R.: Improving alginate printability for biofabrication: establishment of a universal and homogeneous pre-crosslinking technique. Biofabrication. 12 (4), 045004 (2020). https://doi.org/10.1088/1758-5090/AB98E5.
27. Naghieh, S., Chen, D.: Printability – a key issue in extrusion-based bioprinting. J. Pharm. Anal. (2021). https://doi.org/10.1016/J.JPHA.2021.02.001.
28. Pellegrino, A., Bin, Y., Kimberly, D., Williams, F.: Bioprinting alginate structures using the FRESH method. In: Biomedical Engineering and Bioengineering Commons (2021).
29. Chimene, D., Lennox, K.K., Kaunas, R.R., Gaharwar, A.K.: Advanced bioinks for 3D printing: A materials science perspective. Ann. Biomed. Eng. 44 (6), 2090–2102 (2016). https://doi.org/10.1007/S10439-016-1638-Y.
30. Afghah, F., Altunbek, M., Dikyol, C., Koc, B.: Preparation and characterization of nanoclay-hydrogel composite support-bath for bioprinting of complex structures. Sci. Reports . 10 (5257), 1–13 (2020). https://doi.org/10.1038/s41598-020-61606-x.
31. Shiwarski, D.J., Hudson, A.R., Tashman, J.W., Feinberg, A.W.: Emergence of FRESH 3D printing as a platform for advanced tissue biofabrication. APL Bioeng. 5 (1), 10904 (2021). https://doi.org/10.1063/5.0032777.
32. Guo, N., Leu, M.C.: Additive manufacturing: Technology, applications and research needs. Front. Mech. Eng. 8 (3), 215–243 (2013). https://doi.org/10.1007/S11465-013-0248-8.
33. Calzado, M.J., Romero, L., Domínguez, I.A., Espinosa, M.D.M., Domínguez, M.: Additive Manufacturing technologies: An overview about 3D printing methods and future prospects. Complexity. 1047 (2019). https://doi.org/10.1155/2019/9656938.
34. He, Y., Yang, F., Zhao, H., Gao, Q., Xia, B., Fu, J.: Research on the printability of hydrogels in 3D bioprinting. Sci. Reports . 6 (29977), 1–13 (2016). https://doi.org/10.1038/srep29977.
35. Jang, T.S., Jung, H. Do, Pan, H.M., Han, W.T., Chen, S., Song, J.: 3D printing of hydrogel composite systems: Recent advances in technology for tissue engineering. Int. J. Bioprinting. 4 (1), (2018). https://doi.org/10.18063/IJB.V4I1.126.
36. You, F., Eames, B.F., Chen, X.: Application of extrusion-based hydrogel bioprinting for cartilage tissue engineering. Int. J. Mol. Sci. 18 (7), (2017). https://doi.org/10.3390/IJMS18071597.
37. Li, J., Wu, C., Chu, P.K., Gelinsky, M.: 3D printing of hydrogels: Rational design strategies and emerging biomedical applications. Mater. Sci. Eng. R Reports. 140 100543 (2020). https://doi.org/10.1016/J.MSER.2020.100543.

38. Li, Y.-C., Zhang, Y.S., Akpek, A., Shin, S.R., Khademhosseini, A.: 4D bioprinting: the next-generation technology for biofabrication enabled by stimuli-responsive materials. Biofabrication. 9 (1), 012001 (2016). https://doi.org/10.1088/1758-5090/9/1/012001.
39. Aguilar-de-Leyva, Á., Linares, V., Casas, M., Caraballo, I.: 3D printed drug delivery systems based on natural products. Pharm. 12 (7), 620 (2020). https://doi.org/10.3390/PHARMACEUTICS12070620.
40. Vaezi, M., Yang, S.: Extrusion-based additive manufacturing of PEEK for biomedical applications. Virtual Phys. Prototyp. 10 (3), 1–13 (2015). https://doi.org/10.1080/17452759.2015.1097053.
41. Kirchmajer, D.M., Gorkin, R., In Het Panhuis, M.: An overview of the suitability of hydrogel-forming polymers for extrusion-based 3D-printing. J. Mater. Chem. B. 3 (20), 4105–4117 (2015). https://doi.org/10.1039/C5TB00393H.
42. Saarai, A., Kasparkova, V., Sedlacek, T., Saha, P.: A comparative study of crosslinked sodium alginate/gelatin hydrogels for wound dressing | Proceedings of the 4th WSEAS international conference on Energy and development - environment - biomedicine. In: GEMESED'11: Proceedings of the 4th WSEAS international conference on Energy and development - environment - biomedicine. pp. 384–389 (2011).
43. Szekalska, M., Sosnowska, K., Czajkowska-Kośnik, A., Winnicka, K.: Calcium chloride modified alginate microparticles formulated by the spray drying process: A strategy to prolong the release of freely soluble drugs. Mater. 11 (9), 1522 (2018). https://doi.org/10.3390/MA11091522.
44. Li, Y., Kumacheva, E.: Hydrogel microenvironments for cancer spheroid growth and drug screening. Sci. Adv. 4 (4), eaas8998 (2018). https://doi.org/10.1126/SCIADV.AAS8998.
45. Siltanen, C., Diakataou, M., Lowen, J., Haque, A., Rahimian, A., Stybayeva, G., Revzin, A.: One step fabrication of hydrogel microcapsules with hollow core for assembly and cultivation of hepatocyte spheroids. Acta Biomater. 50 428–436 (2017). https://doi.org/10.1016/J.ACTBIO.2017.01.010.
46. Agarwal, S., Saha, S., Balla, V.K., Pal, A., Barui, A., Bodhak, S.: Current developments in 3D bioprinting for tissue and organ regeneration–A review. Front. Mech. Eng. 0 90 (2020). https://doi.org/10.3389/FMECH.2020.589171.
47. O'Brien, F.J.: Biomaterials & scaffolds for tissue engineering. Mater. Today. 14 (3), 88–95 (2011). https://doi.org/10.1016/S1369-7021(11)70058-X.
48. Murr, L.E.: Tissue engineering scaffolds and scaffold materials. In: Handbook of Materials Structures, Properties, Processing and Performance. pp. 597–603. Springer, Cham (2015). https://doi.org/10.1007/978-3-319-01815-7_33.
49. Klontzas, M.E., Drissi, H., Mantalaris, A.: The use of alginate hydrogels for the culture of mesenchymal stem cells (MSCs): in vitro and in vivo paradigms. In: Pereira, L. and Cotas, J. (eds.) Alginates - Recent Uses of This Natural Polymer. IntechOpen (2019). https://doi.org/10.5772/INTECHOPEN.88020.
50. Augst, A.D., Kong, H.J., Mooney, D.J.: Alginate hydrogels as biomaterials. Macromol. Biosci. 6 (8), 623–633 (2006). https://doi.org/10.1002/mabi.200600069.
51. Chaturvedi, K., Ganguly, K., More, U.A., Reddy, K.R., Dugge, T., Naik, B., Aminabhavi, T.M., Noolvi, M.N.: Sodium alginate in drug delivery and biomedical areas. In: Natural Polysaccharides in Drug Delivery and Biomedical Applications. pp. 59–100. Academic Press (2019). https://doi.org/10.1016/B978-0-12-817055-7.00003-0.
52. Lode, A., Krujatz, F., Brüggemeier, S., Quade, M., Schütz, K., Knaack, S., Weber, J., Bley, T., Gelinsky, M.: Green bioprinting: Fabrication of photosynthetic algae-laden hydrogel scaffolds for biotechnological and medical applications. Eng. Life Sci. 15 (2), 177–183 (2015). https://doi.org/10.1002/ELSC.201400205.
53. Neerooa, B.N.H.M., Ooi, L.-T., Shameli, K., Dahlan, N.A., Islam, J.M.M., Pushpamalar, J., Teow, S.-Y.: Development of polymer-assisted nanoparticles and nanogels for cancer therapy: An update. Gels 2021, Vol. 7, Page 60. 7 (2), 60 (2021). https://doi.org/10.3390/GELS7020060.

54. Kondiah, P.P.D., Choonara, Y.E., Marimuthu, T., Kondiah, P.J., du Toit, L.C., Kumar, P., Pillay, V.: Nanotechnological paradigms for neurodegenerative disease interventions. In: Advanced 3D-printed systems and nanosystems for drug delivery and tissue engineering. pp. 277–292. Elsevier (2020). https://doi.org/10.1016/B978-0-12-818471-4.00010-8.
55. Tsou, Y.H., Khoneisser, J., Huang, P.C., Xu, X.: Hydrogel as a bioactive material to regulate stem cell fate. Bioact. Mater. 1 (1), 39–55 (2016). https://doi.org/10.1016/J.BIOACTMAT.2016.05.001.
56. Andersen, T., Auk-Emblem, P., Dornish, M.: 3D cell culture in alginate hydrogels. Microarrays 2015, Vol. 4, Pages 133–161. 4 (2), 133–161 (2015). https://doi.org/10.3390/MICROARRAYS4020133.
57. Kandemir, N., Vollmer, W., Jakubovics, N.S., Chen, J.: Mechanical interactions between bacteria and hydrogels. Sci. Reports 2018 81. 8 (10893), 1–11 (2018). https://doi.org/10.1038/s41598-018-29269-x.
58. Richter, R., Kamal, M.A.M., García-Rivera, M.A., Kaspar, J., Junk, M., Elgaher, W.A.M., Srikakulam, S.K., Gress, A., Beckmann, A., Grißmer, A., Meier, C., Vielhaber, M., Kalinina, O., Hirsch, A.K.H., Hartmann, R.W., Brönstrup, M., Schneider-Daum, N., Lehr, C.M.: A hydrogel-based in vitro assay for the fast prediction of antibiotic accumulation in Gram-negative bacteria. Mater. Today Bio. 8 100084 (2020). https://doi.org/10.1016/J.MTBIO.2020.100084.
59. Krujatz, F., Lode, A., Brüggemeier, S., Schütz, K., Kramer, J., Bley, T., Gelinsky, M., Weber, J.: Green bioprinting: Viability and growth analysis of microalgae immobilized in 3D-plotted hydrogels versus suspension cultures. Eng. Life Sci. 15 (7), 678–688 (2015). https://doi.org/10.1002/ELSC.201400131.
60. Moreno-Garrido, I.: Microalgae immobilization: Current techniques and uses. Bioresour. Technol. 99 (10), 3949–3964 (2008). https://doi.org/10.1016/J.BIORTECH.2007.05.040.
61. Afsana, Jain, V., Haider, N., Jain, K.: 3D printing in personalized drug delivery. Curr. Pharm. Des. 24 (42), 5062–5071 (2018). https://doi.org/10.2174/1381612825666190215122208.
62. Maulvi, F.A., Shah, M.J., Solanki, B.S., Patel, A.S., Soni, T.G., Shah, D.O.: Application of 3D printing technology in the development of novel drug delivery systems. Int. J. Drug Dev. Res. 9 (1), 44–49 (2017).
63. Algahtani, M., Mohammed, A., Ahmad, J.: Extrusion-based 3D printing for pharmaceuticals: contemporary research and applications. Curr. Pharm. Des. 24 (42), 4991–5008 (2018). https://doi.org/10.2174/1381612825666190110155931.
64. Mathew, E., Pitzanti, G., Larrañeta, E., Lamprou, D.A.: 3D printing of pharmaceuticals and drug delivery devices. Pharmaceutics. 12 (3), 266 (2020). https://doi.org/10.3390/PHARMACEUTICS12030266.
65. Khaled, S.A., Burley, J.C., Alexander, M.R., Yang, J., Roberts, C.J.: 3D printing of tablets containing multiple drugs with defined release profiles. Int. J. Pharm. 494 (2), 643–650 (2015). https://doi.org/10.1016/J.IJPHARM.2015.07.067.
66. Awad, A., Fina, F., Trenfield, S.J., Patel, P., Goyanes, A., Gaisford, S., Basit, A.W.: 3D printed pellets (miniprintlets): A novel, multi-drug, controlled r148elease platform technology. Pharmaceutics. 11 (4), 148 (2019). https://doi.org/10.3390/PHARMACEUTICS11040148.
67. Yang, I., Lim, B., Park, J., Hyun, J., Ahn, S.: Effect of orthodontic bonding steps on the initial adhesion of mutans streptococci in the presence of saliva. Angle Orthod. 81 (2), 326–333 (2011). https://doi.org/10.2319/062210-343.1.
68. Aderibigbe, B.A., Buyana, B.: Alginate in wound dressings. Pharm. 2018, Vol. 10, Page 42. 10 (2), 42 (2018). https://doi.org/10.3390/PHARMACEUTICS10020042.
69. Narayanan, L.K., Huebner, P., Fisher, M.B., Spang, J.T., Starly, B., Shirwaiker, R.A.: 3D-bioprinting of polylactic acid (PLA) nanofiber–alginate hydrogel bioink containing human adipose-derived stem cells. ACS Biomater. Sci. Eng. 2 (10), 1732–1742 (2016). https://doi.org/10.1021/ACSBIOMATERIALS.6B00196.

70. López-Marcial, G.R., Zeng, A.Y., Osuna, C., Dennis, J., García, J.M., O'Connell, G.D.: Agarose-based hydrogels as suitable bioprinting materials for tissue engineering. ACS Biomater. Sci. Eng. 4 (10), 3610–3616 (2018). https://doi.org/10.1021/ACSBIOMATERIALS.8B00903.
71. Zou, Q., Grottkau, B.E., He, Z., Shu, L., Yang, L., Ma, M., Ye, C.: Biofabrication of valentine-shaped heart with a composite hydrogel and sacrificial material. Mater. Sci. Eng. C. 108 110205 (2020). https://doi.org/10.1016/J.MSEC.2019.110205.
72. Gu, Q., Tomaskovic-Crook, E., Lozano, R., Chen, Y., Kapsa, R.M., Zhou, Q., Wallace, G.G., Crook, J.M.: Functional 3D neural mini-tissues from printed gel-based bioink and human neural stem cells. Adv. Healthc. Mater. 5 (12), 1429–1438 (2016). https://doi.org/10.1002/ADHM.201600095.
73. Yan, Y., Wang, X., Xiong, Z., Liu, H., Liu, F., Lin, F., Wu, R., Zhang, R., Lu, Q.: Direct construction of a three-dimensional structure with cells and hydrogel. J. Bioact. Compat. Polym. SAGE Journals. 20 (3), 259–269 (2005). https://doi.org/10.1177/0883911505053658.
74. Chung, J.H.Y., Naficy, S., Yue, Z., Kapsa, R., Quigley, A., Moulton, S.E., Wallace, G.G.: Bioink properties and printability for extrusion printing living cells. Biomater. Sci. 1 (7), 763–773 (2013). https://doi.org/10.1039/c3bm00012e.
75. Kesti, M., Eberhardt, C., Pagliccia, G., Kenkel, D., Grande, D., Boss, A., Zenobi-Wong, M.: Bioprinting complex cartilaginous structures with clinically compliant biomaterials. Adv. Funct. Mater. 25 (48), 7406–7417 (2015). https://doi.org/10.1002/ADFM.201503423.
76. Rajaram, A., Schreyer, D., Chen, D.: Bioplotting alginate/hyaluronic acid hydrogel scaffolds with structural integrity and preserved schwann cell viability. https://home.liebertpub.com/3dp. 1 (4), 194–203 (2014). https://doi.org/10.1089/3DP.2014.0006.
77. Antich, C., de Vicente, J., Jiménez, G., Chocarro, C., Carrillo, E., Montañez, E., Gálvez-Martín, P., Marchal, J.A.: Bio-inspired hydrogel composed of hyaluronic acid and alginate as a potential bioink for 3D bioprinting of articular cartilage engineering constructs. Acta Biomater. 106 114–123 (2020). https://doi.org/10.1016/J.ACTBIO.2020.01.046.
78. Markstedt, K., Mantas, A., Tournier, I., Ávila, H.M., Hägg, D., Gatenholm, P.: 3D bioprinting human chondrocytes with nanocellulose–alginate bioink for cartilage tissue engineering applications. Biomacromolecules. 16 (5), 1489–1496 (2015). https://doi.org/10.1021/ACS.BIOMAC.5B00188.
79. Morley, C.D., Ellison, S.T., Bhattacharjee, T., O'Bryan, C.S., Zhang, Y., Smith, K.F., Kabb, C.P., Sebastian, M., Moore, G.L., Schulze, K.D., Niemi, S., Sawyer, W.G., Tran, D.D., Mitchell, D.A., Sumerlin, B.S., Flores, C.T., Angelini, T.E.: Quantitative characterization of 3D bioprinted structural elements under cell generated forces. Nat. Commun. 10 (1), 3029 (2019). https://doi.org/10.1038/S41467-019-10919-1.
80. BB, M., M, G.-F., AG, H., MS, D., RMA, D., RL, R., ME, G.: Human platelet lysate-based nanocomposite bioink for bioprinting hierarchical fibrillar structures. Biofabrication. 12 (1), 015012 (2019). https://doi.org/10.1088/1758-5090/AB33E8.
81. MacKenzie, I.: Development of 3D bioprinting techniques based on supportive media, (2020).
82. Krishnamoorthy, Z., Zhang, Z., Xu, C.: Biofabrication of three-dimensional cellular structures based on gelatin methacrylate-alginate interpenetrating network hydrogel. J. Biomater. Appl. 33 (8), 1105–1117 (2019). https://doi.org/10.1177/0885328218823329.
83. Compaan, A.M., Song, K., Hung, Y.: Gellan fluid gel as a versatile support bath material for fluid extrusion bioprinting. ACS Appl. Mater. Interfaces. 11 5714–5726 (2019).
84. Dávila, J.L., d'Ávila, M.A.: Laponite as a rheology modifier of alginate solutions: Physical gelation and aging evolution. Carbohydr. Polym. 157 1–8 (2017). https://doi.org/10.1016/J.CARBPOL.2016.09.057.
85. Bertassoni, L.E., Cecconi, M., Manoharan, V., Nikkhah, M., Hjortnaes, J., Cristino, A.L., Barabaschi, G., Demarchi, D., Dokmeci, M.R., Yang, Y., Khademhosseini, A.: Hydrogel bioprinted microchannel networks for vascularization of tissue engineering constructs. Lab Chip. 14 (13), 2202–2211 (2014). https://doi.org/10.1039/c4lc00030g.

86. Miller, J.S., Stevens, K.R., Yang, M.T., Baker, B.M., Nguyen, D.-H.T., Cohen, D.M., Toro, E., Chen, A.A., Galie, P.A., Yu, X., Chaturvedi, R., Bhatia, S.N., Chen, C.S.: Rapid casting of patterned vascular networks for perfusable engineered three-dimensional tissues. Nat. Mater. . 11 (9), 768–774 (2012). https://doi.org/10.1038/nmat3357.
87. Lee, V.K., Kim, D.Y., Ngo, H., Lee, Y., Seo, L., Yoo, S.S., Vincent, P.A., Dai, G.: Creating perfused functional vascular channels using 3D bio-printing technology. Biomaterials. 35 (28), 8092–8102 (2014). https://doi.org/10.1016/J.BIOMATERIALS.2014.05.083.
88. Kolesky, D.B., Homan, K.A., Skylar-Scott, M.A., Lewis, J.A.: Three-dimensional bioprinting of thick vascularized tissues. Proc. Natl. Acad. Sci. 113 (12), 3179–3184 (2016). https://doi.org/10.1073/PNAS.1521342113.
89. Elahi, B., Tokaldany, S.A.: Application of Internet of Things-aided simulation and digital twin technology in smart manufacturing. In: Ram, M. (ed.) Advances in Mathematics for Industry 4.0. pp. 335–359. Academic Press (2021). https://doi.org/10.1016/B978-0-12-818906-1.00015-2.
90. Comminal, R., Jafarzadeh, S., Serdeczny, M., Spangenberg, J.: Estimations of intelayer contacts in extrusion additive manufacturing using a CFD model. In: Meboldt, M. and Klahn, C. (eds.) Industrializing additive manufacturing : proceedings of AMPA2020. pp. 241–250. Springer (2021).
91. Golebiowska, A.A., Kim, H.S., Camci-Unal, G., Nukavarapu, S.P.: Integration of technologies for bone tissue engineering. In: Reis, R.L. (ed.) Encyclopedia of Tissue Engineering and Regenerative Medicine. pp. 243–259. Academic Press (2019). https://doi.org/10.1016/B978-0-12-801238-3.11063-3.
92. Wang, Z., Tian, Z., Menard, F., Kim, K.: Comparative study of gelatin methacrylate hydrogels from different sources for biofabrication applications. Biofabrication. 9 044101 (2017). https://doi.org/10.1088/1758-5090/aa83cf.
93. Naghieh, S., Sarker, M.D., Karamooz-Ravari, M.R., McInnes, A.D., Chen, X.: Modeling of the mechanical behavior of 3D bioplotted scaffolds considering the penetration in interlocked strands. Appl. Sci. 8 (9), 1422 (2018). https://doi.org/10.3390/APP8091422.
94. Nazir, R., Bruyneel, A., Carr, C., Czernuszka, J.: Mechanical and degradation properties of hybrid scaffolds for tissue engineered heart valve (TEHV). J. Funct. Biomater. 12 (1), 20 (2021). https://doi.org/10.3390/JFB12010020.
95. Ghasemi-Mobarakeh, L., Prabhakaran, M.P., Tian, L., Shamirzaei-Jeshvaghani, E., Dehghani, L., Ramakrishna, S.: Structural properties of scaffolds: Crucial parameters towards stem cells differentiation. World J. Stem Cells. 7 (4), 744 (2015). https://doi.org/10.4252/WJSC.V7.I4.728.
96. Chou, S.-F., Woodrow, K.A.: Relationships between mechanical properties and drug release from electrospun fibers of PCL and PLGA blends. J. Mech. Behav. Biomed. Mater. 65 733 (2017). https://doi.org/10.1016/J.JMBBM.2016.09.004.
97. Li, D., Yao, W., Meng, X., Jin, G., Liang, H., Yang, R.: Mechanical properties and drug delivery ability of smart hydrogel. In: 9th IEEE International Conference on Cyber Technology in Automation, Control and Intelligent Systems, CYBER 2019. pp. 383–388. Institute of Electrical and Electronics Engineers Inc. (2019). https://doi.org/10.1109/CYBER46603.2019.9066527.
98. Haibo Zhang, Xin Huang, Jianxin Jiang, Shibin Shang, Zhanqian Song: Hydrogels with high mechanical strength cross-linked by a rosin-based crosslinking agent. RSC Adv. 7 (67), 42541–42548 (2017). https://doi.org/10.1039/C7RA08024G.
99. Li, X., Sun, Q., Li, Q., Kawazoe, N., Chen, G.: Functional hydrogels with tunable structures and properties for tissue engineering applications. Front. Chem. 499 (2018). https://doi.org/10.3389/FCHEM.2018.00499.
100. Ahearne, M., Yang, Y., Liu, K.-K.: Mechanical Characterisation of Hydrogels for Tissue Engineering Applications Hydrogels for Tissue Engineering. Tissue Eng. 4 1–16 (2008).

101. Kaklamani, G., Cheneler, D., Grover, L.M., Adams, M.J., Bowen, J.: Mechanical properties of alginate hydrogels manufactured using external gelation Journal Item Mechanical properties of alginate hydrogels manufactured using external gelation. J. Mech. Behav. Biomed. Mater. 36 135–142 (2014). https://doi.org/10.1016/j.jmbbm.2014.04.013.
102. Al-Darkazly, I.A.A.: Optimization of mechanical properties of alginate hydrogel for 3D bioprinting self-standing scaffold architecture for tissue engineering applications. Int. J. Biomed. Biol. Eng. 14 (12), 419–427 (2020). https://doi.org/10.1186/S40824-018-0122-1.
103. Suteja, T.J., Soesanti, A.: Mechanical properties of 3D printed polylactic acid product for various infill design parameters: A review. J. Phys. Conf. Ser. 1569 (4), 042010 (2020). https://doi.org/10.1088/1742-6596/1569/4/042010.
104. Tang, C., Liu, J., Yang, Y., Liu, Y., Jiang, S., Hao, W.: Effect of process parameters on mechanical properties of 3D printed PLA lattice structures. Compos. Part C Open Access. 3 100076 (2020). https://doi.org/10.1016/J.JCOMC.2020.100076.
105. Lee, D., Wu, G.-Y.: Parameters affecting the mechanical properties of three-dimensional (3D) printed carbon fiber-reinforced polylactide composites. Polym. 2020, Vol. 12, Page 2456. 12 (11), 2456 (2020). https://doi.org/10.3390/POLYM12112456.

Coal Fly Ash Utilization in India

Dipankar Das and Prasanta Kumar Rout

1   Introduction

Coal is one of the premier sources of fuel for electricity generation in India and due to the huge reserves of coal in the coming decades, it will be used as a source of fuel [1, 2]. India's electricity demand is expected to surpass 9,50,000 MW by 2030 [3]. The electricity demand around the world is increasing daily. Indian coals are mostly of sub-bituminous rank along with bituminous and lignite. The ash content in Indian coals is very high, which is ranging from 35 to 50% [3–9]. As a result, millions of tons of fly ash is being generated in India and possess various environmental problems i.e.; air and water pollution along with disposal issue [3, 10–15]. One of the industrial waste i.e., fly ash can be defined as the fine solid powders produced by burning of pulverized coal in a coal based thermal power plant [16–19]. The burning of pulverized coal produces 80% of fly ash and 20% of bottom ash. The size of the Indian fly ash powder generally ranges from 1 to 150 μm and it is finer than Portland cement [20]. The fly ash powder mainly consists of aluminosilicate source materials such as alumina and silica as a major element along with some minor elements.

American Society for Testing and Materials standard (ASTM C618) and European standards (EN 450) classified fly ash into two types, namely Class F and Class C, based on chemical analysis and the kind of coal used [21–23]. Class F fly ash is derived from bituminous coals where the percentage of CaO is below 10%, and class C fly ash is derived from the lignite coal having CaO more than 10%. According to IS 3812-1981 codes, fly ash is classified into two types, namely Grade I and Grade II [24–26]. Grade I is derived from bituminous coal where

D. Das (✉) · P. K. Rout
Department of Material Science and Engineering, Tripura University (A Central University), Suryamaninagar, Tripura 799022, India
e-mail: dipankar.msen@tripurauniv.in

the proportions of $SiO_2 + Al_2O_3 + Fe_2O_3$ are greater than 70% and Grade II is derived from lignite coal where $SiO_2 + Al_2O_3 + Fe_2O_3$ is greater than 50%.

The Government of India's Ministry of Environment, Forest, and Climate Change (MoEFCC) is taking lots of initiation for the proper utilization and clearance of fly ash. The global fly ash generation is about 800 million tons per annum and can increase up to 2100 million tons from 2031 to 2032 [27]. India is one of the leading fly ash producer countries along with China and the USA [28–30]. Since 1996 the Central Electricity Authority (CEA), Government of India, has been monitoring the fly ash generation and utilization in India.

The MoEFCC, Government of India, has issued a notification on 25th January 2016, where it is directed to enhance the utilization of fly ash powder and to upload the fly ash availability during the current month on the thermal power plant website along with stock in ash pond [31]. It is also directed to utilize the fly ash powder in the jurisdiction of an area of 100–300 km. The transportation cost of fly ash powder up to 100 km will be borne entirely by the thermal power plants and for more than 100 km up to 300 km; the transportation cost will be borne equally both by the user and the thermal power plants. The fly ash powder should utilize in various Government schemes or programs such as Swachh Bharat Abhiyan (SBA), Pradhan Mantri Gramin Sadak Yojana (PMGSY), Mahatma Gandhi National Rural Employment Guarantee Act (MGNREGA), 2005, etc. [31].

The central electricity authority (CEA), Government of India, published a report where it is reported that the fly ash generation and utilization in India for the first half of the year 2018–2019 were 93.26 million tons and 64.08 million tons, respectively [32]. So, the percentage of utilization is 68.72%, whereas the percentage of utilization for the first half of the year 2017–2018 was 60.38%. The total coal consumed for the first half of the year 2018–2019 is 295.42 million tons, and the average ash content was 31.57%. In the year 2018–2019, 156 thermal power plants reported the fly ash generation and utilization data to the Government of India. Among 156 thermal power plants, 29 power utilities have utilized 100% fly ash, and 17 power utilities have used fly ash in the range of 90% to below 100%. A brief summary of fly ash generation and utilization for the last few years in India is shown in Fig. 1.

If we consider the state-wise fly ash generation than seven states namely, West Bengal, Andhra Pradesh, Madhya Pradesh, Chhattisgarh, Uttar Pradesh, Maharashtra, and Odisha each have generated more than 7 million tons of fly ash for the first half of the year 2018–2019. During the above said period, Chhattisgarh has generated maximum fly ash of more than 13 million tons, and Gujarat, Punjab, and Rajasthan have achieved the highest fly ash utilization of more than 100%. Figure 2 shows the state-wise fly ash generation and utilization for the first half of the year 2018–2019. Few coals based thermal power plants in India such as Mihan TPS (Maharashtra), T.G.S (West Bengal), Chinakuri (West Bengal), Rajghat (Delhi), GEPL TPP (Maharashtra), KWPCL TPS (Chhattisgarh), Bhatinda (Punjab), Swastik power and minerals resources private limited (Chhattisgarh) has no fly ash generation during the first half of the year 2018–2019 due to various reasons such as shutdown condition, long breakdown, permanently closed down, etc.

Coal Fly Ash Utilization in India

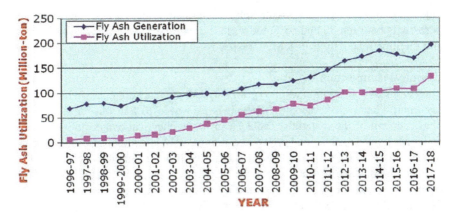

Fig. 1 Fly ash generation and utilization trends for the year 1996–97 to 2017–18 [32]

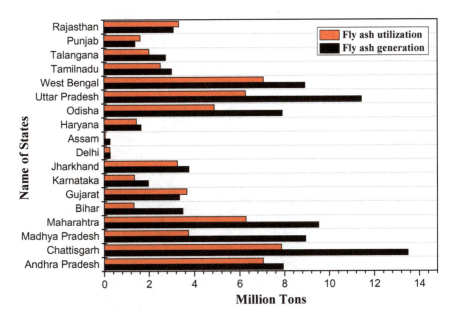

Fig. 2 State-wise fly ash generation and utilization for the first half of the year 2018–2019

1.1 Characterization of Fly Ash

Physical properties

The physical properties of fly ash powders are varied. Naik et al. reported that the bulk density (g/c) of fly ash powders ranges from 0.9 to 1.3, the specific gravity is between 1.6 and 2.6, the plasticity is either lower or non-plastic, the shrinkage limit (volume stability) is higher, the clay content is negligible (%), the free swell index is extremely low, the water holding capacity is between 40 and 60%, the porosity is between 30 and 65%, the surface area ranges from 500 to 5000 m^2/kg, and the lime reactivity is 1–8 MPa, respectively [33].

Chemical composition

The chemical composition of all types of fly ash powders are characterized by X-ray fluorescence spectroscopy (XRF). It is also used to find out the amorphous phase content of the materials. Table 1 shows the chemical compositions of class F fly ash powders. SiO_2 and Al_2O_3 are the major oxide compositions, which are present in the class F fly ash powders along with other minor compositions [34, 35]. The percentage of CaO is less than 10%, i.e., 5.91 wt.%. Table 2 shows the chemical compositions of class C type fly ash powders, where the percentage of SiO_2 and Al_2O_3 is 28.93 and 14.82, respectively [36]. It can be noticed that the % of CaO is high i.e., 39.80%.

Table 1 Oxide composition (wt. %) of fly ash powder [11]

Oxides	Class F powder (%)
SiO_2	55.6
Al_2O_3	29.80
CaO	1.59
Fe_2O_3	5.91
TiO_2	1.63
MgO	1.08
K_2O	1.94
Na_2O	0.23
MnO	0.05
SrO	0.04
ZnO	0.03
SO_3	0.45
LOI	0.47

Table 2 Oxide composition (wt.%) of fly ash powder [36]

Oxides	Class C powder (%)
SiO_2	28.93
Al_2O_3	14.82
CaO	39.80
Fe_2O_3	6.40
MgO	4.86
K_2O	0.56
Na_2O	1.10
Other components	2.63

Particle size distribution

For fly ash powders, one of the crucial parameters is the particle size distribution. A laser diffraction instrument is generally used to measure the size of the particles. The instrument uses the laser diffraction pattern to analyze the interaction between light and particles of raw materials. The analyzed pattern is matched with the mathematical model and the model is calculated by the Fraunhofer or Mie theory. The angle between the diffracted light and incident light is related to the particle size of the powders. Isopropyl alcohol is a recommended dispersive agent due to its high viscosity and reproducibility. The excellent packing density of the mixture and the workability depends on the broader particle size distribution. Another technique to measure the particle size distribution is No. 325 wet sieve analysis as per ASTM C618 and C311/430, respectively. The materials which are coarse are unburnt carbon, and the materials which are fine affect the reactivity. Particle size distribution by using Laser diffraction analysis is done by Das et al. [10]. Figure 3 shows the particle size of distribution of as-received fly ash powders. It was reported that all the particles are below 100 µm, among them 90 volume percent, 50 volume percent, and 30 volume percent particles are 13.62 µm, 2.55 µm 0.65 µm, respectively. The fly ash particles, which are finer in size, show higher surface area [11].

Fourier Transform Infrared Spectroscopy

The structural behavior of the fly ash powders is characterized by the Fourier transform infrared spectroscopy (FTIR). Figure 4 shows the FTIR spectra of as received fly ash powders [11]. The band presence in fly ash powders at 778 and 1047 cm^{-1} is associated with the Si–O–Si and Si–O–Al asymmetric stretching vibration [37–39].

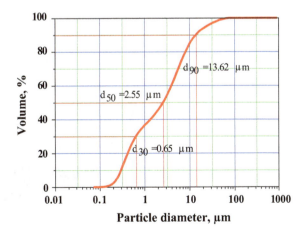

Fig. 3 Particle size distribution of fly ash powder [10]

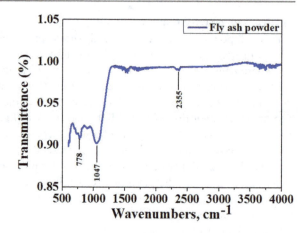

Fig. 4 FTIR spectra of fly ash powder [11]

X-ray diffraction study

The qualitative analyses of fly ash-based powders are done by x-ray diffraction (XRD) study, where various phases are identified. Here in the XRD instrument, the samples are hit by electrons and an x-ray is emitted. The instrument collects all the diffraction angle, scattered x-rays and their intensities are also collected to form the pattern. The x-ray diffraction pattern of fly ash powders is shown in Fig. 5 [40]. Nath et al. [40] reported that quartz and mullite are the major crystalline phases present in the fly ash powders collected from the Tata power company limited, Jamshedpur, India. Ismail et al. [41] observed that along with quartz and mullite, hematite and maghemite are also observed. The presence of hump in the range between 15 and 35° of 2Θ indicates the amorphous phase of the fly ash powders [40–45].

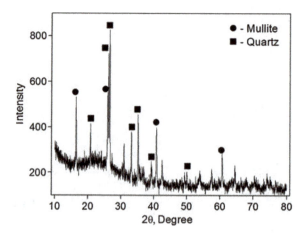

Fig. 5 XRD pattern of fly ash powder [40]

Morphological study

The morphology of fly ash powders is shown in Fig. 6a, b [37]. It was studied by using scanning electron microscopy. From the figure, it can be observed that most of the powders are spherical in size with various size particles. Few angular particles can also observe from the micrograph. The energy dispersive x-ray spectroscopy (EDS) analyses the elemental composition of fly ash powders. Figure 6c shows the EDS spectra analysis of fly ash powder, where the presence of various elements such as silicon, calcium, aluminium, iron, oxygen etc. can be observed along with other minor elements.

The objective of the present study is to report the fly ash generation in coal-based thermal power plants in India and its utilization in various industries for practical applications.

Organization of Chapter

Section 2 elaborates the fly ash utilization in multiple fields such as cement and concrete industries, mine filling and agriculture, construction of

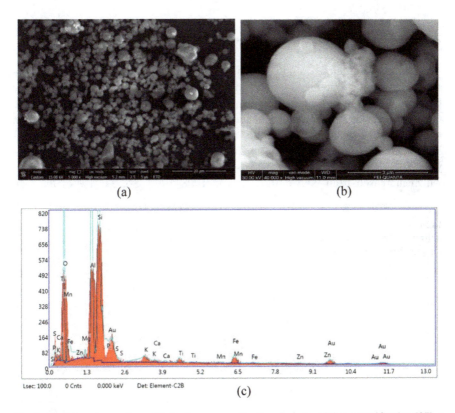

Fig. 6 Coal fly ash powder: **a** Under lower magnification, **b** under higher magnification [37], **c** EDS spectra of fly ash powder [10]

roads/embankments/flyovers, raising of ash dykes, reclamation of low laying area and building materials, geopolymer materials, and additive manufacturing for sustainable materials, respectively. Section 3 concludes the chapter with future scope.

2 Fly Ash Utilization in India

India is one of the world's leading countries for fly ash generation, and every year it is increasing. So, there is an urgent need for the utilization of this fly ash powder to save the environment.

Table 3 and Fig. 7 show the major modes of fly ash utilization for the first half of the year 2018–2019 in India. The table and graph show that a maximum of 26.85% fly ash was utilized in the cement industries, and the lowest 0.77% of total fly ash was used in the agriculture sector. The remaining 31.28% of fly ash was unutilized for the first half of the year 2018–2019. A few important fly ash utilization areas in India are described below.

2.1 Cement and Concrete Industries

According to estimates, 1 tonne of carbon dioxide (CO_2) is released into the atmosphere during the manufacture of 1 tonne of cement [46]. Fly ash has pozzolanic property, which results in the application of fly ash in cement industries for the manufacturing of Portland pozzolana cement (PPC). By utilizing fly ash powder, both coal and limestone can be saved. Figure 8a shows the trends of fly ash utilization in cement industries. From the figure, it can be noticed that from the year 2018–2019, the cement industries used 60.11 million tons of fly ash powder, which increased from 2.45 million tons in 1998–1999. So, 27.71% of the total

Table 3 Modes of fly ash utilization for the first half of the year 2018–2019 [32]

Description	Quantity of fly ash utilized	
	Million tons	%
Cement	25.0370	26.85
Mine filling	4.8014	5.15
Bricks and tiles	8.0691	8.65
Concrete	0.9660	1.04
Ash dyke raising	8.5285	9.15
Hydro power sector	0.0000	0.00
Roads and flyovers	2.5199	2.70
Agriculture	0.7155	0.77
Others	4.4346	4.76
Unutilized fly ash	29.17	31.28

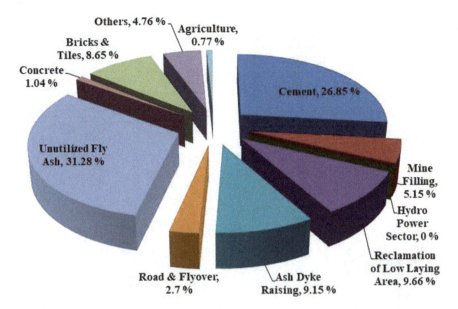

Fig. 7 Fly ash utilization graph (First half of the year 2018–2019 in India) [32]

fly ash generation is used by the cement industries for the year 2018–2019. The substitution of fly ash in cement can reduce the overall CO_2 and greenhouse gas emissions. Figure 8b shows the utilization of fly ash in the field of roads and highways including flyovers.

2.2 Mine Filling and Agriculture

Fly ashes are being used for mine filling and agriculture fields. Figure 9a shows the fly ash utilization trends in mine filling and agriculture fields from 1998–1999 to 2018–2019. Due to mine filling, top fertile soil and precious river sand can be saved. From the Fig. 9a it can be seen that 0.65 million tons of fly ash were used from 1998 to 1998, but if we are looking to the progressive graph, it can also be observed that fly ash utilization increased in 2017–2018, but after that, in the year 2018–2019 the graph has reduced slightly. The fly ash utilization for the year 2018–2019 in the field of mine filling is 10.10 million tons.

Fly ash powder contains various micronutrients such as potassium (K), calcium (Ca), magnesium (Mg), iron (Fe), zinc (Zn), molybdenum (Mo), sulfur (S), etc., improves water holding capacity and soil aeration which results in the potential growth of crops and vegetable [47]. The trends of fly ash utilization from the year 1998–1999 to 2018–2019 are shown in Fig. 9b. In the year 1998–1999, 0.13 million tons of fly ash were used in the agriculture field, which increased to 1.38 million tons from 2018 to 2019. The percentage of utilization of fly ash in the agriculture field from the total fly ash generation for the year 2018–2019 is 0.63%

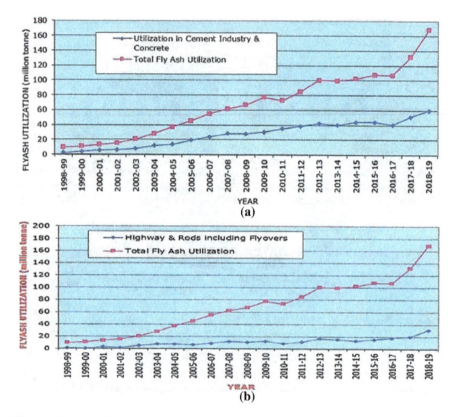

Fig. 8 Utilization of fly ash utilization: **a** Cement and concrete industries, **b** construction of roads/Embankments/highway and including flyovers [31]

2.3 Construction of Roads/Embankments/Flyovers and Raising of Ash Dykes

Fly ash powders have tremendous potential for utilization in the area of construction of roads, Embankments, Flyovers, and raising of ash dykes, etc. From Fig. 8b, it can be observed that in the year 1998–1999, 1.055 million tons of fly ash is used in the field of construction of roads, Embankments, Flyovers, and raising of ash dykes, etc. In 2018–2019, the utilization was 31.30 million tons, and the percentage of usage is 14.42% of total fly ash generation for the above said year.

2.4 Reclamation of Low Laying Area and Building Materials

One of the potential areas of fly ash utilization is the reclamation of low laying areas. Fly ash powder can be used as a substitute for soil or sand. Figure 10a shows the fly ash utilization trends in the reclamation of low laying areas during

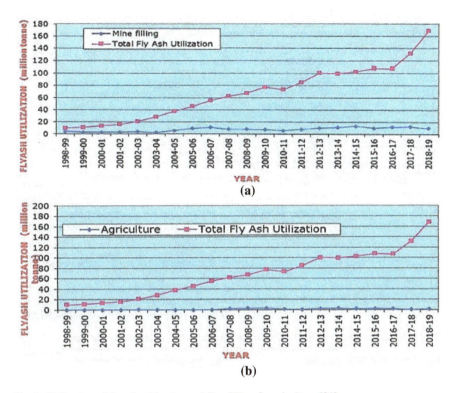

Fig. 9 Utilization of fly ash utilization: **a** Mine filling, **b** agriculture [31]

the period of 1998–99 to 2018–19. In the year 1998–1999, 4.17 million tons of fly ash was used in the reclamation of low laying areas. Later it was increased to 29.31 million tons during the year 2018–2019. Figure 10b shows the fly ash utilization trends in the area of construction materials such as Bricks, Blocks, and Tiles. These fly ash-based construction materials are good as conventional materials. From the Fig. 10b, it can be observed that during the years 1998–1999, the fly ash utilization in the field of construction materials was 0.70 million tons and from 2018 to 2019, it increased to 21.61 million tons. The percentage of fly ash utilization in the reclamation of low laying areas and construction materials is 13.51% and 9.96%, respectively, for the years 2018–2019.

2.5 Geopolymer Materials

A French scientist Joseph Davidovits in 1978, first used the term "Geopolymer" [37]. An aluminosilicate source materials and an alkaline solution can react to generate geopolymers [37]. As a starting raw material, a variety of byproducts including coal fly ash, slag, rice husk, red mud, metakaolin, etc. were used

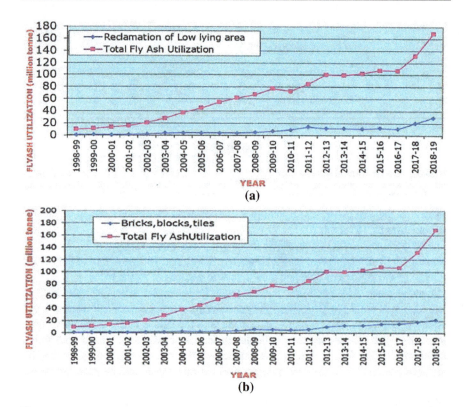

Fig. 10 Fly ash utilization in **a** Reclamation of low laying area, **b** construction materials [31]

and sodium hydroxide (NaOH), potassium hydroxide (KOH) or a combination of NaOH and Na_2SiO_3 or KOH and Na_2SiO_3 are used as alkaline solutions for geopolymer synthesis [10, 48, 49]. Mn $[-(SiO_2)z-AlO_2]$n. wH_2O is the general empirical formula for geopolymer, where M = alkaline element, (−) denotes the presence of a bond, n = degree of polycondensation, and z = 1, 2, or 3 [37, 50–53]. A few authors suggested a three-step mechanism for geopolymerization, where in the first step the dissolution of Si and Al starts due to the high alkaline activation, the second stage involves the transportation or orientation, and the next one is polycondensation [54, 55]. Figure 11 shows the conceptual model of the geopolymer mechanism which was given by Duxson et al. [56]. Due to better mechanical strength, superior properties, the geopolymeric materials becoming more significant as a substitute construction material for several industries [15, 57, 58]. Due to all these properties, geopolymer has gained much more interest in various industries such as cement alternative, coating material, aggregates, insulation materials, soil stabilization, marine structure, composite materials, etc.

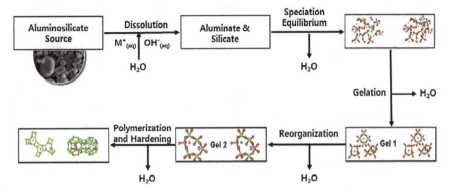

Fig. 11 Conceptual reaction mechanism of fly ash based geopolymer [56]

2.6 Additive Manufacturing for Sustainable Materials

Nowadays, 3D printing technology is one of the rapid technologies that have many advantages over subtractive technologies, such as energy efficiency, resource-saving, low labor cost, speed, flexible design, etc. [59, 60]. It deposits the materials layer by layer, wherever required. There are numerous uses of 3D printing technology in the fields of transportation, aerospace, biomaterials, automobiles, ceramics, electronics, and healthcare [61–67]. The application of 3D printing technology in the building sector is still in its infancy. Due to this, the application of this technology has gained much more attention in the construction industries and academia. Yuan et al. [59] studied the various factors which influence the extrusion-based geopolymer. It was reported by Alghamdi et al. that the printability of 3D printed geopolymer products are affected by various parameters such as concentration of chemical compositions of fly ash powders, type and the alkaline solutions, respectively [68]. The printability and rheological parameters of fly ash based geopolymers were studied by Chougan et al. [69]. Panda et al. [70] studied the mechanical properties of 3D printed geopolymer products and reported that the products are quite variable. The performance depends on printing quality, material open time, layers, and the direction of loadings [70]. Figure 12 shows the pictorial representation of the 3D mortar printing setup.

3 Conclusion and Future Scope

The primary source of electricity in India is generated by thermal power plants using coal as fuel. India produces millions of tons of fly ash every year. Fly ash powders, a solid thermal industrial waste, are a matter of concerns for today's world because it raises various environmental issues. Fly ashes are now one of the important raw materials for various industries. This study has been done to investigate the fly ash generation in India and its utilization in various industries. The fly

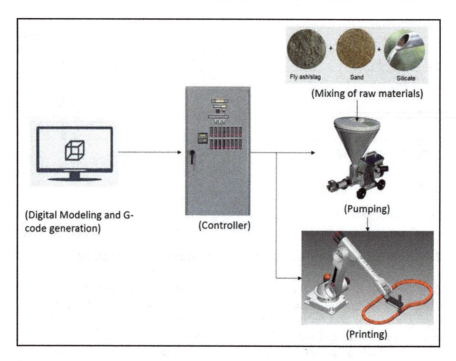

Fig. 12 Shows the pictorial representation of 3D mortar printing set up

ash powder generation and utilization in the year 2018–2019, 217.04 million tons and 168.40 million tons, respectively. The fly ash utilization is increased as compared to the last year, but it's required a lot of effort to achieve the target of 100% utilization as per the notification from the Government of India, which can influence to save the environment along also the economy of the country. Few states such as Gujarat, Punjab, and Rajasthan, have achieved the highest fly ash utilization of more than 100%. More application areas and more technologies should be adopted to utilize fly ash in India.

It has been observed that there are lots of future scopes to utilize the fly ash in various fields apart from the above-mentioned area, such as zeolites, aerogels, lightweight aggregates, composite materials, soil stabilization, absorbents for wastewater treatment, rare earth elements recovery, carbon nanotubes, catalysis, metallurgy, mine fills, precast fly ash based concrete units, etc.

References

1. K.V. Narayanan, E. Natarajan: Experimental Studies on Cofiring of Coal and Biomass Blends in India. Renewable Energy 32(15), 2548–2558 (2007). https://doi.org/10.1016/j.renene.2006.12.018.
2. S.K. Guttikunda, P. Jawahar: Atmospheric Emissions and Pollution from the Coal-Fired Thermal Power Plants in India. Atmospheric Environment. 92, 449–460 (2014). https://doi.org/10.1016/j.atmosenv.2014.04.057.
3. V.C. Pandey, J.S. Singh, R.P. Singh, N. Singh, M. Yunus: Arsenic Hazards in Coal Fly Ash and Its Fate in Indian Scenario. Resources, Conservation and Recycling. 55(9–10), 819–835 (2011). https://doi.org/10.1016/j.resconrec.2011.04.005.
4. S. Chakravarty, A. Mohanty, A. Banerjee, R. Tripathy, G.K. Mandal, M.R. Basariya, M. Sharma: Composition, Mineral Matter Characteristics and Ash Fusion Behavior of Some Indian Coals. Fuel. 150, 96–101(2015). https://doi.org/10.1016/j.fuel.2015.02.015.
5. M. Penchala Reddy, A. Shankar Singh, V. Mahendra Reddy, A. Elwardany, H. Reddy: Computational Analysis of Influence of Particle Size, Oxygen Concertation, and Furnace Temperature on the Ignition Characteristics of Pulverized High Ash and High Moisture Coal Particle. Alexandria Engineering Journal. 61(8), 6169–6180 (2022). https://doi.org/10.1016/j.aej.2021.11.047.
6. T. Rudra Paul, H. Nath, V. Chauhan, A. Sahoo: Gasification Studies of High Ash Indian Coals Using Aspen plus Simulation, Materials Today Proceedings. 46, 6149–6155 (2021). https://doi.org/10.1016/j.matpr.2020.04.033.
7. V. Khanpit, S.P. Tajane, S.A. Mandavgane: Experimental Studies on Coal-Water Slurry Fuel Prepared from Pretreated Low-Grade Coal, International Journal of Coal Preparation and Utilization. 42(3), 831–845 (2022). https://doi.org/10.1080/19392699.2019.1666830.
8. P.S. Parhi, U. Balunaini, S.M. Sravanam, Seismic Site Characterization of a Few Indian Coal Ash Deposits Using Multichannel Analysis of Surface Waves. Soil Dynamics and Earthquake Engineering. 155, 107192 (2022). https://doi.org/10.1016/j.soildyn.2022.107192.
9. U. Bhattacharjee, T.C. Kandpal: Potential of Fly Ash Utilisation in India, Energy, 27(2), 151–166 (2002).
10. D. Das, P.K. Rout: Synthesis, Characterization and Properties of Fly Ash Based Geopolymer Materials, Journal of Materials Engineering and Performance. 30(5), 3213–3231 (2021). https://doi.org/10.1007/s11665-021-05647-x.
11. D. Das, P.K. Rout: Synthesis and Characterization of Fly Ash and GBFS Based Geopolymer Material. Biointerface Research in Applied Chemistry. 11(6), 14506–14519 (2021). https://doi.org/10.33263/BRIAC116.1450614519.
12. P.K.R. Dipankar Das: Industrial Solid Wastes and Their Resources, Emerging Trends in Science and Technology, D.N.S. Dr. Mukesh Kumar Kumawat, Ed., 1st Editio, Bhumi Publishing, 96–101 (2021).
13. R. Roy, D. Das, P.K. Rout: Fabrication of Mullite Ceramic by Using Industrial Waste. Smart Cities, (Boca Raton), CRC Press, 285–291 (2022). https://doi.org/10.1201/9781003287186-13.
14. R. Roy, D. Das, P.K. Rout: A Review of Advanced Mullite Ceramics. Engineered Science. 18, 20–30 (2022). https://doi.org/10.30919/es8d582
15. D. Panias, I.P. Giannopoulou, T. Perraki: Effect of Synthesis Parameters on the Mechanical Properties of Fly Ash-Based Geopolymers, Colloids and Surfaces A: Physicochemical and Engineering Aspects. 301(1–3), 246–254 (2007). https://doi.org/10.1016/j.colsurfa.2006.12.064.
16. D. Das, A.P. Das, P.K. Rout: Effect of Slag Addition on Compressive Strength and Microstructural Features of Fly Ash Based Geopolymer. Circular Economy in the Construction Industry (Boca Raton), CRC Press, 61–68 (2021). https://doi.org/10.1201/9781003217619-9.

17. R. Shanmugam, S. Vetrivel, G. Sri Aathava, M. Kirubhakaran: Investigation of Engineering Properties of Sun-Dried Bottom Ash Based Eco Clay Blocks, Materials Today Proceedings. (2022). https://doi.org/10.1016/j.matpr.2022.06.345.
18. P. He, X. Zhang, H. Chen, Y. Zhang: Waste-to-Resource Strategies for the Use of Circulating Fluidized Bed Fly Ash in Construction Materials: A Mini Review. Powder Technology. 393, 773–785 (2021). https://doi.org/10.1016/j.powtec.2021.08.035.
19. Hitesh, R. Wattal, S. Lata: Development and Characterization of Coal Fly Ash through Low-Energy Ball Milling. Materials Today Proceedings. 47, 2970–2975 (2021). https://doi.org/10.1016/j.matpr.2021.05.204.
20. D.P. Mishra, S.K. Das: A Study of Physico-Chemical and Mineralogical Properties of Talcher Coal Fly Ash for Stowing in Underground Coal Mines., Materials Characterization. 61(11), 1252–1259 (2010). https://doi.org/10.1016/j.matchar.2010.08.008.
21. T. Kim, J.M. Davis, M.T. Ley, S. Kang, P. Amrollahi: Fly Ash Particle Characterization for Predicting Concrete Compressive Strength. Construction and Building Materials. 165, 560–571 (2018). https://doi.org/10.1016/j.conbuildmat.2018.01.059.
22. Z. Li, G. Xu, X. Shi: Reactivity of Coal Fly Ash Used in Cementitious Binder Systems: A State-of-the-Art Overview. Fuel. 301, 121031 (2021). https://doi.org/10.1016/j.fuel.2021.121031.
23. P. Yoosuk, C. Suksiripattanapong, P. Sukontasukkul, P. Chindaprasirt: Properties of Polypropylene Fiber Reinforced Cellular Lightweight High Calcium Fly Ash Geopolymer Mortar. Case Studies in Construction Materials. 15, e00730 (2021). https://doi.org/10.1016/j.cscm.2021.e00730.
24. P.S. Reddy, D. Rajitha: The International Journal of Analytical and Experimental Modal Analysis An Experimental Study On High Performance Concrete By Using Admixture Metakaolin, Slag, Silica Fume, Fly Ash On M80 Grade Concrete ISSN NO : 0886-9367 Page No : 961 ISSN NO: 0886-2020, XII(0886), 961966.
25. G.S. Priya, L.S. Prasath: Experimental Investigation on Flyash Bricks By Using Granite Saw Dust. Technical Research Organization India. 4(6), 2394–0697 (2019).
26. L. Wu, X. Xu, H. Wang, J.-Q. Yang: Experimental Study on Bond Properties between GFRP Bars and Self-Compacting Concrete. Construction and Building Materials. 320, 126186 (2022). https://doi.org/10.1016/j.conbuildmat.2021.126186.
27. G.L. Golewski: Energy Savings Associated with the Use of Fly Ash and Nanoadditives in the Cement Composition. Energies. 13(9), 1–20(2020). https://doi.org/10.3390/en13092184.
28. A. Dwivedi, M.K. Jain: Fly Ash – Waste Management and Overview: A Review. Recent Research and Technology. 6(1), 30–35 (2014).
29. A. Rastogi, V. Kumar Paul: A Critical Review of the Potential for Fly Ash Utilisation in Construction-Specific Applications in India. Environ. Research, Engineering and Management. 76(2), 65–75 (2020). https://doi.org/10.5755/j01.erem.76.2.25166.
30. A. Bhatt, S. Priyadarshini, A. Acharath Mohanakrishnan, A. Abri, M. Sattler, S. Techapaphawit: Physical, Chemical, and Geotechnical Properties of Coal Fly Ash: A Global Review. Case Studies in Construction Materials. 11, e00263 (2019). https://doi.org/10.1016/j.cscm.2019.e00263.
31. G. of I. Central Electricity Authority, CEA Annual Report 2018–19, Annu. Rep., 2019, p 248. www.cea.nic.in, last accessed 2022/08/24.
32. G. of I. Central Electricity Authority, "Flyash Utilization for First Half of the Year 2018–2019.Pdf," CEA, Government of India, n.d. http://www.cea.nic.in/reports/others/thermal/tcd/flyash_201819-firsthalf.pdf. last accessed 2022/08/24.
33. B. Naik, D. Kumar Bagal, S.S. Pradhan: Mechanical Characterization Based on Partial Replacement Analysis of Portland Pozzolana Cement with Industrial Waste in M30 Grade Concrete Partial Replacement of Concrete View Project Parametric Optimization Dry Turning of TItanium Alloy (Grade 5), 2019, (June). http://www.ripublication.com.
34. D. Das, P.K. Rout: Synthesis of Inorganic Polymeric Materials from Industrial Solid Waste. Silicon (2022). https://doi.org/10.1007/s12633-022-02116-5

35. R. Roy, D. Das, P.K. Rout: Mullite Ceramics Derived from Fly Ash Powder by Using Albumin as an Organic Gelling Agent. Biointerface Research in Applied Chemistry. 13, 339 (2022). https://doi.org/10.33263/BRIAC134.339
36. P. Jani, A. Imqam: Class C Fly Ash-Based Alkali Activated Cement as a Potential Alternative Cement for CO2 Storage Applications. Journal of Petroleum Science and Engineering. 201, 108408 (2021). https://doi.org/10.1016/j.petrol.2021.108408.
37. D. Das, P.K. Rout, Utilization of Thermal Industry Waste: From Trash to Cash, Carbon – Science and Technology. 11(2), 43–48 (2019).
38. P. Rożek, M. Król, W. Mozgawa: Spectroscopic Studies of Fly Ash-Based Geopolymers. Spectrochimica Acta - Part A Molecular and Biomolecular Sepctroscopy. 198, 283–289 (2018). https://doi.org/10.1016/j.saa.2018.03.034.
39. T. Bakharev: Geopolymeric Materials Prepared Using Class F fly Ash and Elevated Temperature Curing. 35(6), 1224–1232 (2005). https://doi.org/10.1016/j.cemconres.2004.06.031.
40. S.K. Nath, S. Maitra, S. Mukherjee, S. Kumar: Microstructural and Morphological Evolution of Fly Ash Based Geopolymers. Construction and Building Materials. 111, 758–765 (2016). https://doi.org/10.1016/j.conbuildmat.2016.02.106.
41. I. Ismail, S.A. Bernal, J.L. Provis, R. San Nicolas, S. Hamdan, J.S.J. Van Deventer: Modification of Phase Evolution in Alkali-Activated Blast Furnace Slag by the Incorporation of Fly Ash. Cement and Concrete Composites. 45, 125–135 (2014). https://doi.org/10.1016/j.cemconcomp.2013.09.006.
42. T. Sakthivel, D.L. Reid, I. Goldstein, L. Hench, S. Seal: Hydrophobic High Surface Area Zeolites Derived from Fly Ash for Oil Spill Remediation. Environmental Science and Technology. 47(11), 5843–5850 (2013). https://doi.org/10.1021/es3048174.
43. H. Assaedi, T. Alomayri, C.R. Kaze, B.B. Jindal, S. Subaer, F. Shaikh, S. Alraddadi: Characterization and Properties of Geopolymer Nanocomposites with Different Contents of Nano-CaCO3. Construction and Building Materials. 252, 119137 (2020). https://doi.org/10.1016/j.conbuildmat.2020.119137.
44. H. Li, Y. Chen, Y. Cao, G. Liu, B. Li: Comparative Study on the Characteristics of Ball-Milled Coal Fly Ash, Journal of Thermal Analysis and Calorimetry. 124(2), 839–846 (2016). https://doi.org/10.1007/s10973-015-5160-5.
45. S.K. Nath, S. Kumar: Role of Particle Fineness on Engineering Properties and Microstructure of Fly Ash Derived Geopolymer. Construction and Building Materials. 233, 117294 (2020). https://doi.org/10.1016/j.conbuildmat.2019.117294.
46. P.K. Sarker: Bond Strength of Reinforcing Steel Embedded in Fly Ash-Based Geopolymer Concrete. Materials and Structures. 44(5), 1021–1030 (2011). https://doi.org/10.1617/s11527-010-9683-8.
47. R. Kaur, D. Goyal: Mineralogical Studies of Coal Fly Ash for Soil Application in Agriculture Mineralogical Studies of Coal Fly Ash for Soil Application in Agriculture. Particulate Science and Technology. 33(1), 76–80 (2015). https://doi.org/10.1080/02726351.2014.938378.
48. M.T. Ghafoor, Q.S. Khan, A.U. Qazi, M.N. Sheikh, M.N.S. Hadi: Influence of Alkaline Activators on the Mechanical Properties of Fly Ash Based Geopolymer Concrete Cured at Ambient Temperature. Construction and Building Materials. 273, 121752 (2021). https://doi.org/10.1016/j.conbuildmat.2020.121752.
49. H. Castillo, H. Collado, T. Droguett, S. Sánchez, M. Vesely, P. Garrido, S. Palma: Factors Affecting the Compressive Strength of Geopolymers: A Review. Minerals 11(12), 1317 (2021). https://doi.org/10.3390/min11121317.
50. M.C.M. Nasvi, P.G. Ranjith, J. Sanjayan, H. Bui: Effect of Temperature on Permeability of Geopolymer: A Primary Well Sealant for Carbon Capture and Storage Wells. Fuel 117(PART A), 354–363 (2014). https://doi.org/10.1016/j.fuel.2013.09.007.
51. A.C. Derrien, H. Oudadesse, J.C. Sangleboeuf, P. Briard, A. Lucas-Girot: Thermal Behaviour of Composites Aluminosilicate-Calcium Phosphates, Journal of Thermal analysis and Calorimetry 75(3), 937–946 (2004). https://doi.org/10.1023/B:JTAN.0000027187.14921.86.

52. S. Nagajothi, S. Elavenil, S. Angalaeswari, L. Natrayan, W.D. Mammo: Durability Studies on Fly Ash Based Geopolymer Concrete Incorporated with Slag and Alkali Solutions, Advances in Civil Engineering (2022), 1–13 (2022). https://doi.org/10.1155/2022/7196446.
53. E. ARIÖZ, G.B. BÜKE: Removal of Methylene Blue from Aqueous Solutions with Fly Ash Based Geopolymer Foam. European Journal of Science and Technology. (28), 1437–1441 (2021). https://doi.org/10.31590/ejosat.1016237.
54. H. Xu, J.S.J. Van Deventer: The Geopolymerisation of Alumino-Silicate Minerals. International Journal of Mineral Processing. 59(3), 247–266 (2000). https://doi.org/10.1016/S0301-7516(99)00074-5.
55. J.G.S. Van Jaarsveld, J.S.J. Van Deventer, L. Lorenzen: Factors Affecting the Immobilization of Metals in Geopolymerized Flyash. Metallurgical and Materials Transactions B. 29(1), 283–291 (1998). https://doi.org/10.1007/s11663-998-0032-z.
56. P. Duxson, A. Fernández-Jiménez, J.L. Provis, G.C. Lukey, A. Palomo, J.S.J. Van Deventer: Geopolymer Technology: The Current State of the Art. Journal of Material Science 42(9), 2917–2933 (2007). https://doi.org/10.1007/s10853-006-0637-z.
57. P. Cong and Y. Cheng: Advances in Geopolymer Materials: A Comprehensive Review. Journal of Traffic and Transportation Engineering. 8(3), 283–314 (2021). https://doi.org/10.1016/j.jtte.2021.03.004.
58. K. Debnath, D. Das, P.K. Rout: Effect of mechanical milling of fly ash powder on compressive strength of geopolymer. Mater Today Proc. (2022). https://doi.org/10.1016/j.matpr.2022.08.321.
59. Q. Yuan, C. Gao, T. Huang, S. Zuo, H. Yao, K. Zhang, Y. Huang, J. Liu: Factors Influencing the Properties of Extrusion-Based 3D-Printed Alkali-Activated Fly Ash-Slag Mortar. Materials (Basel) 15(5), 1969 (2022). https://doi.org/10.3390/ma15051969.
60. K. Korniejenko, M. Łach, S. Chou, W. Lin, J. Mikuła, D. Mierzwiński, A. Cheng, M. Hebda. A Comparative Study of Mechanical Properties of Fly Ash-Based Geopolymer Made by Casted and 3D Printing Methods. IOP Conference Series: Materials Science and Engineering 660(1), 012005 (2019). https://doi.org/10.1088/1757-899X/660/1/012005.
61. B. Khoshnevis: Automated Construction by Contour Crafting—Related Robotics and Information Technologies. Automation in Construction 13(1), 5–19 (2014). https://doi.org/10.1016/j.autcon.2003.08.012.
62. C.K. Chua, K.F. Leong: 3D Printing and Additive Manufacturing. World Scientific, 2014. https://doi.org/10.1142/9008.
63. A. Goulas, R.A. Harris, R.J. Friel: Additive Manufacturing of Physical Assets by Using Ceramic Multicomponent Extra-Terrestrial Materials. Additive Manufacturing. 10, 36–42 (2016). https://doi.org/10.1016/j.addma.2016.02.002.
64. D. Powell, A.E.W. Rennie, L. Geekie, N. Burns: Understanding Powder Degradation in Metal Additive Manufacturing to Allow the Upcycling of Recycled Powders. Journal of Cleaner Production. 268, 122077 (2020). https://doi.org/10.1016/j.jclepro.2020.122077.
65. M. Govindharaj, U.K. Roopavath, S.N. Rath: Valorization of Discarded Marine Eel Fish Skin for Collagen Extraction as a 3D Printable Blue Biomaterial for Tissue Engineering. Journal of Cleaner Production. 230, 412–419 (2019). https://doi.org/10.1016/j.jclepro.2019.05.082.
66. F. Giudice, R. Barbagallo, G. Fargione: A Design for Additive Manufacturing Approach Based on Process Energy Efficiency: Electron Beam Melted Components. Journal of Cleaner Production 290, 125185 (2020). https://doi.org/10.1016/j.jclepro.2020.125185.
67. A. Meurisse, A. Makaya, C. Willsch, M. Sperl: Solar 3D Printing of Lunar Regolith. Acta Astronautica 152, 800–810 (2018). https://doi.org/10.1016/j.actaastro.2018.06.063.
68. H. Alghamdi, S.A.O. Nair, N. Neithalath: Insights into Material Design, Extrusion Rheology, and Properties of 3D-Printable Alkali-Activated Fly Ash-Based Binders. *Materials* and *Design* 167, 107634 (2019). https://doi.org/10.1016/j.matdes.2019.107634.
69. M. Chougan, S. Hamidreza Ghaffar, M. Jahanzat, A. Albar, N. Mujaddedi, R. Swash: The Influence of Nano-Additives in Strengthening Mechanical Performance of 3D Printed Multi-Binder Geopolymer Composites. Construction and Building Materials. 250, 118928 (2020). https://doi.org/10.1016/j.conbuildmat.2020.118928.

70. B. Panda, N.A.N. Mohamed, M.J. Tan: Effect of 3D Printing on Mechanical Properties of Fly Ash-Based Inorganic Geopolymer. In: International Congress on Polymers in Concrete (ICPIC 2018), pp 509–515, Springer International Publishing, (2018). https://doi.org/10.1007/978-3-319-78175-4_65.

3D Printing Pathways for Sustainable Manufacturing

Granville Embia, Bikash Ranjan Moharana, Aezeden Mohamed, Kamalakanta Muduli, and Noorhafiza Binti Muhammad

1 Introduction

Additive manufacturing has been defined as laying down materials in successive layers to form a desired 3D object using computer integrated manufacturing technology. Additive manufacturing with its advantage over traditional manufacturing has become a widely used manufacturing technology in the recent decade. The ongoing research and improvement on additive manufacturing has improved manufacturing costs by reducing waste, reducing parts weights, energy savings, printing complex geometries in consolidated assembly, and reducing supply chain and logistic costs [1, 2].

The newest manufacturing technology is 3D printing or Additive manufacturing. Additional technology enables digital data to be transformed directly into a physical product. The additive process chain may look very short and straightforward at first sight: software for the description of the physical object is computer-aided design. The digital information is transmitted to a special 3D printer that produces the designed part directly.

G. Embia · A. Mohamed · K. Muduli (✉)
Mechanical Engineering Department, PNG University of Technology, Lar, Papua New Guinea
e-mail: kamalakantam@gmail.com

A. Mohamed
e-mail: aezeden.mohamed@pnguot.ac.pg

B. R. Moharana
Mechanical Engineering Department, C V Raman Global University, Bhubaneswar, Odisha, India

N. B. Muhammad
Faculty of Mechanical Engineering Technology, Universiti Malaysia Perlis, Kangar, Malaysia
e-mail: noorhafiza@unimap.edu.my

Additive manufacturing is a widely used technology and has a broader range of applications as it can use it to do anything everywhere. 3D printing is cheaper and economical, which can be used locally for personal use. At school, in factories, aerospace, automotive, medical and health care, logistics, etc., 3D printing is announced as one of the intelligent and sustainable manufacturing technologies. In striving for a sustainable future, companies are looking for ways to optimize their production processes to reduce energy consumption and waste. 3D printing has two advantages, mainly because of its facilities which design more efficiently and create less waste, as one of the significant sustainable technologies [2, 3].

3D printing during the last decay has improved significantly in producing lightweight parts with complex geometries and consolidated features in assemblies. This has enhanced costs in aerospace, automotive, biomedical, factories, etc., and contributes to the world economy. However, 3D printing has many potentials to improve further post-work and surface finished, printing more extensive and complete products like aircraft, and skill gap in operating and maintaining 3D printers for personal use.

Time to the market is the decisive factor in evaluating a brand's success in highly competitive industries today. 3D printing has significantly reduced the time to the market by dodging conventional tooling strategies, cutting down lead times on models, and making the ultimate item. In addition, 3D printing has helped a lot of industries to survive in the competitive world today. Additive manufacturing continues to grow and has become more successful in making manufacturing technology more sustainable in recent years [1].

There are several additive manufacturing processes used to print 3D products. These processes are classified based on the physical condition of raw material; solid-based, powder-based, and fluid-based systems. They can also be classified based on the fusion of the material. They all build 3D objects by adding layer-wise material or storing it on a path. The various printing processes are based on different material types (metal, ceramic, polymer, composites) and other layer construction methods (e.g., stereolithography).

The advancement of any subject depends on the logical synthesis of previous studies based on their revelations [4]. This has been found to be helpful for both academicians as well as practitioners in formulating a response tailored to the socio-environmental requirements [5]. The literature review has been suggested as a research methodology by several researchers as it is capable of contributing significantly to methodological, thematic and conceptual development in different domains [6–8]. Hence, review methodology was preferred in this research for extracting and analyzing literature pertinent to 3D printing and Sustainable Manufacturing practices.

The objectives of this chapter:

- To review the current status of 3D printing in various industrial applications.
- To explore how 3D printing leads to sustainable manufacturing technology.
- To reveal the potential roadblocks to the 3D printing adoption.

Organization of Chapter

The chapter is organized as: Sect. 2 deals with a literature review on Sustainable Manufacturing and 3D printing. The benefits of additive manufacturing over traditional manufacturing are described in Sect. 3. The different types of materials used in 3D printing dealt with Sect. 4. 3D printing types in Sect. 5 and its applications in Sect. 6. Section 7 is concerned with the contribution of 3D printing towards sustainable manufacturing. The limitations of this additive manufacturing process are explained in Sect. 8. The concluding remarks and recommendations of this additive manufacturing process for the sustainable growth of the nation are described in Sect. 9.

2 Literature Review

This section discusses the literature on Sustainable Manufacturing and 3D printing.

2.1 Sustainable Manufacturing

Sustainable manufacturing covers practices of manufacturing that result in responsible, environmentally safe and eco-friendly products as well as process and stimulate green growth simultaneously stabilizing a competitive economic growth [9]. Sustainable manufacturing is a win–win practice in manufacturing because its innovations are based on reduced greenhouse gas emission, waste reduction, re-use and recycling of virgin material. The biggest advantage or achievement of sustainable manufacturing is bringing together all the manufacturing practices together as one sharing one ambition and goal. The goal is to extract, produce, transport and dispose of in a responsible manner that will lower the use of virgin material, reduce waste generation safeguard the interest of future generations and lower environmental hazards [10–14].

2.2 3D Printing

Three-dimensions printing is unique and sustainable in current manufacturing technology. It differs from the traditional manufacturing of reducing more significant workpieces to the desired part, which creates a lot of waste, loss of energy, loss of time, and labor loss. It is also unique as it is based on a digital blueprint that is easier to refine. Printing is becoming more valuable as its pricing is lowered and affordable since the technology was first designed and realized by Charles W. Hull in 1984.

Three main phases of 3D printing are the modelling, the printing, and the finishing of products as shown in Fig. 1. The advantage of 3D printing is; design freedom, fast product development cycle, ease to produce, low startup cost for production, local production, and on-demand manufacturing. It also provides a simple, effective, low-cost supply chain or logistics and no need for mass production in factories. 3D printing offers to manufacture of low waste and less energy usage. 3D printing can print metals, ceramics, plastics, composites, foods, and even living organic cells, making it more significant and sustainable [1, 3, 15].

Additive production is considered an additive. The material added sequentially is deduced from a solid block until the final part is produced, contrary to more subtractive traditional output. For example, turning/winding/milling, numeric computer control, laser trimming processes, water jet cutting, machine cutting, etc. are subtracting or convection processes [15]. The difference between additive manufacturing and traditional manufacturing as shown in Fig. 2.

The advantage of 3D printing is; design freedom, fast product development cycle, ease to produce, low startup cost for production, local production, and on-demand manufacturing. It also provides a simple, effective, low-cost supply chain

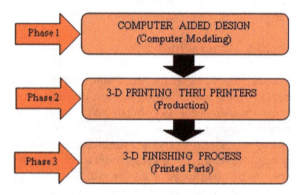

Fig. 1 Three phases in additive manufacturing

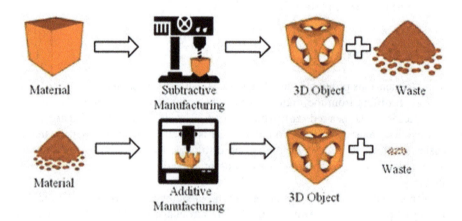

Fig. 2 Showing the difference between additive manufacturing and traditional manufacturing [15, 16]

or logistics and no need for mass production in factories. 3D printing offers to manufacture of low waste and less energy usage. 3D printing can print metals, ceramics, plastics, composites, foods, and even living organic cells, making it more significant and sustainable [3, 15].

Time to market is the crucial determinant of the success of many industries and their brands in today's highly competitive industry. 3D printing mass production can significantly reduce market time by preventing conventional tooling methods, cutting lead times for prototyping and end-use parts [15, 16].

For instance, Jake Fallon, founder and chairman of the 3D printing club Penn State Behrend and Braskem's additive manufacturing team had an enormous problem with COVID-19 distribution, which required thousands of facial straps for covers to secure their worldwide workforce. The group turned to a 3D printer and infusion forming machine for 48 h. Without getting to 3D printing, Braskem offered an exorbitant metal form and took a toll on significant time and cash for the staff. Instep, the group delivered thousands of straps and arranged them for conveyance to workplaces worldwide within a week of the beginning of this extend [15].

Time to market is the crucial determinant of the success of many industries and their brands in today's highly competitive industry. 3D printing mass production can significantly reduce market time by preventing conventional tooling methods, cutting lead times for prototyping and end-use parts.

For instance, Jake Fallon, founder and chairman of the 3D printing club Penn State Behrend and Braskem's additive manufacturing team had an enormous problem with COVID-19 distribution, which required thousands of facial straps for covers to secure their worldwide workforce. The group turned to a 3D printer and infusion forming machine for 48 h. Without getting to 3D printing, Braskem offered an exorbitant metal form and took a toll on significant time and cash for the staff. Instep, the group delivered thousands of straps and arranged them for conveyance to workplaces worldwide within a week of the beginning of this extend [15, 16].

The objective of this paper is to identify how 3D printing works in achieving sustainable manufacturing. With a broader range of materials used in 3D printing, an analysis of how 3D printing works in current industries and its application will achieve sustainability in 3D printing [17].

2.3 Related Studies

AL-Maliki et al. [18] stressed on the technological aspect and process of additive or 3D manufacturing. The types of 3D manufacturing and their limitations to confined space and large volume productions. In another study Kubáč and Kodym [19], explained in depth the understanding of 3D or additive manufacturing in accordance to supply chain. Joshi and Sheikh [15] explored the drawbacks of printed components. The impact additive manufacturing can have on the supply chain and changing the perspective of it through cost, environment and emission especially with spare parts manufacturing. Study by Bourell et al. [17] focused on material usage in AM. Westerweel et al. [20] discussed on how additive manufacturing can be introduced as a major aspect in preventive Maintenance strategies. Parupelli and Desai [21], discussed the future of additive manufacturing and the impact it can have on sustainable manufacturing. Stock and Seliger [22] elaborated the opportunity that comes along with additive manufacturing, in the aspects of sustainable development bringing forth economic, social and environment development. Royo et al. [23] clarified the approach of Additive manufacturing towards sustainable manufacturing with waste management of products. Lee et al. [24] elaborated further on the promising future additive manufacturing has with the current trend in terms of sustainable manufacturing in science and engineering. Kai et al. [25] elaborated on the progress of sustainable manufacturing with additive manufacturing in small and large scale manufacturing and what they are capable of with their current impacts on manufacturing.

3   Additive Manufacturing Over Traditional Manufacturing

Additive manufacturing is more efficient and reduces the need for extensive transport, storage, and travel required by traditional manufacturing. It works in the inverse of conventional manufacturing methods (Table 1).

The waste material in metal application of additive manufacture is 40% less than subtractive or traditional manufacturing procedures. In this process, approximately 95–98% of 3D printed waste can be recycled. The Airbus innovation

Table 1 Showing advantage of additive manufacturing compared to traditional manufacturing with their outcomes

Outcome manufacturing	Traditional manufacturing	Additive manufacturing
Features	Only produce simple parts flat & round surface	Produce parts & assembles in different thicknesses with deep channels & cut
Geometries	Only produce complex parts flat & round surfaces	Produce various & complex parts, topological forms, blind holes, geometry & design with a high specific weight ratio
Parts consolidation	Each part of an assembly is manufactured separately from the traditional method	Traditionally manufactured parts formed thru joining together to an assembly can now be coordinated into a single printed part with three-dimensions printing
Fabrication to consolidation	Parts machined separately & fabricated to form one assembly	Traditionally manufactured components require assembly & can be printed at the same time in many steps

group's study in the United Kingdom and its partners has shown that additive production can decrease up to 75% of raw material consumption. Certain studies show that additive manufacturing, with its nature of production, has 70% less environmental effect than traditional manufacturing [17, 26].

Because of its production, most parts produced by 3D printing have reduced weight considerably. For instance, an aircraft manufacturing bracket has a weight reduction of 50–80%, which can save fuel annually by about 2.5 million, as the fuel consumption is directly related to the aircraft's weight [17, 26].

With additive manufacturing, there is a high possibility of producing high-quality products in terms of the complex geometry of the part, good surface finish (in Stereolithography), consolidation of different elements into one component, reduction of maintenance alignment, and improving maintenance labor utilization.

The additional cost associated with the creation or assembly of complex parts is eliminated by additive manufacturing. A highly complicated feature of traditional production typically requires much more production costs than a simple part. In addition, however, the production process is the same in additive production, despite the complexity [27, 28]. Additional costs are, therefore, no longer a feature of modern additive production technology. The fact that additive production can be used in easy production to create custom products is also a significant functional difference between additive manufacturing and the traditional production method. The injection molding, for example, is one of many examples of expensive three-dimensions printing tools; it is another to produce custom 3D printed jigs and devices. Jigs and devices are used to simplify, make reliable and efficient manufacturing and assembly processes while reducing cycle times and improving worker's safety [26–28].

Additive production varies in terms of technology used to produce three dimensions from traditional manufacturing methods. Traditional processes often involve many steps, each with a different machine. For example, for turning, milling, and drilling, a finished metal product is frequently used in concert. A single 3D printer handles all the processes and functionality of creation in additive production. Additionally, there is less work involved in the production of additives than in traditional production. 3D printing has produced products fully automated and requires little control from an operator. In comparison to conventional manufacturing, each machine is operated with different skills [28].

4 Materials Used in 3D Printing

With additive manufacturing, a wide range of materials (almost all types of materials) are used in producing additive manufactured products. In this case, the selection of materials to be made is essential as any other manufacturing process. Material selection is based on a wide variety of printable materials, the applications of which are different depending on a certain number of factors, such as material properties, cost, color, appearance, and thickness of the layer [26–29].

Achieving a high-quality product requires a careful selection of materials with a suitable additive manufacturing process. The properties of an additively manufactured component are highly dependent on its underlying microstructure. This intern depends on the property of the raw materials and process conditions used. The temperature, pressure, and disease under which the product is made can affect the property of the final product produced [29].

Additive manufacturing can use metallic, ceramics, polymers, and their combination in hybrid, composites, or functionally graded materials. 3D printing to be successful requires careful characterization of raw material components such as metal powder, polymer powders, wires, sheets [29].

4.1 Metallic Materials

As a unique material, metal is a popular alternative material of choice since it can be printed directly using a variety of 3D printing techniques such as powder sintering and direct melting. The most popular material is metals like aluminum, titanium and stainless steel because they have high strength and good metallic finish. Metallic materials in 3D printing have gained a lot of space in recent manufacturing technology because of their unique character in producing the final product [29–31].

Metal materials in the form of sheet metal, powder, or wire can be used in 3D printing. The advantage of 3D printers has been that the aerospace, car, medical, and manufacturing industries have gained a lot of attention. Studies have shown that there is a considerable saving made through the reduction of waste by recycling it. Additive manufacturing has produced metallic parts of the exact specification as traditional manufacturing parts but has less weight with standard stiffness and fewer materials [29–32].

4.2 Ceramics Materials

The ceramics are an incredibly diverse range of materials covering traditional ceramics (e.g., pottery and refractory) to modern ceramics (such as alumina and silicon nitride). These are in the fields of electronic equipment, aviation, and cutting instruments. Using pottery and concrete without large pores or any cracks, 3D printing technology can produce 3D print objects by optimizing parameters and setting up suitable mechanical properties. Ceramic is resistant to fire, solid, and durable. Their unique characteristics, such as heat resistance, allow them to be used in many applications in which materials like metals and polymer are unsuitable. Ceramics can be used in all geometries and forms and are very sustainable for creating future buildings due to the ceramic fluid before the setting. Ceramic materials include alumina, biological glass, and zirconia [26, 27, 29].

Fast ceramics production is a proprietary additive manufacturing technology developed by 3D Ceram, who were amongst the first to produce ceramic parts from direct 3D printing. Fast ceramics production is a stereolithography photopolymerization technology; whereby a laser is used to polymerize a paste made of photosensitive resin and ceramic material. The most effective application of 3D with ceramic material is in the pioneering use of printing bone substitutes and cranial implants. Additive manufacturing allows for total control over the geometry of porous structures compared to traditional methods of producing implants. This promotes osteointegration and increases the compressive mechanical strength between living bone and the implant, reducing inflammation and implant rejection. The fact that ceramic is light, durable, and biocompatible makes it a more excellent material for the medical and surgical industries used for implants, surgical tools and guides, and diagnostic equipment [28, 29].

4.3 Polymers Materials

For most additive manufacturing processes, polymers are founded materials. Polymers may be used for additive manufacturing for more complex designs in hours, resulting from the accelerating prototyping process. Polymers are advantageous because they're lighter than their metal counterparts. In three-dimensions printing, the weight of polymer products has saved 20–40% on fuel consumption for aerospace parts. In sectors like automotive and aviation, where light-weighting is a problem, this is particularly important. Polymer components may also have unique characteristics such as heat, high resistance, or water repellence [3, 29–32].

The production of polymer 3D printing requires the material needed for the part itself and can produce less or no waste as the waste can be recycled. The polymer in additive manufacturing has become environmentally friendly by reducing waste and improving fuel efficiency (i.e., less CO_2 released into the atmosphere). The polymer has been a better choice with additive manufacturing than metals in many areas: consumer products, sustainable applications, and advanced manufacturing and biomedical devices.

Several polymeric materials are available for additive manufacturing, and selections of a material depend on the method used and the mechanical properties to be achieved. Some of the most important 3D print polymers include acrylonitrile butadiene styrene, polycarbonate, multi-colored polylactides, polystyrene, polyamide. These materials are mainly used for low performance or prototyping components, but demand for new higher performance polymers and composites is increasing. The last few years have been investigated for using 3D printing with high temperature and chemical resistant polymers such as polyphenylene sulphides, polyetherimides, polyphenylesulphones, and polyether ether ketones [3, 33–37]. Although plastic materials are not involved in developing structural, mechanical components like metals, they still represent near to 80% of the market referring to Fig. 3 [37–39].

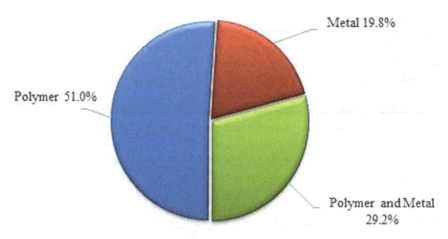

Fig. 3 Material used in additive manufacturing [39]

4.4 Composites Materials

Composite materials being a unique material that can adapt in many different functions with their low weight and tailorable properties has become a driver in transforming high-performance industries of the world. Composite material providing high quality and long service life, with high strength, less maintenance, and the ability to recycle has enabled it to replace metals and ceramic in most applications [27, 31].

Examples of composite materials are carbon-fiber enhanced polymers and glass-fiber enhanced polymer composites, for example. Composite structures in carbon fiber enhanced polymers are widely used in the aerospace industry due to their high specific stiffness, strength, resistance to corrosion, and good tiredness [3].

Reinforced polymer composites are widely used for various applications in additive production due to the cost-efficiency, and great performance of glass fibers. Reinforced composites are mostly used in electronics, automobile and aviation applications. It has excellent characteristics such as high strength, flexibility, rigidity and chemical hazard resistance [3, 27].

4.5 Smart Materials

A clever material may change the geometry and the form of the components under external conditions such as water and heat. Self-evolving and soft robotic systems are excellent examples of 3D printed parts manufactured using intelligent materials. Another example of intelligent group material is memory alloys and polymers. Certain memory alloys such as nickel-titanium can be used in implant biomedicine for the application of micro-electro-mechanical devices. Shape memory polymers are a type of functional material that answers stimuli such as light and electric heat.

The complex shape of the memory shape polymer can easily and conveniently be manufactured with 3D printing technology. The quality assessment is based on the precise measurement, smoothness, and density of the components of this material [16, 29, 36–39, 48].

5 Types of 3D Printing

The production of 3D printed products can be accomplished through a variety of additive manufacturing procedures. These procedures are categorized according to the feedstock that will be used (powder, sheet, wire, or liquid) and the manner in which components will be fused together. It is important to keep in mind that the benefits and drawbacks associated with an additive manufacturing process are directly proportional to the manner in which it is utilized to create the final product. So, picking the right 3D printing method is important if one wants to get the part or assembly needed for a certain application.

5.1 Fuse Deposition Modelling

Fuse deposition modelling was developed by Scott and Lisa Crump in 1988 as an additive manufacturing technology. It is commonly used to model, prototype, and manufacture mechanical systems. Modelling of fuse deposition uses additive production to print parts by laying the layers of materials. A plastic filament or metal wire is released from a spiral and delivers material to an extruder to activate and decrease the flow. The dust is heated to molten the material and is moved horizontally and vertically by a numerical control mechanism directly controlled through a software package of computer-aided production.

The model is manufactured by extruding tiny thermoplastic beads into layers when the material immediately hardened after the extruder of the dust. The printer shifts the extrusion head to a particular place melted material, where it refreshes and solidifies. When a layer is completed, it moves down the building platform and repeats it until 3D are completed [40–43].

For quick production, fuse deposition modelling is a prominent form of fast prototyping. Fast prototyping enables iterative testing, and fast production can be a comparatively inexpensive alternative for every short period. Plastics being the cheap material used in fuse deposition modelling makes the final product produced in fuse deposition modelling cheaper. Still, because it leaves the support and marks after production, a surface finish process such as sanding is required. The behavior of the melt and mechanical properties of the finished parts are related to the molecular structure and molecular weight distribution of the raw material [40].

Usually, the part is ready to use after printing, but there may be some post-processing requirements, such as removing support structures or surface lining. Modelling of fused deposition can be fully functional in the form of standard plastics. They, however, have Z-directions and step structures on the surface. They have anisotropy (in the vertical direction). The most common polymer used in fuse deposition modelling is acrylonitrile butadiene styrene, but polyetherimide is also used. For example, modelling fused deposition works with standard materials like acrylonitrile butadiene styrene, polyurethane, polyamide, and polycarbonate. Hence, these parts are more reliable with good mechanical properties and can last for an extended period. Like any other plastic part produced in traditional production, components and assemblies made in fuse deposition modelling can be post-processed.

5.2 Direct Energy Deposition

Direct energy deposition was first used in 1997 to produce 3D aerospace components with Titanium. A more complex additive manufacturing process for aircraft parts is direct energy deposition. Typically, it is used for repairing existing features and adding additional material. Parts can be completely manufactured using the technology from scratch. This method can be used to do maintenance and repair damaged parts like impeller blades or propellers for turbines. A focused energy supply, such as a laser or electric beam, uses direct energy deposition to melt materials. Materials are melted and deposited at the same time through a nozzle to form a 3D part. Direct energy deposition thus became the boundary of extrusion and fusion of the powder bed.

Laser engineered net forming, direct metal deposition, electron beam additive manufacturing are also known as direct energy deposition. The name of the technology is given according to the specific application used.

The advantage of direct energy deposition are as follows; it does not require starting on a flat plane. Complete axis control with deposition head, direct energy deposition can be done either with powder or wire, the deposition rate is almost up to 3 kg/h, the power deposition rate is 2 kg/h, a wide range of materials is used in this method [41, 42].

5.3 Selective Laser Sintering

Selective laser sintering has the capability, because it requires no support during the fabrication process, to precisely and precisely manufactures different design elements and complex geometry. Particular laser sintering components have good isotropic mechanical properties that make them ideal for functional parts and prototypes. This 3D printing technology is used mainly in biotechnology in the field of medical care and therapy worldwide. Selective laser sintering uses a powerful laser to synthesize polymer powder particles into a solid three-dimension structure based on a three-dimension model. For more than 20 years, particular laser sintering methods have been popular with engineers. Low part cost, high production productivity, and established materials make the technology perfect for various applications, forming rapid prototyping for small-batch, bridge, or custom production. Selective laser sintering (SLS) is trusted to produce vital functional components by engineers and manufacturers in different industries [35, 40, 41].

5.4 Digital Light Processing

Digital light processing is a type of 3D vat polymerization technology which uses a liquid photopolymer resin that can be solidified from a light source. The 3D of light-based printing includes digital light processing, laser-assisted printing, selective laser melt, and selective laser sinters [42–45]. The polymerization reaction is applied through digital light processing and laser-supported printing methods. The material powder is melted and transformed at high temperatures generated by laser during selective laser melting and laser selection sintering. This 3D of platforms differs significantly in the printing mechanism, speed, selection, and product resolution.

3D printing technology for digital light processing comes from Texas instruments' image projection technology of the 1980s. This method uses a set of chipsets based on micro-electromechanical optical technology to process photosensory working light sources.

A digital micro-mixer comprising a group of small, controllable mirrors is the system's primary function—the digital micro-mixer. The mirrors rotate in order to control the path of light and then project onto the resin during work. The ordinary arrays have many mirrors, from one million to over two million mirrors.

5.5 Stereolithography

One of the additive manufacturing methods employed in creating 3D printed objects is Stereolithography. This process converts liquid plastic into solid 3D parts by the unique, stereo-lithographed 3D printing machine. 3D stereo printer Stereolithography has become extremely popular due to its ability to produce high-precision, isotropic and waterproof prototypes [45]. It can manufacture components in a range of advanced materials with a smooth finish and delicate features. But 3D stereo printers are more than prototyping, and they can easily be made into mass production powerhouses by managing and performing them correctly.

Form's laboratories Standard resins and laboratory engineering resins offer a powerful alternative to comprehensive tests and stress-related performance. The generation of robust plastic jigs and appliances printed in-house on a 3D stereolithographic printer can be supported and facilitated by many [45].

6 Applications of 3D Printing

In recent years, the 3D printing industry has been expanding and succeeding. The technology has been widely used in aerospace, automotive, medical and other fields. Below are a few of the primary industries that use 3D printing products to improve operational cost and performance for each sector, enabling them to be competitive and survive in the challenging world [45].

6.1 Aerospace Industry

Metal 3D printing is now used in aerospace or aviation industries. Metal 3D printing produces different heavy-duty and functional motor parts. The aerospace industry began to take advantage of the convenience of the flexibility in design and reliability of 3D metal parts. In relatively low-term areas such as the fan pipe, the engine access panel, the compressor vanes, the by-pass fan statistics, and acoustic motor line, polymer or polymer composites are used. Polyether-ketone-ketone (PEKK) and composite carbon fiber materials are highly thermal-resistant, chemically inert, mechanically resistant. 3D PEKK composite printing parts have a potential application and are highly applicable in the aerospace industry [18]. Composites of 3D printed polyetherimide families also provide good mechanical resistance to high temperatures and pressure for high-performance applications.

Polymer additive production (commonly referred to as drawings) in unused aerial vehicles has gained popularity due to light weight-prints, easy-to-make aerodynamic forms, and flexible design options, which can be modified and reported at remote locations [34].

6.2 Automotive Industries

The major advances in additive manufacturing for the automotive industry opened up the door for new designs, clear, lighter and safer products, shorter lead times and low costs. The automotive manufacturer uses mainly additive production for fast prototyping. The additive manufacturing process produces ingredients that are less restricted in design, and which are often restricted by traditional manufacturers. In eliminating the need for new tools and producing end parts directly, additive production reduces the overall lead time and improves reactivity on the market. By manufacturing light weight parts that are less costly to handle, on requirement and at the location. Manufacturing additives reduce inventory and supply chain costs. Additive production has a high impact in changing the business model of automotive companies, with product innovation and supply chain transformation. Examples of automobile components produced by selective laser sintering processes are such as car bumpers, windbreakers, interior/exterior trims, and other accessories [31–34].

6.3 Medical Sector

The significant advances in additive manufacturing for the automotive industry opened the door for new designs, clear, lighter, and safer products, shorter lead times, and low costs. The automotive manufacturer uses mainly additive production for fast prototyping. The additive manufacturing process produces fewer local ingredients in design and is often restricted by traditional manufacturers. In eliminating the need for new tools and creating end parts directly, additive production reduces the overall lead time. It improves reactivity on the market by manufacturing lightweight components that are less costly to handle, on the requirement, and at the location. Manufacturing additives reduce inventory and supply chain costs. Additive production has a high impact in changing the business model of automotive companies, with product innovation and supply chain transformation. Examples of automobile components produced by selective laser sintering processes include car bumpers, windbreakers, interior/exterior trims, and other accessories [36, 38].

7 Contribution of 3D Printing Towards Sustainable Manufacturing

The 3D printing industry was rapidly expanding in the last decay, and in 2013 it researched the market size of some 2,6 billion dollars. Increased commercial 3D are printing from 355 in 2008 to approximately 23,000 in 2013, a rapid increase of 6400% [2, 36, 37, 46]. The rise in product design freedom has allowed a broader range of designs and materials in additive production. In contrast, the use of different designs and styles by traditional production has been restricted. Second, there are no extra costs of complexity; in conventional products, complex products lead to higher manufacturing costs, while manufacturing complex products does not have to be changed by 3D printing. Third, the manufacture in one lot; traditionally, single, single goods are more expensive. There is a big difference in the cost of printing a single unit and mass production in 3D printing. However, increased customization; additive manufacture requires the production of products layer by layer, making it easier to adapt to the output without changing the entire manufacturing design. The manipulative process simplification: 3D printers are not as concerned about operator involvement and the level of expertise as in traditional manufacturing, as they use an already designed digital module. The elimination of the supplies and the production lines; the additive manufacturing enables the manufacture of the entire product in one process where it involves many (100–100) steps from the beginning of the production cycle, as is the case in conventional manufacturing. Fourth, instant production at the global level; because most 3D print designs are accessible via the internet, production can be launched anywhere in the world where an internet connection is available [36, 37]. The reduction of waste and emissions; the material necessary for production, which almost does not lead to destruction, is used in additive manufacturing only. The need for local transport is also decreased by the printing or display of the goods [43, 44].

In today's highly competitive industries, time is a significant factor in success in the market and creating customer confidence in your brand by delivering customer satisfaction. By avoiding traditional tooling methods, reducing lead times on prototypes, and manufacturing final parts and components, significantly reduces market time with mass three dimensions printing [2].

3D mass production does not always mean the direct printouts of end-use components but the creation of affordable, personalized tools. The custom 3D molds for injection mounting are a perfect example of this. 3D printed molds save much time and save a good deal of money on cost for low volume manufacturing in 10–100 parts. It also enables a more flexible and sustainable manufacturing approach whereby design engineers can easily modify molds and iterate on the final design or assembly [33].

Automotive continuously improves their vehicle fuel usage efficiency through production for lightweight parts and assemblies with additive manufacturing. This has resulted in business growing revenues by delivering more excellent value to the customers and complying with the current increasing standard for fuel compliance such as corporate average fuel economy. The fact that 3D printing is more economical in lightweight production for aircraft and automotive industries has gained a lot of confidence to be sustainable in the future. With its flexibility to produce parts of any geometry, 3D printing has saved many materials to make the same part as produced in traditional manufacturing with the same material. An excellent example of this is the 2015 Ford F-150. The Ford F-150s s body, unveiled in January 2014, is almost entirely made of aluminum, weighing about 317 kg less [46].

Working in innovation with EADA and EOS, a direct metal laser sintering manager, replacing the cast steel hanging bracket with an additive titanium part on the Airbus A320 is optimal in placing metals where loads exist. This reduction in raw material consumption by 75% reduces energy and emissions from generation, operation, and life-cycle recycling by 10 kg per chipset.

3D printing is a highly influential technology with the potential to make clinical care more affordable, more accessible, and personalized and improve the field of medicine and medical attention. As technology and printers evolve, biomaterial printing is regulated in terms of safety, and the general public has a sense of how 3D printing works. Many successful cases demonstrate the potential of additive production in pediatric planning. In the medical field, the role of 3D print is relevant in five patients with a double-decker ventricular septal defect, ranging from 7 months to 11 years old. The 3D model printed on computer tomography (CT) or MRI data has contributed to a complete anatomy assessment, leading to a successful surgical repair of three of the five patients. 3D printing's main pillars in healthcare have made the ability of physicians to treat more patients than previous cases sustainable, obtaining results from patients and taking less time to handle the medical specialists [2, 37].

3D printing in the recent decay has evolved with a lot of improvements in a wide range of applications from aerospace industries, automotive industries, medical institutions, and so on [38]. 3D printing is more sustainable and has a lot of room for improvement in the future; research is ongoing to further make life easier worldwide and throughout the universe. Additive production tends to save more resources and allow for more effective, new designs that contribute to lower production standards [44–46]. It is also greener than traditional production techniques, cleaner, and more durable, with much scope for improvement [37, 47, 48].

8 Limitation of 3D Printing

Additive manufacturing with unique technology can use any materials to print a variety of design parts/products for different applications [41, 45]. However, 3D printed materials must carefully examine their various characteristics, including dimensional stability, strength, viscosity, and resistance to heat and moisture. The temperature cannot be controlled sufficiently by all metals and polymers for 3D printing. Delamination and stress breakage may occur through weak links between layers.

In addition, there are small printing chambers for 3D printers, which are not suitable for printing large parts. This can increase the cost and time for more significant components with increased work to mount the components. Most 3D printed components need cleaning to remove supports and smooth the surface to obtain the necessary finish with the production and connectivity of 3D printing parts, layer by layer. This also allows them to delaminate during use under specific stresses or guidance. When producing items using modelling fuse deposition, this issue is more significant.

9 Conclusion and Future Scope

From several findings, additive manufacturing is the most reliable and sustainable manufacturing technology in the challenging world today. Additive manufacturing has done a lot of savings in the automobile and aerospace industry. It has saved fuel costs for Aeroplan's as the weight of the part is reduced by 55%. With the recycling and reduction of waste by 90%, 3D printing can produce high-quality products with less cost on the market than traditional manufacturing.

Since 3D printing uses computer-aided design software to print products, it is highly recommended to save supply chain costs by printing the part locally where it will be used. Additive manufacturing can reduce maintenance and installation time through their ability to print complex geometry with consolidated parts into one assembly or component. Additive manufacturing is more economical in which it can save 90% of energy and uses only one machine to altogether produce a finished product. These provide an excellent opportunity for additive manufacturing to be competitive and help industries survive and sustain their operations.

There are some areas where additive manufacturing has some limitations and needs to improve. This technology is still being developed in the medical industries with the potential risk of bio-printing the harmful substance in print. There are also challenges to solve and eliminate, for example, printing patterns, the accumulation of porosity, variable printing flows, etc. With 3D printing becoming more popular and accessible, people have a more significant opportunity to create fake and falsified products that cannot tell the difference. This makes the final component differ from the original design when some 3D printers have a low tolerance. This can be fixed after processing, but the time and cost of production will increase further. However, it is just a matter of time. As research is ongoing, 3D printing would improve and fill up the gaps, improving the technology and making life easier in today's competitive world.

Based on prior research, the impact of 3D printing on environmentally responsible manufacturing practices is investigated in this study. The main goal of future research will be to figure out how to measure sustainable performance in manufacturing industries, especially as it relates to the use of 3D printing processes.

References

1. Shahrubudin, N., Lee, T. C., Ramlan, R.: An overview on 3D printing technology: Technological, materials, and applications. Procedia Manufacturing, 35, 1286–1296 (2019).
2. Attaran, M.: The rise of 3-D printing: The advantages of additive manufacturing over traditional manufacturing. Business horizons, 60(5), 677–688 (2017).
3. Nath, S. D., Nilufar, S.: An overview of additive manufacturing of polymers and associated composites. Polymers, 12(11), 2719 (2020).
4. Kumar, A., Paul, J., Unnithan, A. B.: 'Masstige' marketing: A review, synthesis and research agenda. Journal of Business Research, 113, 384–398 (2020).
5. Muduli, K., Barve, A., Tripathy, S., & Biswal, J. N.: Green practices adopted by the mining supply chains in India: a case study. International Journal of Environment and Sustainable Development, 15(2), 159–182 (2016).
6. Palmatier, R. W., Houston, M. B., Hulland, J.: Review articles: Purpose, process, and structure. Journal of the Academy of Marketing Science, 46(1), 1–5 (2018).
7. Snyder, H.: Literature review as a research methodology: An overview and guidelines. Journal of Business Research, 104, 333–339 (2019).
8. Peter, O., Swain, S., Muduli, K., Ramasamy, A.: IoT in Combating COVID-19 Pandemics: Lessons for Developing Countries. In Assessing COVID-19 and Other Pandemics and Epidemics using Computational Modelling and Data Analysis, pp. 113–131, Springer, Cham. (2022).
9. Muduli, K., Kusi-Sarpong, S., Yadav, D. K., Gupta, H., Jabbour, C. J. C.: An original assessment of the influence of soft dimensions on implementation of sustainability practices: implications for the thermal energy sector in fast growing economies. Operations Management Research, 14(3), 337–358 (2021).
10. Moharana, B. R., Sahu, S. K., Maiti, A., Sahoo, S. K., Moharana, T. K.: An experimental study on joining of AISI 304 SS to Cu by Nd-YAG laser welding process. Materials Today: Proceedings, 33, 5262–5268 (2020).
11. Biswal, J. N., Muduli, K., Satapathy, S., Yadav, D. K.: A TISM based study of SSCM enablers: an Indian coal-fired thermal power plant perspective. International Journal of System Assurance Engineering and Management, 10(1), 126–141 (2019).

12. Biswal, J. N., Muduli, K., Satapathy, S.: Critical analysis of drivers and barriers of sustainable supply chain management in Indian thermal sector. International Journal of Procurement Management, 10(4), 411–430 (2017).
13. Moharana, B. R., Sahoo, S. K.: An ANN and RSM Integrated Approach for Predict the Response in Welding of Dissimilar Metal by Pulsed Nd: YAG Laser. Universal Journal of Mechanical Engineering, 2(5), 169–173 (2014).
14. Moharana, B. R., Patro, S. S.: Multi objective optimization of machining parameters of EN-8 carbon steel in EDM process using GRA method. International Journal of Modern Manufacturing Technologies, 11(2), 50–56 (2019).
15. Joshi, S. C., Sheikh, A. A.: 3D printing in aerospace and its long-term sustainability. Virtual and Physical Prototyping, 10(4), 175–185 (2015).
16. Sriram, V., Shukla, V.: Additive Manufacturing Using Metal Powder-An Insight, Technology Information Forecasting and Assessment Council (TIFAC), New Delhi (2016).
17. Bourell, D., Kruth, J. P., Leu, M., Levy, G., Rosen, D., Beese, A. M., Clare, A.: Materials for additive manufacturing. CIRP annals, 66(2), 659–681 (2017).
18. AL-Maliki, J. Q., AL-Maliki, A. J. Q.: The Processes and Technologies of 3D Printing, International Journal of Advances in Computer Science and Technology, 4(10), 161–165 (2015).
19. Kubáč, L., Kodym, O.: The impact of 3D printing technology on supply chain. In MATEC web of conferences. vol. 134, p. 00027. EDP Sciences (2017).
20. Westerweel, B., Basten, R. J., Van Houtum, G. J.: Preventive maintenance with a 3D printing option. Available at SSRN 3355567 (2019).
21. Parupelli, S., Desai, S. A.: comprehensive review of additive manufacturing (3d printing): processes, applications and future potential. American journal of applied sciences, 16(8), 1–29 (2019).
22. Stock, T., Seliger, G.: Opportunities of sustainable manufacturing in industry 4.0. procedia CIRP, 40, 536–541 (2016).
23. Royo, J. D., Soldevila, L. M., Kayser, M., Oxman, N.: Modelling Behaviour for Distributed Additive Manufacturing. In Modelling Behaviour, pp. 295–302, Springer, Cham (2015).
24. Lee, H. T., Song, J. H., Min, S. H., Lee, H. S., Song, K. Y., Chu, C. N., & Ahn, S. H.: Research trends in sustainable manufacturing: a review and future perspective based on research databases. International Journal of Precision Engineering and Manufacturing-Green Technology, 6(4), 809–819 (2019).
25. Kai, D. A., De Lima, E. P., Cunico, M. W. M., Da Costa, S. E. G.: Measure Additive Manufacturing for Sustainable Manufacturing. In ISPE TE, pp. 186–195 (2016).
26. Zocca, A., Colombo, P., Gomes, C. M., Günster, J.: Additive manufacturing of ceramics: issues, potentialities, and opportunities. Journal of the American Ceramic Society, 98(7), 1983–2001 (2015).
27. Regassa, Y., Lemu, H. G., Sirabizuh, B.: Trends of using polymer composite materials in additive manufacturing. In IOP conference series: materials science and engineering, vol. 659(1) p. 012021, IOP Publishing. (2019).
28. Craig, A., Gangula, G. B., Illinda, P.: 3D Opportunity in the Automotive Industry, A Deloitte Series on Additive Manufacturing, (2014).
29. Moritz, T., Maleksaeedi, S.: Additive manufacturing of ceramic components. In Additive Manufacturing, pp. 105–161, Butterworth-Heinemann (2018).
30. Richards, F.: Aerospace today: composites, 3d printing, & a shot of espresso. Advanced Materials & Processes, 173(5), 4–5 (2015).
31. Tadjdeh, Y.: 3D printing promises to revolutionize defense, aerospace industries. National Defense, 98(724), 20–23 (2014).
32. Gibson, I., Rosen, D. W., Stucker, B., Khorasani, M., Rosen, D., Stucker, B., Khorasani, M.: Additive manufacturing technologies, vol. 17, Cham, Switzerland: Springer (2021).
33. Liu, Z., Jiang, Q., Zhang, Y., Li, T., Zhang, H. C.: Sustainability of 3D printing: A critical review and recommendations. International Manufacturing Science and Engineering Conference, vol. 49903, p. V002T05A004. American Society of Mechanical Engineers (2016).

34. Njuguna, J., Pielichowski, K.: The role of advanced polymer materials in aerospace. Res. Gate, 148 (2013).
35. Mierzejewska, Z. A., Markowicz, W.: Selective laser sintering-binding mechanism and assistance in medical applications. Advances in materials science, 15(3), 5 (2015).
36. Ben-Ner, A., Siemsen, E.: Decentralization and localization of production: The organizational and economic consequences of additive manufacturing (3D printing). California Management Review, 59(2), 5–23 (2017).
37. Kim, H.: Marketing Analysis and Future of Sustainable Design Using 3D Printing Technology Archives of Design Research 31(1), 23–35 (2018).
38. Aimar, A., Palermo, A., Innocenti, B.: The role of 3D printing in medical applications: a state of the art. Journal of healthcare engineering, 2019, 1–10 (2019).
39. Sireesha, M., Lee, J., Kiran, A. S. K., Babu, V. J., Kee, B. B., Ramakrishna, S.: A review on additive manufacturing and its way into the oil and gas industry. RSC advances, 8(40), 22460–22468. (2018).
40. McAlister, C., Wood, J.: The potential of 3D printing to reduce the environmental impacts of production. European Council for an Energy Efficient Economy, ECEEE Industrial Summer Study Proceedings, 213–221 (2014).
41. Kearney, A.: 3D Printing: Ensuring Manufacturing Leadership in the 21st Century, HP Confidential, (2017).
42. Stratasys.: Will 3D Printing Eliminate the Warehouse? The 3D Printing Solutions Company, (2017). www.stratasys.com.
43. Wing, I., Gorham, R., Sniderman, B.: 3D opportunity for quality assurance and parts qualification, Deloitte University, (2015). DUPress.com.
44. Nasr, N.: OECD Sustainable Manufacturing Toolkit. OECD Directorate for Science, Technology and Industry (DSTI) (2011). www.oecd.org/innovation/green/toolkit.
45. Nayyar, A., & Kumar, A. (Eds.).: A roadmap to industry 4.0: smart production, sharp business and sustainable development, pp. 1–21, Berlin: Springer (2020).
46. Pathak, P., Singh, M. P., Sharma, P.: Sustainable Manufacturing: An Innovation and Need for Future, 2017. Proceedings of International Conference on Recent Innovations in Engineering and Technology, pp 21–26, Jaipur, India (2017).
47. Kumar, A., Nayyar, A.: si 3-Industry: A Sustainable, Intelligent, Innovative, Internet-of-Things Industry. In A Roadmap to Industry 4.0: Smart Production, Sharp Business and Sustainable Development, pp. 1–21, Springer, Cham. (2020).
48. Nayyar, A., Puri, V., Le, D. N.: Internet of nano things (IoNT): Next evolutionary step in nanotechnology. Nanoscience and Nanotechnology, 7(1), 4–8 (2017).

Role of 3D Printing in Pharmaceutical Industry

Rajeshwar Kamal Kant Arya, Dheeraj Bisht, Karuna Dhondiyal, Meena Kausar, Hauzel Lalhlenmawia, Pem Lhamu Bhutia, and Deepak Kumar

1 Introduction

Industry 4.0 is a revolutionary road toward using intelligent automation technology in manufacturing [1]. In this fourth generation of the industrial revolution (Industry 4.0), the latest manufacturing expertise is integrated with novel technologies to support and enhance the manufacturing of complex systems with some intelligent technologies, including artificial intelligence and the international internet of things [2, 3]. The pharmaceutical industry is a fast-paced, growing industry. The demand for novel patient-customized drugs, drug delivery systems, and pharmaceutical devices is increased in the last few years [4]. The digitalization of the pharmaceutical industry can reduce product costs with improved efficiency and production capacity with the help of automatic and computerized systems. Still, the pharmaceutical industry restricts digitalization because of the complex manufacturing processes. However, there is an excellent requirement for the digitalization of pharmaceutical industries for both novel and conventional delivery systems [5].

R. K. K. Arya (✉) · D. Bisht · K. Dhondiyal · M. Kausar
Department of Pharmaceutical Sciences, Bhimtal Campus, Kumaun University Nainital, Bhimtal, Uttarakhand, India
e-mail: rajeshwrarya@gmail.com

H. Lalhlenmawia
Department of Pharmacy, Regional Institute of Paramedical and Nursing Sciences, Aizawl, Mizoram, India

P. L. Bhutia
Govt Pharmacy College, Sajong, Rumtek, Sikkim, India

D. Kumar
Department of Pharmaceutical Chemistry, School of Pharmaceutical Sciences, Shoolini University, Solan, Himachal Pradesh, India

© The Author(s), under exclusive license to Springer Nature Switzerland AG 2023
A. Nayyar et al. (eds.), *New Horizons for Industry 4.0 in Modern Business*, Contributions to Environmental Sciences & Innovative Business Technology,
https://doi.org/10.1007/978-3-031-20443-2_13

In the pharmaceutical sector, industry 4.0 can help improve the manufacturing of personalized medication, and additive manufacturing, by collaborating with three Dimension Printing (3DP) [6]. Artificial intelligence should also be added to the 3DP for manufacturing personalized medication systems with artificial intelligence and automation in production. Some infrastructural support systems such as cloud and back chain are also required, which would give some pace to 3DP-based clinical research and other formulations and navigate researchers towards the personalized medication system and industry 4.0 [7].

Recently three 3DP gained popularity, and the concept of 3D printing has opened new avenues in the scientific community and pharmaceutical industries. 3D printing was designed to help the industry, and it has shown encouraging results in the last few years [8], the 3D printing revolutionized manufacturing techniques for developing novel drug delivery system, personalized medicine, tissue modification, and disease modeling. 3D printing can emerge as a milestone in the pharmaceutical industry with a novel multipurpose drug delivery system, tissue modification, and tunable customized dosage form per patient need [4]. 3D printing shows broad applicability in tissue engineering, prosthetic organ fabrication, diagnostic and medical device and, novel drug delivery fabrication [9]. A computed Tomography scan and M.R.I. are used to generate the anatomical and medical scaffold per the patient's requirement. The prosthetic organ can be fabricated to imitate the function same as the original organs [9].

The 3D printing concept came to light in the 1980s when Charles Hull used photopolymers to fabricate plastic devices. [10]. 3D printing may be defined as the process of making any possible object three-dimensional (3D), utilizing material deposition in layers above one another from digital designs [10]. This technology has also been used in different fields such as aerospace, consumer goods, robotics, and automotive. [11]. In 2015, F.D.A. approved Spritam (an antiepileptic drug), the first 3D-printed pill. After this approval, 3D printing-based manufacturing of medicines attracted the interest of pharmaceutical industries in manufacturing 3D printing-based formulations [12].

In pharmaceutical manufacturing, additive manufacturing, or especially 3D printing, is an innovative era because it opens the opportunity of making unconstrained dosage forms that are equal to the traditional drug fabrication approach [13]. As 3-dimensional printing is the twenty-first-century production technology, 3D printing has become one of the most multipurpose production technologies, in which sequential layers are formed by the accumulation of components along with polymers and metals within the X and Y direction to give a 3D object [14]. 3D printing can make objects of any shape and size by accumulating materials layer by layer. For the last few years, 3D printing has become a promising technology in industrial utility functions [15]. 3D printing can develop a wide range of drug delivery systems, i.e., oral controlled release systems and drug implants to multiphase-release dosage forms; consequently, 3D printing technology is highlighted as a vital device for designing both easy and complicated design drug delivery systems [13]. With the help of 3D printing, we can fabricate various designs that are impossible to create with our traditional manufacturing system

[16]. M.I.T. (Massachusetts Institute Technology, Cambridge, MA, U.S.A.) was the one that introduced 3D printing technology in the pharmaceutical industry [17].

The main objectives of the chapter are:

1. To understand the basics of 3D printing technology,
2. To elaborate the role of 3D printing in the field of pharmaceuticals,
3. To have better understanding of the complications and limitations of 3D printing in the pharmaceutical industry,
4. To provide information on novel formulations manufactured with help of 3D printing, and
5. And, to understand the importance of 3D printing in the manufacturing of prosthetics and customized drug delivery system.

Organization of Chapter

The chapter is organized as : Sect. 2 elaborates procedure for 3D Printing. Section 3 enlightens advantages and limitations of 3D Printing. Section 4 details about Fabrication of 3D Medical Product. Section 5 illustrates challenges of 3D printing. Section 6 stresses on Application of 3D Printing in the Pharmaceutical Field. Section 7 illustrates advantages of 3D printing in the pharma industry. Section 8 concludes the chapter with future scope.

2 Procedure for 3D Printing

The basic process of 3D printing is given here

Initially, digital design software such as Creo parametric, Autodesk, Solidworks, AutoCAD, Onshape, etc., is used to form the 3D image of an object.

After creating the 3D image, a Standard tessellation language (S.T.L.)—a digital file format has to create.

Triangulated facets are used to gain detailed information about the surface of the 3D model (S.T.L. file).

Now digital file (S.T.L.) is transformed into G-file with the help of slicer software (previously installed in the 3D printer).

With the help of print head movement along with the X-Y axis, the base of the 3D object is created.

By allowing the movement of the print head along with the z-axis, material deposition in layers can be done to create a complete 3D object [9].

3   Advantages and Limitations of 3D Printing

3D printing is more advantageous than conventional manufacturing techniques as elaborated in the following points:

1. 3D printing fabricates personalized medication based on the patient's age and disease.
2. Medicine manufacturing is expedited with more rapid analysis than conventional, and it can be easily customized by the pharmacist or manufacturer [18].
3. It does not require expensive tools, so there is a reduction in cost, time, and labor. The product with an excellent surface finish can be manufactured with 3D printing [19].
4. The printer occupies very little space; that's why 3D printing manufacturing is affordable in terms of space requirements. A small batch can be easily manufactured in a single run. High drug loading can be achieved with 3D printing than the conventional dosage form [20].
5. 3D printing also minimizes material wastage [21].
6. 3D printing also improves formulation properties [18].

The following the limitations of 3D Printing.

Being most advantageous, it also suffers from disadvantages such as the industry cannot monitor all product manufacturing operations [17]. The following are some of the key limitations of 3D printing:

1. The 3D blueprint is shared with pharmaceutical companies for drug manufacturing. If any defect arises, the blueprint provider is not solely responsible for the defects, and the vendor, software maker, and manufacturing unit are also accountable for this. Regulatory policies are required for blueprint certification to protect the product from legal and financial conflicts [17].
2. The law infringement by multiple persons is there because many people are involved in the manufacturing, and this system cannot identify the main culprit indulging in the act or patent infringement, and this is the biggest disadvantage of 3D printing [22].

4 Fabrication of 3D Medical Product

The 3D medical product is fabricated using digital software, and a 3D virtual design of the drug product is prepared. The software involved in this process is Onshape, Solid works, Creo parametric Autodesk, etc. Then the virtual design is converted into a readable system S.T.L. (standard tessellation language) or stereolithography [8]. After this, the 3D is sliced horizontally into a 2D design called a G file with the help of Slicer software. Then the printing head moved along the X and Y axis and constructed a plinth of the final product, and then the printing head moved along the Z-axis to create a 3D product by accumulating layers of desired material repetitively and fabricating the final 3D product [9] Fig. 1.

Various techniques are available for manufacturing products with the help of 3D printing. The best techniques are selected based on speed, precision, and raw material. The following 3D printing technology is widely employed in pharmaceutical production, e.g., Fused Deposition Modeling, Vat Photo-Polymerization, Selective Laser, Binder Jetting, and Material Jetting.

4.1 Fused Deposition Modeling or Material Extrusion

It is a widely used 3D printing method; in this, the final product is formed by melting the polymeric material with the help of heat or using the soft material [20]; it is also known as the material extrusion technique [22], the FDM technique is inexpensive technique than selective laser sintering technique [18]. In this technique, the materials used for making the product are heated previously and then passed through the nozzle, moving along all three axes and placing the material layer by layer. After this, the other layers are added to the previous layer [21], and after solidification, the exact 3D product designed by the computer-aided design is

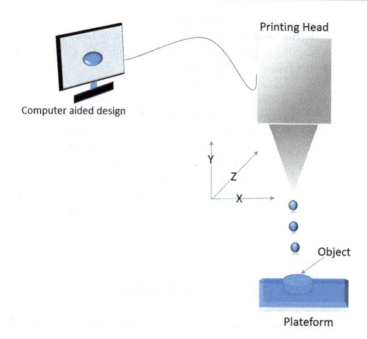

Fig. 1 Basic diagram of 3D printer

fabricated [9]. The product shows good mechanical strength [15]. FDM uses various software for creating virtual designs, e.g., AutoCAD, Inventor®, Solidworks®, CATIA™, etc. There is some freeware also available [23]. A better understanding of running software is required for designing complex designs. The amount of material extruded from the nozzle can be controlled through the software and saved in G-file.

Polylactic acid is a widely used material because it is a bioplastic, produced by a green synthesis, and shows good tensile strength. Other materials include poly-caprolactone polyethylene oxide, polypropylene, cellulose derivative, eudragits, polyethylene, polyvinyl alcohol polybutylene terephthalate, ethyl vinyl acetate wood, nylon, metals, sorbitol, graphene-doped P.L.A., etc. [24]. This system suffers drawbacks, e.g., limited availability of suitable material and fabricated weak layers [25]. Various kinds of dosages form can be prepared by using FDM, e.g., tablets, transdermal patches, microneedles, oral or buccal patches, vaginal inserts, and parenteral [23]. Esposito Corcione et al. [26] prepared hydroxyapatite-loaded microspheres in a polylactic acid scaffold, they prepared microspheres by using spray drying techniques, and then the scaffold was prepared by dispersing the prepared microspheres in P.L.A. by using the FDM technique for enhancing the bone healing process [26] Some of the dosage form prepared by fused deposition modeling are listed in Table 1.

Table 1 Various formulations prepared by fused deposition modeling

Sl. no	Drug name	Formulation type	References
1	Theophylline	Floating tablet	[27]
2	Prednisolone	Extended-release tablet	[28]
3	Captopril	Immediate release tablets	[29]
4	Budesonide	Tablet	[30]
5	Hydroxyapatite	Microporous implant	[25]
6	Domperidone	Floating tablet	[31]
7	Ibuprofen	Tablet	[32]
8	Salicylic acid	Face mask	[33]
9	Deflazacort	Nanoparticles	[34]

4.2 Vat Photo-Polymerization (V.P.P.)

Vat photopolymerization (V.P.P.) is a 3D printing technique. It uses U.V. light or other high-intensity light resources to create polymerization of liquid photograph-polymerizable resins that form the complex polymer and eventually solidify into the final product [13, 35]; using this method product with high accuracy and good surface finish are prepared. For these monomers, oligomers and initiators are required to form a complex polymer. In this technique, various polymers are utilized, e.g., Soybean oil epoxidized acrylate, Irgacure 819, Lignin, methacrylic anhydride, SR494 (Sartomer), Ebecryl 8210 (Allnex), Genome 1122, vinyl terminated, polydimethylsiloxane, etc. [36]. V.P.P. technique generally employs method top-down and bottom-up methods, and this technique includes stereolithography (S.L.A.), digital light processing (D.L.P.), continuous liquid interface production (CLIP), and two-photon polymerization (2PP).

Stereolithography (S.L.A.) Stereolithography is a widely employed method of vat photopolymerization. It uses both top-down and bottom-up methods. In the top-down method, the laser beam is positioned just below the resin reservoir, and in the case of the top-down laser source is put above the resin reservoir.

Digital light processing (D.L.P.) This technique uses U.V. light as a light source and provides a faster speed than S.L.A. [37]. Dynamic mask is used to illuminate the light on photosensitive resin, it uses digitally projected light, and the image is composed of pixels that make layers [35].

Continuous Liquid Interface Production (CLIP)
Continuous Liquid Interface Production (CLIP) uses a liquid emitting diode and an oxygen-permeable window, also known as the dead zone, that permits the liquid resin to flow betwixt the printing surface and oxygen-permeable window. The flowing resin enhances the resolution of the printing object. In this technique, the printing object continuously rotates, improving the printing object's quality. This

Table 2 Various formulations prepared by Vat photo-polymerization

Sl. no	Drug name	Formulation type	Reference
1	Paracetamol	Oral modified-release tablet	[41]
2	4-Aminosalicylic acid and Paracetamol	Oral modified-release tablet	[42]
3	Salicylic acid	Anti-acne patch	[33]

technique shows better results over objects made through layer-by-layer techniques and removes the staircase effect [35]. However, this technique is enormously expensive and takes more time to process [38].

Two-photon polymerization 2PP: In this technique, the monomer is converted into the polymer with the help of a photochemical reaction; two photons near I.R. are utilized to generate high energy that initiates a photochemical reaction [39]. For this method, the radical is generated by using U.V. light, the electron transition occurs by absorbing two photons simultaneously and generates the radicals that cause the polymerization process where the light is focused, and eventually, the solidification process occurs near the focal point and a 3D structure is created [40]. With this technique, a porous product can be prepared. Some of the dosage forms prepared by Vat photopolymerization are listed below in Table 2.

4.3 Selective Laser or Powder Bed Fusion

This technique is also known as powder bed fusion. In this technique, the powdered raw material is utilized to prepare drug products by melting and fusing the material with a high-energy laser, which is eventually converted into a solid product. In 1984 Carl Deckard invented this technique and utilized neodymium-doped yttrium 74 aluminium garnet (Nd: YAG) laser [43]. A carbon dioxide laser is widely used to fuse the powder layer because it is economical and generates very high energy. If the powder material is thermoplastic, the S.L.S. technique is best suited for making drug products without requiring a binder or drying process [16]. In this process, the thermoplastic agents are generally utilized, e.g., Poly tetrafluoro ethylenes, polycaprolactone, polymethyl acrylate, epoxy polyethers, polyethylene oxide, Eudragit-L100, Eudragit-55, and Eudragit-RL [44]. With the help of a roller, the powder material is first spread as a thin layer on the formulation area and then allowed to melt and fuse (sintering) by using laser radiation [38]. This technique is lesser-used due to the opportunity for chemical degradation [15]. This technique is most effective in formulating tablets containing guaifenesin as an expectorant [18]. Generally, this technique is used with thermo-stable drugs and raw materials, but for thermolabile drugs and materials, the temperature and beam intensity can be modified for better results [44]. Based on the type of material, types, and intensity of light employed, this technique is categorized into four subsets, Selective Laser Melting (S.L.M.), Electron Beam Melting (EBM), and Multijet Fusion (M.J.F.). The S.L.M. and EBM are used when the feed material is metals or alloys, for

Table 3 Various formulations prepared by selective laser technique

Sl. no	Drug name	Formulation type	References
1	Ondansetron	Orally disintegrating tablet	[45]
2	Fluorouracil	Implantable tablet	[46]
3	Paracetamol	Printlets	[47]
4	Ibuprofen	Sustained release tablet	[48]
5	Salicylic acid	Mask	[33]
6	Paracetamol and Ibuprofen	Printlets	[49]

this high-intensity electron beam is utilized [16]. For nylon, M.J.F. is employed, and the infrared lamp is used for melting the feed. In M.J.F., fusion agents and detailing agents are added to enhance efficacy and printing resolution. The fusion agent enhances the method's efficacy, and the detailing agent minimizes thermal bleeding. There are various formulations prepared by the selective laser technique given the Table 3.

4.4 Binder Jetting

Binder jetting is also known as powder bed 3D printing, and Sachs and colleagues invented it in 1993 [21]. This technique is very popular and successful in the pharma sector, and the first FDA-approved tablet Spritam was prepared by this technique [50]. This method does not require a high-intensity beam to create the 3D product [51]. The liquid binder is jetted on the powder bed, the powder gets aggregated, densification occurs, the formed bed is pulled below, and a new layer is allowed to form over this layer [13, 52]. In 2015 F.D.A. accredited the first 3D-printed drug product. This technique is similar to the S.L.S. technique [38]. Among other 3D printing, this technique provides the rapid formation of complex structures [47]. The combination of active pharmaceutical ingredients and excipient is prepared and sprayed over the non-powder substrate. This method can be achieved by utilizing Drop on Demand (D.O.D.) or Continuous Jet Method techniques [13]. The formulation which contains the starting material as the liquid is best suitable for this approach [8]. Its excessive accuracy in creating 3D drug products is the predominant benefit in the pharmaceutical field. Generally, low-density materials are used as binders, and various polymers are used as binders e.g., polyglycolic acid, polylactic acid, polycaprolactone, polyethylene oxide, calcium phosphates, and calcium silicate [21]. Huang et al. [53] prepared levofloxacin implant using the binder jetting technique and found a better release than the conventional dosage form. Scoutaris et al. [54] prepared felodipine solid dispersion using inkjet printing to enhance the aqueous solubility. Some of the dosage forms prepared by inkjet printing are listed below in Table 4.

Table 4 Various formulations prepared by the binder jetting technique

Sl. no	Drug name	Formulation type	References
1	Felodipine	Microdots	[54]
2	5-Fluorouracil, curcumin, and cisplatin	Microneedles	[55]
3	Paclitaxel	Microparticles	[56]
4	Levofloxacin	Implant	[53]
5	Nitroglycerin	Injection	[57]
6	Enalapril maleate and propranolol hydrochloride	Orodispersible film	[58]

4.5 Material Jetting

Material jetting is a very rapid and accurate 3D printing, in this technique, the liquid photopolymer is sprayed through the printer head and get deposited on the object platform after solidifying with the help of U.V. light, and making the product by forming layers repetitively. This technique is very similar to the 2D printing technique. This system uses a D.O.D. 3D printing head for viscous polymer solution. There are two types of D.O.D. printing head thermal and piezoelectric. The thermal D.O.D. generates heat that creates a vapor bubble in the polymer reservoir, and the polymer is in the form of droplets released from the printing head [59]. This technique is unsuitable for heat-sensitive polymers, and solvents with high vapor pressure are generally utilized. In a piezoelectric system, the head contains a piezoelectric element that creates mechanical motion in the presence of electric current, and high pressure is generated that elutes the solution from the printing head [60]. There is no need for high volatile solvents, and this technique can be utilized for heat-sensitive materials. Therefore, it is widely used in the pharma sector. Solutions and nano-suspension are used for making pharmaceutical dosage forms [61]. Biuanz et al. [62] prepared salbutamol sulfate-loaded oral film for personalized medication using the thermal inkjet. The salbutamol solution was dispersed upon the potato starch-based scaffold. Some formulations prepared by the material jetting technique are given in Table 5.

4.6 Direct Energy Deposition

The direct energy deposition technique combines powder bed fusion and material extrusion technique. In this technique, the raw material is deposited on the platform through the nozzle, then melted with an energy source's help and converted into a 3D product. Generally, the wire or powdered material is used as raw material, and the laser beam, plasma arc, or electron beam is used as the energy source. In this method, the powder is melted being deposited on the substrate, whereas in the powder fusion method, the raw material is previously deposited on the substrate, and then the melting process is done [70]. The printer consists of a nozzle head that moves on a different angle over the stationary object and deposits the

Table 5 Various formulations prepared by the material jetting technique

Sl. no	Drug name	Formulation type	References
1	Enalapril and hydrochlorothiazide	Orodispersible film	[63]
2	Cidofovir and paclitaxel	Nanoparticles	[64]
3	Propranolol hydrochloride riboflavin sodium phosphate	Orodispersible film	[65]
4	Liothyronine sodium) and T4 (levothyroxine sodium)	Orodispersible film	[66]
5	Ibuprofen	Mucoadhesive buccal films	[67]
6	Lidocaine hydrochloride and piroxicam	Oromucosal	[68]
7	Indomethacin	Microneedles for transdermal delivery	[69]

raw material on the preferred space [20], and the laser beam then falls onto the powdered bed and melts the material that gets hard on the surface, and a layer is formed, and nozzle moves over the layer repetitively and forms the final product. This technique can also be used for repairing purposes but is limited in the pharmaceutical field [70]. The laser beam-assisted technique provides the précised object with the minimum material loss [71].

5 Challenges of 3D Printing

3D printing has shown an important role in the pharmaceutical sciences, and this technology has shown favourable results. But 3D printing faces several challenges while manufacturing. Therefore, these challenges must be controlled before manufacturing the personalized dosage form [20]. The important challenges are the selection of raw material, physical and chemical properties of materials, the ability of printing, and thermal conductivity have to be cautiously evaluated for the utilization of the product for human welfare [69, 72].

i. In powder-based 3D technologies, friability is a major challenge in the manufacturing of 3D dosage forms; to overcome this problem, fine particulate polymers should be used [21].
ii. The 3D printer has a high cost that enhances the production cost that can be reduced by making a low-cost printer.
iii. Being an automatic technology, the jobs within the production area would be considerably lower, so the economy would have a greater effect.
iv. The selection of excipient is very important because the material used, is non-pharmaceutical grade therefore the toxicity can obstruct the utilization of these materials in the pharmaceutical dosage form. Hence to advance the utilization of printing in the pharmaceutical field there should be a proper study of non-toxic biocompatible and stable excipients [17].

6 Application of 3D Printing in the Pharmaceutical Field

In the field of pharmaceutical 3D printing have many applications. In the medical field, dental implants and customized prosthetics are formulated with the help of 3D printing technology. Oral solid dosage forms and transdermal delivery systems are two dosage form in which 3D technology have shown a greater application in the pharmaceutical field. These formulations have shown good progress [73].

6.1 Customized Implant and Prostheses

3D printing has shown revolutionary steps in making various kinds of implants and prosthetics. It allows the making of customized individual-specific morphological prosthetic organs for a controlled release. Zuniga et al. [74] developed a prosthetic hand for children with deformed upper limbs by using the economic 3D printing technique; with the advent of 3D printing technology, nearly any imaginable structure could be made for implants and prostheses by translating X-Rays, M.R.I., and C.T. scan into a digital form of 3D print. This approach has formulated hip implant, spinal and dental, and hearing [18]. 3D printing techniques have emerged as a boon in the hearing devices industry. In 2013 about 10 million hearing aid was sold out globally [75]. 3D printing-based technology has also shown promising results in the dental health system; dental crowns and braces are very popular. In a study, dental implants were placed to support the restoration process, and the titanium implant has shown a positive response with a high success rate in 3 yearlong study [76]

6.2 Bio-printing Tissues and Organs

Damaging of Tissue or organ due to old age, trauma, or some congenital problem causes organ dysfunction, and organ replacement is the only line of treatment. We require human donors for organ replacement, which is a big issue [77]. In 2019, approximately 113,000 patients needed organ transplants in the U.S., but there is a huge scarcity of organs for transplant, and it takes an average of 213–370 days to receive the organs from the donor [78], and the demand increases day by day. Tissue engineering has changed the scenario of organ replacement, the isolated cell or tissue was regenerated in-vitro by using a degradable scaffold, but this method has its limitations [17]. 3D inject bio-printing of tissue and organ has shown promising results. Bio-printing product is safe for the receptor. Generally, bio-printing is based on the layering of biomaterials, and there is three steps preparatory phase fabrication of computer-aided 3D design, supplying the design for 2D layer formation, and then processing into it a 3D structure with the help of a 3D printer. In inkjet printing, biomaterial and the cell suspension are used as bio-ink sprayed over a hydrogel base polymer scaffold. With the help of a nozzle, the bio-ink is sprayed over the base; generally, the D.O.D. technique is the best-suited technique because it uses the piezoelectric mechanism, in which the biomaterial is vaporized by the electric current and sprayed over the scaffold [79]. Kim et al. [80] developed an artificial tubular tracheal graft by using polycaprolactone as a base material with the help of 3d printing techniques; the graft was blended with mesenchymal (M.S.C.s) and chondrocyte stem cells that were activated by pluripotent stem cell (iPSC), for regenerating the mucosal cell line and cartilage of trachea in rabbit [80].

6.3   Tablets

Regarding the preferred form in the pharmaceutical field, oral dosage forms are the best-preferred form. Tablet and capsule are examples of oral dosage form. There are two groups in which tablets are categorized: single API tablets and multiple API tablets. FDM technology is the preferred technology for the formulation of single API tablets. Giri et al. [27] developed gastro-retentive floating tablets of theophylline by using the hot-melt extrusion 3-D printing technique; they used hydroxypropyl cellulose as an excipient. In a study, a laser sintering 3D printing technique was used to develop an orally disintegrating tablet of ondansetron, they used mannitol and cyclodextrin as an excipient, and the prepared tablet showed a low disintegration time of 15s [45]. In another study, the haloperidol tablets were developed using a 3D technique; they used hot-melt extrusion techniques for formulating the tablets using Kollidon® VA64, Kollicoat® IR, Affinsiol™15 cP, and HPMCAS in the various ratio [81].

3D printed pediatric tablets Scoutaris and his colleagues prepared 3D printed indomethacin tablets for pediatrics under the name of Starmix® by extrusion method in which indomethacin and a thermoplastic polymer, hypromellose acetate succinate, compressed to form multiple layers of indomethacin tablet [82].

6.4   Microneedles

Microneedles are the type of transdermal drug delivery system. As the name indicates, microneedles are micro-sized needles. These micro-sized needles are present on the matrix's surface to enhance biologically active molecules to penetrate the skin. As 3D printing technology help in the formulation of tiny and very small structures, it increases the application of 3D printing in the fabrication of microneedles. Mansor et al. [83] developed microneedles using a 3D printing technique; they used various polymers for the study e.g., polyvinyl alcohol, polylactic acid, polyester resin, and acrylonitrile butadiene styrene. The study reveals that polyvinyl alcohol has high strength and durability. Johnson et al. [84] prepared low-cost techniques for formulating microneedles by using a 3D printing technique, and used photo photopolymerizable resin.

6.5 Transdermal Patch

With the help of 3D printing technology, a patch that has a complex structure can be manufactured for the transdermal delivery system. The patch prepared through 3D printing technology has a great benefit compared to traditional patches [73]. In a study, the montelukast containing transdermal patches was developed using 3D printing technology with low melting polymeric filament printed onto the packaging material. In the study, the researcher used various combinations of Kollidon 12PF, PEG 4000, and P.E.O. 900 k to achieve low melting polymers, and the transdermal has shown good drug release [85].

6.6 Wound Dressing

Wound dressing scaffolds of polyethylene glycol with pericardium were prepared using 3D printing to support the healing process of the lesion created during the vascular graft that helps replace the damaged vessel during surgery [4]. Andriotis et al. [86] developed chitosan and cyclodextrin loaded with propolis extract-based dressing material for healing lesions using the 3D printing technique.

6.7 Buccal Film

Oral delivery of peptides or proteins is one of the major pharmaceutical challenges because of inadequate bioavailability (due to instability in G.I.T. fluid) and efficacy. Montenegro et al. [87] formulated buccal film via thermal inkjet printing technique, using two different types of proteins, including lysozyme (obtained from egg chicken) and ribonuclease-A (obtained from bovine pancreas).

6.8 Abuse-Deterrent or Discouragement Dosage Forms

The major concept behind making these dosage forms is to make the drug unpleasant for drug abusers by compressing the drug in layers and slowing down the drug release from a dosage form or by creating hindrance in drug release. Turk et al. [88] focused on this on this concept initially; after that, it was also supported by others. In 2019, Nukala [89] made an abuse-deterrent dosage form, in which P.V.A. (Polyvinyl alcohol) was used and a hydrophilic polymer, to make the egg-shaped coat, filled Metformin HCl in the core, and the formulation also fulfilled the F.D.A. requirement of drug release [89].

7 Advantages of 3D Printing in the Pharma Industry

The use of 3D printing technology in the pharma industry shows various advantages. It can be used to fabricate site-specific dosage forms, e.g., colon targeting. The tablet ingredient suffers from oxidation problems when broken down. 3D printing technology overcomes this problem by using the micro-dosing technique. With the help of 3D printing techniques, the solubility of various insoluble drugs can also be increased, e.g., thermal inkjet printing enhances the solubility of miconazole [90], and Zip dosing is also used to improve the disintegration behaviour of various drugs [91]. The 3D printing technology can also be utilized for a rapid clinical trial. It also provides non-contact processing of pharmaceuticals with high accuracy and reproducibility [92].

3D printing has shown great success in manufacturing various dosage forms and prosthetics or artificial organs. Many researchers believe that 3D printing could also play a crucial role in the future of pharmacy. A hypothesis is developed which defines the fabrication of a customized 'polypill' that could potentially include all the drugs in a single pill. The only resistance to this hypothesis's success is the local pharmacy level's regulation and safety assurance. Some big players in 3D printed drug manufacturing include Aprecia Pharmaceuticals, Hewlett Packard Caribe, B.V., L.L.C., Galaxo SmithKline Plc, Rab RX Ltd, Cycle Pharmaceuticals, and others. Aprecia Pharmaceuticals was the first pharmaceutical industry to manufacture the tablet through the 3D printing technique and was approved by the F.D.A. in 2015 [93]. In 2017, Aprecia Pharmaceuticals and Cycle Pharmaceuticals made a partnership agreement to develop and sell several orphan drugs using 3D printing technology. Later after 2 years, one more tablet for Parkinson's disease was manufactured. Glaxo Smith Klein used inkjet 3D Printing and the U.V. curing technique to form the drug through 3D printing [94]. Due to economic reasons, several orphan drugs are not formulated by the pharmaceutical industry, so 3D printing technology promised to formulate the drug and treat rare diseases [95]—the U.S.F. Health Radiology and Northwell Health System developed a COVID-19 testing kit using 3D-printing technology. They designed testing 3 D printed nasopharyngeal swabs to meet the global demand for testing kits [96]. With the help of 3D printing, companies like Sanova can produce millions of hearing aid products annually [97]. Not just mass production, but using this technology, they can customize the product as per their customer's needs. It's been predicted by many research companies like Markets and Markets.com that the valuation of the market for 3D printed medical devices alone can reach up to $3.2 billion in the coming 5 years [98].

8 Conclusion and Future Scope

Industry 4.0 plays a pivotal role in manufacturing personalized dosage forms with the help of 3D printing technology. 3D printing technology uses software to create 3D objects. 3D printing technology has flexibility in the formulation and has a very innovative way of creating new medical products. 3D printing made manufacturing easy and customized the formulation according to the patient's requirements with enhanced stability, patient compliance, and cost-effectiveness. From the above discussion, we can conclude that the 3D printing technique can play an important role in personalized drug delivery systems and the making of various kinds of organs, prosthetics, and formulations, and in the upcoming years, we can expect a large amount of 3D printing based medical and pharmaceutical product in the market. 3D printing technology can easily transform the manufacturing technique and the manufacturing style of the pharmaceutical industry. As we know, with traditional methods, complex dosage forms cannot be formulated, whereas 3D printing technology made it easy to formulate highly advanced, complex dosage forms with small structures and shapes. 3D printing has several advantages over conventional manufacturing techniques. Still, some technological challenges restrict the commercialization of 3D printing-based products, including selecting suitable material that affects the stability or process reproducibility, product quality, and regulatory approvals that must be answered [73]. Researchers continuously trying to discover the answer to overcome these technological and regulatory problems. Industries are working to evolve the clinical trials and G.M.P. solutions in the near future [99]. In the future, 3D printing technology must be included in the syllabus of pharmacy education to train health care professionals for preparing medicine [100]. The global market for 3D printing-based pharmaceutical products is increasing daily; in the future, it is estimated to reach US$ 2064.8 million by 2027 [101]. The implication of industry 4.0 in 3D printing is required in a more aggressive way to address the big data, cyber security, cloud computing, A.I., blockchain, machine learning, internet of things, and deep learning to make the pharmaceutical industry into a smart or futuristic industry [102].

References

1. Nayyar, A., & Kumar, A.: A roadmap to industry 4.0: smart production, sharp business and sustainable development. Berlin: Springer (2020).
2. Andreadis, I.I., Gioumouxouzis, C.I., Eleftheriadis, G.K., Fatouros, D.G.: The Advent of a New Era in Digital Healthcare: A Role for 3D Printing Technologies in Drug Manufacturing? Pharmaceutics 14, (2022). https://doi.org/10.3390/pharmaceutics14030609.
3. Kumar, A., Nayyar, A.: si3-Industry: A Sustainable, Intelligent, Innovative, Internet-of-Things Industry, in: Nayyar, A., Kumar, A. (Eds.), A Roadmap to Industry 4.0: Smart Production, Sharp Business and Sustainable Development. Springer International Publishing, Cham, pp. 1–21 (2020) https://doi.org/10.1007/978-3-030-14544-6_1.
4. Jamróz, W., Szafraniec, J., Kurek, M., & Jachowicz, R.: 3D Printing in Pharmaceutical and Medical Applications – Recent Achievements and Challenges. Pharmaceutical Research, 35(9), 176 (2018). https://doi.org/10.1007/s11095-018-2454-x.
5. Hole, G., Hole, A.S., McFalone-Shaw, I.: Digitalization in pharmaceutical industry: What to focus on under the digital implementation process? International Journal of Pharmaceutics: X 3 (2021). 100095. https://doi.org/10.1016/j.ijpx.2021.100095.
6. Reinhardt, I.C., Oliveira, J., Ring, D.: Industry 4.0 and the Future of the Pharmaceutical Industry. Pharmaceutical Engineering 41 (2021).

7. Elbadawi, M., McCoubrey, L.E., Gavins, F.K.H., Ong, J.J., Goyanes, A., Gaisford, S., Basit, A.W.: Harnessing artificial intelligence for the next generation of 3D printed medicines. Advanced Drug Delivery Reviews 175, 113805 (2021). https://doi.org/10.1016/j.addr.2021.05.015.
8. Samiei, N.: Recent trends on applications of 3D printing technology on the design and manufacture of pharmaceutical oral formulation: A mini review. Beni-Suef University Journal of Basic and Applied Sciences, 9(1), 12 (2020). https://doi.org/10.1186/s43088-020-00040-4.
9. Ali, A., Ahmad, U., & Akhtar, J.: 3D Printing in Pharmaceutical Sector: An Overview. In Pharmaceutical Formulation Design—Recent Practices. IntechOpen. (2020). https://doi.org/10.5772/intechopen.90738.
10. Pucci, Josephine U., Christophe, Brandon R., Sisti, Jonathan A., Connolly, Edward S.: Three-dimensional printing: technologies, applications, and limitations in neurosurgery. Biotechnol. Adv. 35 (5), 521–529 (2017). 10.1016/j. biotechadv.2017.05.007.
11. Azad, M.A., Olawuni, D., Kimbell, G., Badruddoza, A.Z.M., Hossain, M.S., Sultana, T.: Polymers for extrusion-based 3D printing of pharmaceuticals: A holistic materials–process perspective. Pharmaceutics. 12(2), 124 (2020). https://doi.org/10.3390/pharmaceutics12020124.
12. Warsi, M.H., Yusuf, M., Al Robaian, M., Khan, M., Muheem, A., Khan, S.: 3D Printing Methods for Pharmaceutical Manufacturing: Opportunity and Challenges. Curr. Pharm. Des. 24 (42), 4949–4956 (2018). https://doi.org/10.2174/1381612825666181206121701.
13. Pandey, M., Choudhury, H., Fern, J. L. C., Kee, A. T. K., Kou, J., Jing, J. L. J., Her, H. C., Yong, H. S., Ming, H. C., Bhattamisra, S. K., & Gorain, B.: 3D printing for oral drug delivery: A new tool to customize drug delivery. Drug Delivery and Translational Research, 10(4), 986—1001 (2020). https://doi.org/10.1007/s13346-020-00737-0.
14. Wallis, M., Al-Dulimi, Z., Tan, D. K., Maniruzzaman, M., & Nokhodchi, A.: 3D printing for enhanced drug delivery: Current state-of-the-art and challenges. Drug Development and Industrial Pharmacy, 46(9), 1385–1401 (2020). https://doi.org/10.1080/03639045.2020.1801714.
15. Jose, P. A., & Gv, P. C.: 3D printing of pharmaceuticals – a potential technology in developing personalized medicine. *Asian Journal of Pharmaceutical Research and Development*, 6(3), 46–54 (2018). https://doi.org/10.22270/ajprd.v6i3.375.
16. Awad, A., Fina, F., Goyanes, A., Gaisford, S., & Basit, A. W.: 3D printing: Principles and pharmaceutical applications of selective laser sintering. International Journal of Pharmaceutics, 586, 119594 (2020). https://doi.org/10.1016/j.ijpharm.2020.119594.
17. Cui, X., Boland, T., D'Lima, D. D., & Lotz, M. K.: Thermal Inkjet Printing in Tissue Engineering and Regenerative Medicine. *Recent Patents on Drug Delivery & Formulation*, 6(2), 149–155 (2012).
18. Chen, A. (n.d.). Benefits vs drawbacks of 3D printing in the Pharmaceutical industry. Retrieved August 14, (2021). https://www.cmac.com.au/blog/benefits-drawbacks-3d-printing-pharmaceutical-industry.
19. Gujrati, A., Sharma, A., & Mahajan, S. C.: Review on Applications of 3D Printing in Pharmaceuticals. 25, 7 (2019).
20. Ponni, R., Swamivelmanickam, M., & Sivakrishnan, S. (2020). 3D Printing in Pharmaceutical Technology – A Review. International journal of pharmaceutical investigation, 10, 8–12.
21. Islam, R., & Sadhukhan, P.: An Insight of 3d Printing Technology in Pharmaceutical Development and Application: An Updated Review. Current trends in Pharmaceutical Research 7, 55–80 (2021).
22. Wong, S: 3D printing: Risks vs. benefits for the pharma industry (United Kingdom) [Text]. PharmaTimes; PharmaTimes Media Limited. (2018, May 29). http://www.pharmatimes.com/web_exclusives/3d_printing_risks_vs._benefits_for_the_pharma_industry_1237380.
23. Mwema, F. M., & Akinlabi, E. T.: Basics of Fused Deposition Modelling (FDM). Fused Deposition Modeling, 1–15 (2020). https://doi.org/10.1007/978-3-030-48259-6_1.

24. Melocchi, A., Parietti, F., Maroni, A., Foppoli, A., Gazzaniga, A., & Zema, L.: Hot-melt extruded filaments based on pharmaceutical grade polymers for 3D printing by fused deposition modeling. *International Journal of Pharmaceutics*, 509(1), 255–263 (2016). https://doi.org/10.1016/j.ijpharm.2016.05.036.
25. Gao, X., Yu, N., & Li, J.: Influence of printing parameters and filament quality on structure and properties of polymer composite components used in the fields of automotive. In K. Friedrich, R. Walter, C. Soutis, S. G. Advani, & I. H. B. Fiedler (Eds.), Structure and Properties of Additive Manufactured Polymer Components. Woodhead Publishing. pp. 303–330 (2020). https://doi.org/10.1016/B978-0-12-819535-2.00010-7.
26. Esposito Corcione, C., Gervaso, F., Scalera, F., Padmanabhan, S. K., Madaghiele, M., Montagna, F., Sannino, A., Licciulli, A., & Maffezzoli, A.: Highly loaded hydroxyapatite microsphere/P.L.A. porous scaffolds obtained by fused deposition modelling. *Ceramics International*, 45(2, Part B), 2803–2810 (2019). https://doi.org/10.1016/j.ceramint.2018.07.297.
27. Giri, B. R., Song, E. S., Kwon, J., Lee, J.-H., Park, J.-B., & Kim, D. W.: Fabrication of Intragastric Floating, Controlled Release 3D Printed Theophylline Tablets Using Hot-Melt Extrusion and Fused Deposition Modeling. Pharmaceutics, 12(1), 77(2020). https://doi.org/10.3390/pharmaceutics12010077.
28. Skowyra, J., Pietrzak, K., & Alhnan, M. A.: Fabrication of extended-release patient-tailored prednisolone tablets via fused deposition modelling (FDM) 3D printing. European Journal of Pharmaceutical Sciences: Official Journal of the European Federation for Pharmaceutical Sciences, 68, 11–17 (2015). https://doi.org/10.1016/j.ejps.2014.11.009.
29. Hussain, A., Mahmood, F., Arshad, M. S., Abbas, N., Qamar, N., Mudassir, J., Farhaj, S., Nirwan, J. S., & Ghori, M. U.: Personalised 3D Printed Fast-Dissolving Tablets for Managing Hypertensive Crisis: In-Vitro/In-Vivo Studies. *Polymers*, 12(12), 3057 (2020). https://doi.org/10.3390/polym12123057.
30. Goyanes, A., Chang, H., Sedough, D., Hatton, G., Wang, J., Buanz, A., Gaisford, S., & Basit, A.: Fabrication of controlled-release budesonide tablets via desktop (FDM) 3D printing. *International Journal of Pharmaceutics*, 496 (2015). https://doi.org/10.1016/j.ijpharm.2015.10.039.
31. Chai, X., Chai, H., Wang, X., Yang, J., Li, J., Zhao, Y., Cai, W., Tao, T., & Xiang, X.: Fused Deposition Modeling (FDM) 3D Printed Tablets for Intragastric Floating Delivery of Domperidone. *Scientific Reports*, 7(1), 2829 (2017). https://doi.org/10.1038/s41598-017-03097-x.
32. Thakkar, R., Pillai, A. R., Zhang, J., Zhang, Y., Kulkarni, V., & Maniruzzaman, M.: Novel On-Demand 3-Dimensional (3-D) Printed Tablets Using Fill Density as an Effective Release-Controlling Tool. Polymers, 12(9), 1872 (2020). https://doi.org/10.3390/polym12091872.
33. Goyanes, A., Det-Amornrat, U., Wang, J., Basit, A. W., & Gaisford, S.: 3D scanning and 3D printing as innovative technologies for fabricating personalized topical drug delivery systems. Journal of Controlled Release: Official Journal of the Controlled Release Society, 234, 41–48 (2016). https://doi.org/10.1016/j.jconrel.2016.05.034.
34. Beck, R. C. R., Chaves, P. S., Goyanes, A., Vukosavljevic, B., Buanz, A., Windbergs, M., Basit, A. W., & Gaisford, S.: 3D printed tablets loaded with polymeric nanocapsules: An innovative approach to produce customized drug delivery systems. International Journal of Pharmaceutics, 528(1), 268–279 (2017). https://doi.org/10.1016/j.ijpharm.2017.05.074.
35. Pagac, M., Hajnys, J., Ma, Q.P., Jancar, L., Jansa, J., Stefek, P., & Mesicek, J.: A Review of Vat Photopolymerization Technology: Materials, Applications, Challenges, and Future Trends of 3D Printing. Polymers, 13(4), 598 (2021). https://doi.org/10.3390/polym13040598.
36. Xu, X., Awad, A., Robles-Martinez, P., Gaisford, S., Goyanes, A., & Basit, A. W.: Vat photopolymerization 3D printing for advanced drug delivery and medical device applications. Journal of Controlled Release, *329*, 743–75 (2021). https://doi.org/10.1016/j.jconrel.2020.10.008.
37. Davoudinejad, A., Péreza, L. C. D., Quagliotti, D., Pedersen, D. B., Garcíaa, J. A. A., Yagüe-Fabra, J. A., & Tosello, G.: Geometric and Feature Size Design Effect on Vat Photopolymerization Micro Additively Manufactured Surface Deatures. Proceedings of the Joint Special

Interest Group Meeting: Additive Manufacturing. 2018 ASPE and euspen Summer Topical Meeting: Advancing Precision in Additive Manufacturing. (2018). https://orbit.dtu.dk/en/publications/geometric-and-fature-size-design-effect-on-vat-photopolymerizatio.

38. Rahman, Z., Barakh Ali, S. F., Ozkan, T., Charoo, N. A., Reddy, I. K., & Khan, M. A.: Additive Manufacturing with 3D Printing: Progress from Bench to Bedside. The AAPS Journal, 20(6), 101 (2018). https://doi.org/10.1208/s12248-018-0225-6.

39. Nguyen, A.K., Narayan, R.J.: Two-photon polymerization for biological applications. Materials Today 20, 314–322 (2017). https://doi.org/10.1016/j.mattod.2017.06.004.

40. Correa, D. S., De Boni, L., Otuka, A. J. G., Tribuzi, V., & Mendonça, C. R.: Two-Photon Polymerization Fabrication of Doped Microstructures. In Polymerization. (2012). https://doi.org/10.5772/36061.

41. Abaci, A., Gedeon, C., Kuna, A., & Guvendiren, M.: Additive Manufacturing of Oral Tablets: Technologies, Materials and Printed Tablets. Pharmaceutics, 13(2), 156 (2021). https://doi.org/10.3390/pharmaceutics13020156.

42. Wang, J., Goyanes, A., Gaisford, S., & Basit, A. W.: Stereolithographic (S.L.A.) 3D printing of oral modified-release dosage forms. *International Journal of Pharmaceutics*, *503*(1), 207–212 (2016). https://doi.org/10.1016/j.ijpharm.2016.03.016.

43. George, M., Aroom, K., Hawes, H. G., Gill, B. S., & Love, J.: 3D Printed Surgical Instruments – The Design and Fabrication Process. *World Journal of Surgery*, *41*(1), 314–319 (2017). https://doi.org/10.1007/s00268-016-3814-5.

44. Charoo, N. A., Barakh Ali, S. F., Mohamed, E. M., Kuttolamadom, M. A., Ozkan, T., Khan, M. A., & Rahman, Z.: Selective laser sintering 3D printing – an overview of the technology and pharmaceutical applications. Drug Development and Industrial Pharmacy, 46(6), 869–877(2020). https://doi.org/10.1080/03639045.2020.1764027.

45. Allahham, N., Fina, F., Marcuta, C., Kraschew, L., Mohr, W., Gaisford, S., Basit, A. W., & Goyanes, A.: Selective Laser Sintering 3D Printing of Orally Disintegrating Printlets Containing Ondansetron. Pharmaceutics, *12*(2), 110 (2020). https://doi.org/10.3390/pharmaceutics12020110

46. Salmoria, G., Vieira, F. E., Ghizoni, G. B., Marques, M. S., & Kanis, L.: 3D printing of P.C.L./Fluorouracil tablets by selective laser sintering: Properties of implantable drug delivery for cartilage cancer treatment (2017). https://doi.org/10.15761/ROM.1000121.

47. Gueche, Y. A., Sanchez-Ballester, N. M., Bataille, B., Aubert, A., Leclercq, L., Rossi, J.-C., & Soulairol, I.: Selective Laser Sintering of Solid Oral Dosage Forms with Copovidone and Paracetamol Using a CO2 Laser. Pharmaceutics, *13*(2), 160 (2021). https://doi.org/10.3390/pharmaceutics13020160.

48. Madzarevic, Medarevic, Vulovic, Sustersic, Djuris, Filipovic, & Ibric.: Optimization and Prediction of Ibuprofen Release from 3D D.L.P. Printlets Using Artificial Neural Networks. Pharmaceutics, 11(10), 544 (2019). https://doi.org/10.3390/pharmaceutics11100544.

49. Awad, A., Fina, F., Trenfield, S. J., Patel, P., Goyanes, A., Gaisford, S., & Basit, A. W.: 3D Printed Pellets (Miniprintlets): A Novel, Multi-Drug, Controlled Release Platform Technology. Pharmaceutics, 11(4), 148 (2019). https://doi.org/10.3390/pharmaceutics11040148.

50. Trenfield, S. J., Madla, C. M., Basit, A. W., & Gaisford, S.: Binder Jet Printing in Pharmaceutical Manufacturing. In A. W. Basit & S. Gaisford (Eds.), 3D Printing of Pharmaceuticals Springer International Publishing. pp. 41–54 (2018). https://doi.org/10.1007/978-3-319-90755-0_3.

51. Mostafaei, A., Elliott, A. M., Barnes, J. E., Li, F., Tan, W., Cramer, C. L., Nandwana, P., & Chmielus, M.: Binder jet 3D printing—Process parameters, materials, properties, modeling, and challenges. *Progress in Materials Science*, 119(C) (2021). https://doi.org/10.1016/j.pmatsci.2020.100707.

52. Kjar, A., & Huang, Y.: Application of Micro-Scale 3D Printing in Pharmaceutics. Pharmaceutics, 11 (8) 390. (2019). https://doi.org/10.3390/pharmaceutics11080390.

53. Huang, W., Zheng, Q., Sun, W., Xu, H., & Yang, X.: Levofloxacin implants with predefined microstructure fabricated by three-dimensional printing technique. International Journal of Pharmaceutics, 339(1–2), 33–38 (2007). https://doi.org/10.1016/j.ijpharm.2007.02.021.

54. Scoutaris, N., Alexander, M.R., Gellert, P.R., Roberts, C.J.: Inkjet printing as a novel medicine formulation technique. Journal of Controlled Release 156, 179–185 (2011). https://doi.org/10.1016/j.jconrel.2011.07.033.
55. Shi, K., Tan, D. K., Nokhodchi, A., & Maniruzzaman, M.: Drop-On-Powder 3D Printing of Tablets with an Anti-Cancer Drug, 5-Fluorouracil. Pharmaceutics, 11(4), 150 (2019). https://doi.org/10.3390/pharmaceutics11040150.
56. Radulescu, D., Trost, H. J., Taylor, D. T., Antohe, B., Silva, D., & Schwade, N. D.: 3D Printing of Biological Materials for Drug Delivery and Tissue Engineering Applications. Digital Fabrication, 4 (2005).
57. Daly, R., Harrington, T. S., Martin, G. D., & Hutchings, I. M.: Inkjet printing for pharmaceutics – A review of research and manufacturing. *International Journal of* Pharmaceutics, 494(2), 554–567(2015). https://doi.org/10.1016/j.ijpharm.2015.03.017.
58. Vakili, H., Nyman, J. O., Genina, N., Preis, M., & Sandler, N.: Application of a colorimetric technique in quality control for printed pediatric orodispersible drug delivery systems containing propranolol hydrochloride. *International Journal of* Pharmaceutics, 511(1), 606–618(2016). https://doi.org/10.1016/j.ijpharm.2016.07.032.
59. Zikulnig, J., & Kosel, J.: Flexible Printed Sensors—Overview of Fabrication Technologies. In Reference Module in Biomedical Sciences. Elsevier. (2021). https://doi.org/10.1016/B978-0-12-822548-6.00010-8.
60. Ozbolat, I. T.: Bioprinter Technologies With contributions by Hemanth Gudupati and Kazim Moncal, The Pennsylvania State University. In I. T. Ozbolat (Ed.), *3D Bioprinting* Academic Press. pp. 199–241). (2017). 7 https://doi.org/10.1016/B978-0-12-803010-3.00007-X.
61. Vithani, K., Goyanes, A., Jannin, V., Basit, A. W., Gaisford, S., & Boyd, B. J.: An Overview of 3D Printing Technologies for Soft Materials and Potential Opportunities for Lipid-based Drug Delivery Systems. Pharmaceutical Research, 36(1), 4 (2019). https://doi.org/10.1007/s11095-018-2531-1.
62. Buanz, A. B. M., Saunders, M. H., Basit, A. W., & Gaisford, S.: Preparation of personalized-dose salbutamol sulphate oral films with thermal inkjet printing. Pharmaceutical Research, 28(10), 2386–2392 (2011). https://doi.org/10.1007/s11095-011-0450-5.
63. Thabet, Y., Lunter, D., & Breitkreutz, J.: Continuous inkjet printing of enalapril maleate onto orodispersible film formulations. International Journal of Pharmaceutics, 546(1–2), 180–187 (2018). https://doi.org/10.1016/j.ijpharm.2018.04.064.
64. Varan, C., Wickström, H., Sandler, N., Aktaş, Y., & Bilensoy, E.: Inkjet printing of antiviral P.C.L. nanoparticles and anticancer cyclodextrin inclusion complexes on bioadhesive film for cervical administration. International Journal of Pharmaceutics, *531*(2), 701–713 (2017). https://doi.org/10.1016/j.ijpharm.2017.04.036.
65. Sandler, N., Määttänen, A., Ihalainen, P., Kronberg, L., Meierjohann, A., Viitala, T., & Peltonen, J.: Inkjet printing of drug substances and use of porous substrates-towards individualized dosing. In Journal of Pharmaceutical Sciences (Vol. 100). (2012).
66. Alomari, M., Vuddanda, P. R., Trenfield, S. J., Dodoo, C. C., Velaga, S., Basit, A. W., & Gaisford, S.: Printing T3 and T4 oral drug combinations as a novel strategy for hypothyroidism. International Journal of Pharmaceutics, 549(1), 363–369 (2018). https://doi.org/10.1016/j.ijpharm.2018.07.062.
67. Eleftheriadis, G. K., Katsiotis, C. S., Andreadis, D. A., Tzetzis, D., Ritzoulis, C., Bouropoulos, N., Kanellopoulou, D., Andriotis, E. G., Tsibouklis, J., & Fatouros, D. G.: Inkjet printing of a thermolabile model drug onto FDM-printed substrates: Formulation and evaluation. Drug Development and Industrial Pharmacy, 46(8), 1253–1264 (2020). https://doi.org/10.1080/03639045.2020.1788062.
68. Palo, M., Kogermann, K., Laidmäe, I., Meos, A., Preis, M., Heinämäki, J., & Sandler, N.: Development of Oromucosal Dosage Forms by Combining Electrospinning and Inkjet Printing. Molecular Pharmaceutics, 14(3), 808–820 (2017). https://doi.org/10.1021/acs.molpharmaceut.6b01054.
69. Arshad, M. S., Shahzad, A., Abbas, N., AlAsiri, A., Hussain, A., Kucuk, I., Chang, M.-W., Bukhari, N. I., & Ahmad, Z.: Preparation and characterization of indomethacin loaded films

by piezoelectric inkjet printing: A personalized medication approach. Pharmaceutical Development and Technology, 25(2), 197–205 (2020). https://doi.org/10.1080/10837450.2019.1684520.
70. Ashish, Ahmad, N., Gopinath, P., & Vinogradov, A.: Chapter 1 - 3D Printing in Medicine: Current Challenges and Potential Applications. In N. Ahmad, P. Gopinath, & R. Dutta (Eds.), 3D Printing Technology in Nanomedicine (pp. 1–22 (2019). Elsevier. https://doi.org/10.1016/B978-0-12-815890-6.00001-3.
71. Majumdar, J. D., Madapana, D., & Manna, I.: 3-D Printing by Laser-Assisted Direct Energy Deposition (LDED): The Present Status. Transactions of the Indian National Academy of Engineering (2021). https://doi.org/10.1007/s41403-021-00252-9.
72. Ngo, T.D., Kashani, A., Imbalzano, G., Nguyen, K.T.Q., Hui, D.: Additive manufacturing (3D printing): A review of materials, methods, applications and challenges. Composites Part B: Engineering 143, 172–196 (2018). https://doi.org/10.1016/j.compositesb.2018.02.012.
73. Park, B. J., Choi, H. J., Moon, S. J., Kim, S. J., Bajracharya, R., Min, J. Y., & Han, H.-K.: Pharmaceutical applications of 3D printing technology: Current understanding and future perspectives. Journal of Pharmaceutical Investigation, 49(6), 575–585 (2019). https://doi.org/10.1007/s40005-018-00414-y.
74. Zuniga, J., Katsavelis, D., Peck, J., Stollberg, J., Petrykowski, M., Carson, A., & Fernandez, C.: Cyborg beast: A low-cost 3d-printed prosthetic hand for children with upper-limb differences. B.M.C. Research Notes, 8, 10 (2015). https://doi.org/10.1186/s13104-015-0971-9.
75. Dodziuk, H. (2016). Applications of 3D printing in healthcare. *Kardiochirurgia i Torakochirurgia Polska Polish Journal of Cardio-Thoracic Surgery, 13*(3), 283–293. https://doi.org/10.5114/kitp.2016.62625.
76. Tunchel, S., Blay, A., Kolerman, R., Mijiritsky, E., & Shibli, J. A.: 3D Printing/Additive Manufacturing Single Titanium Dental Implants: A Prospective Multicenter Study with 3 Years of Follow-Up. *International Journal of Dentistry, 2016*, e8590971(2016). https://doi.org/10.1155/2016/8590971.
77. Ventola, C. L.: Medical Applications for 3D Printing: Current and Projected Uses. Pharmacy and Therapeutics, 39(10), 704–711 (2014).
78. Lewis, A., Koukoura, A., Tsianos, G.-I., Gargavanis, A. A., Nielsen, A. A., & Vassiliadis, E.: Organ donation in the U.S. and Europe: The supply vs demand imbalance. Transplantation Reviews, 35(2), 100585 (2021). https://doi.org/10.1016/j.trre.2020.100585.
79. Agarwal, S., Saha, S., Balla, V.K., Pal, A., Barui, A., Bodhak, S.: Current Developments in 3D Bioprinting for Tissue and Organ Regeneration–A Review. Frontiers in Mechanical Engineering 6 (2020). https://doi.org/10.3389/fmech.2020.589171.
80. Kim, I. G., Park, S. A., Lee, S.-H., Choi, J. S., Cho, H., Lee, S. J., Kwon, Y.-W., & Kwon, S. K.: Transplantation of a 3D-printed tracheal graft combined with iPS cell-derived M.S.C.s and chondrocytes. Scientific Reports, 10(1), 4326(2020). https://doi.org/10.1038/s41598-020-61405-4.
81. Solanki, N. G., Tahsin, M., Shah, A. V., & Serajuddin, A. T. M.: Formulation of 3D Printed Tablet for Rapid Drug Release by Fused Deposition Modeling: Screening Polymers for Drug Release, Drug-Polymer Miscibility and Printability. Journal of Pharmaceutical Sciences, 107(1), 390–401(2018). https://doi.org/10.1016/j.xphs.2017.10.021.
82. Scoutaris, N., Ross, S.A., Douroumis, D.: 3D Printed "Starmix" Drug Loaded Dosage Forms for Paediatric Applications. Pharm Res 35, 34 (2018). https://doi.org/10.1007/s11095-017-2284-2.
83. Mansor, N. H. A., Markom, M. A., Tan, E. S. M. M., & Adom, A. H.: Design and Fabrication of Biodegradable Microneedle Using 3D Rapid Prototyping Printer. Journal of Physics: Conference Series, 1372, 012053 (2019). https://doi.org/10.1088/1742-6596/1372/1/012053.
84. Johnson, A. R., & Procopio, A. T.: Low-cost additive manufacturing of microneedle masters. 3D Printing in Medicine, 5(1), 2 (2019). https://doi.org/10.1186/s41205-019-0039-x.
85. Azizoğlu, E., Özer, Ö.: Fabrication of Montelukast sodium-loaded filaments and 3D printing transdermal patches onto packaging material. International Journal of Pharmaceutics 587, 119588 (2020). https://doi.org/10.1016/j.ijpharm.2020.119588.

86. Andriotis, E. G., Eleftheriadis, G. K., Karavasili, C., & Fatouros, D. G.: Development of Bio-Active Patches Based on Pectin for the Treatment of Ulcers and Wounds Using 3D-Bioprinting technology. *Pharmaceutics*, *12*(1), 56 (2020). https://doi.org/10.3390/pharmaceutics12010056.
87. Montenegro-Nicolini, M., Miranda, V., Morales, J.O.: Inkjet Printing of Proteins: an Experimental Approach. The AAPS Journal 19, 234–243 (2017). https://doi.org/10.1208/s12248-016-9997-8.
88. Turk, D.C., O'Connor, A.B., Dworkin, R.H., Chaudhry, A., Katz, N.P., Adams, E.H., Brownstein, J.S., Comer, S.D., Dart, R., Dasgupta, N., Denisco, R.A., Klein, M., Leiderman, D.B., Lubran, R., Rappaport, B.A., Zacny, J.P., Ahdieh, H., Burke, L.B., Cowan, P., Jacobs, P., Malamut, R., Markman, J., Michna, E., Palmer, P., Peirce-Sandner, S., Potter, J.S., Raja, S.N., Rauschkolb, C., Roland, C.L., Webster, L.R., Weiss, R.D., Wolf, K.: Research design considerations for clinical studies of abuse-deterrent opioid analgesics: IMMPACT recommendations. Pain 153(10), 1997–2008 (2012). https://doi.org/10.1016/j.pain.2012.05.029.
89. Nukala, P.K., Palekar, S., Patki, M., Patel, K.: Abuse Deterrent Immediate Release Egg-Shaped Tablet (Egglets) Using 3D Printing Technology: Quality by Design to Optimize Drug Release and Extraction. AAPS PharmSciTech. 20(2), 80 (2019a). https://doi.org/10.1208/s12249-019-1298-y.
90. Boehm, R.D., Miller, P.R., Daniels, J., Stafslien, S., Narayan, R.J.: Inkjet printing for pharmaceutical applications. Materials Today 17, 247–252 (2014) https://doi.org/10.1016/j.mattod.2014.04.027.
91. Zip Dose® Technology [WWW Document], n.d. URL https://www.pharmaceuticalonline.com/doc/zipdose-technology-0001 (accessed 11.20.21).
92. Overview of Pharmaceutical 3D printing, Pharma Excipients. URL https://www.pharmaexcipients.com/pharmaceutical-3dp-overview (n.d.) (accessed 11.20.21).
93. Osouli-Bostanabad, K., & Adibkia, K.: Made-on-demand, complex and personalized 3D-printed drug products. BioImpacts : BI, 8(2), 77–79 (2018). https://doi.org/10.15171/bi.2018.09.
94. Martin J. 3D Printing In The Pharmaceutical Industry — Where Does It Currently Stand? https://www.pharmaceuticalonline.com/doc/3d-printing-in-the-pharmaceutical-industry-where-does-it-currently-stand-0002.
95. Brewer, G. J.: Drug development for orphan diseases in the context of personalized medicine. Translational Research: The Journal of Laboratory and Clinical Medicine, 154(6), 314–322 (2009). https://doi.org/10.1016/j.trsl.2009.03.008.
96. Ford, J., Goldstein, T., Trahan, S. et al.: A 3D-printed nasopharyngeal swab for COVID-19 diagnostic testing. 3D Print Med 6, 21 (2020). https://doi.org/10.1186/s41205-020-00076-3.
97. https://3dprintingindustry.com/news/ucl-researchers-fabricate-3d-printed-antibacterial-hearing-aids-to-prevent-ear-infections-178558/.
98. https://www.prnewswire.com/news-releases/global-3d-printing-medical-devices-market-to-reach-3-2-billion-by-2026--301323882.html.
99. Huels, C.: 3D printing or Additive Manufacturing of Tablets. Presentation presented at PHARMAP 2021 2021; Virtual (2021)].
100. Melnyk, L.A., Oyewumi, M.O.: Integration of 3D printing technology in pharmaceutical compounding: Progress, prospects, and challenges. Annals of 3D Printed Medicine 4, 100035 (2021). https://doi.org/10.1016/j.stlm.2021.100035.
101. Mohapatra, S., Kar, R.K., Biswal, P.K., Bindhani, S.: Approaches of 3D printing in current drug delivery. Sensors International 3, 100146 (2022). https://doi.org/10.1016/j.sintl.2021.100146.
102. The Importance of 3D Printing in Industry 4.0. https://www.3dnatives.com/en/3d-printing-in-industry-4-0-150220215.

3D Printing: A Game Changer for Indian MSME Sector in Industry 4.0

Nidhi U. Argade and Hirak Mazumdar

Abbreviations

ABS	Acrylonitrile Butadiene Styrene
AI	Artificial Intelligence
AM	Additive Manufacturing
CAGR	Compound Annual Growth Rate
CNC	Computer Numerical Control
CAD	Computer Aided Design
CDPL	Continuous Digital Light Processing
DED	Directed Energy Deposition
DLP	Digital Light Processing
EBM	Electron Beam Melting
FGMs	Functionally Graded Materials
FDM	Fused Deposition Modelling
GeM	Government e-marketplace
LOM	Laminated Object Manufacturing
ML	Machine Learning
MSME	Micro Small and Medium Enterprises
PLA	Polylactic Acid
PP	Polypropylene

N. U. Argade (✉)
Department of Accounting and Financial Management, Faculty of Commerce, The Maharaja Sayajirao University of Baroda, Vadodara, Gujarat, India
e-mail: nidhiu.argade-afm@msubaroda.ac.in

H. Mazumdar
College of Natural & Applied Science, Department of Computer Science, University of Houston-Victoria, 3007 N. Ben Wilson St, Victoria, TX 77901, USA
e-mail: mazumdarh@uhv.edu

© The Author(s), under exclusive license to Springer Nature Switzerland AG 2023
A. Nayyar et al. (eds.), *New Horizons for Industry 4.0 in Modern Business*, Contributions to Environmental Sciences & Innovative Business Technology,
https://doi.org/10.1007/978-3-031-20443-2_14

SAF	Selective Absorption Fusion
SLM	Selective Laser Melting
SHS	Selective Heat Sintering
SLS	Selective Laser Sintering
SLA	Stereo Lithography
TC	Technology Centres
UAM	Ultrasound Additive Manufacturing

1   Introduction

Energy Technology by its very nature possesses capacity to make things appealing while maintaining its utility. It is that subset of knowledge that take efforts from better to its best regime. The intention of a mankind may decide its destiny as to whether technology will have value addition it terms of innovation, comfort, utility or otherwise may be fatal by not using it for a noble cause and using the same destructively. The recent spread of new dimension of technological innovation is in the form of Digital Fabrication Technology which is also known as 3-D Printing or Additive Manufacturing (AM). It mainly used in creating physical parts through a geometrical image by adding material in layers, that may have appealing demonstration in final outcome of creations, in general, may be in objects. The recently surfaced version of 3D printing is a technology that has grown up with the passage of time and given vivid shapes in past couple of decades. It is an apt illustration of changing dimension of knowledge which obsoletes faster than anything else on the earth, and is changing and growing at an astonishing speed. Additive Manufacturing, with its flexibility, plays a critical role in Industry 4.0, saving time and cost, determining process efficiency and decreasing complexity, and enabling for quick prototyping and highly decentralised production processes [1].

The historical perspective of the said revolutionary journey is not too old, in general, but has gone through different layers of innovations since its inception. The journey of 3D printing started in 1983, when Charles Hull, popularly known as 'Chuck' developed Stereo Lithography Apparatus (SLA-1). In 1986, the first patent of this system was granted to him and first 3D printer was realised on commercial basis by, 3D Systems Corporation, company of Charles Hull. According to the author, this process is a system which generates 3D objects through developing cross-sectional design of objects. The author could give shape to the innovation, through creativity and untiring efforts, with which he could register around 60 U.S. patents in rapid Prototyping and ion optics. Traditionally, in late 1980s, the conventional technology was reckoned suitable for developing aesthetic prototypes known as 'Rapid Prototyping'. In the beginning of 21st century, the term Additive Manufacturing was regarded as a synonym for 3D printing as it was possible to have addition of the material in layers to give shape to 3D object in more appealing manner. Since it has inherent precision in terms of giving desirable output on account of capacity to use wide range of material, 3D printing becomes popular and scaled reasonable heights in 2019–20. In the recent times, it is considered quite apt and relevant in industrial production, as it has enriched versatility. The most mesmerising advantage of this technology is its strength to develop composite shapes as well as geometric designs which were not possible to produce otherwise manually, together with hollow objects and shapes designed to reduce weight through internal truss structure. Furthermore, it has added glory to the field by converting mass production in to mass customisation, meeting need and aspiration of individual customers. This unique aspect has unfolded immense room to have customised marketing and thereby almost made ordinary type of competition completely irrelevant.

The initial purchase and installation involve capital but later it could be easily recovered through reduction of wastage and time saving as the technology is very cost effective. The technology of 3D printing has evolved from the drawings of Computer Aided Design (CAD), which uses the technology which does fabrication layer after layer for 3D structures. 3D printing is an innovative method which has emerged as a very important step in designing technology. It unlocks various new opportunities and possibilities for usage of metal as well as many other materials for 3D printing. It will revolutionise industries and improve the production line by enhancing the production speed and lowering the cost of the same. Due to accessibility of production tools and formation of a supportive online community that shares their work publicly, customers increasingly have the power to create and make the products they use, many of which are advanced technology and design, as a result of these two influences [2]. The adoption rate of consumer 3d printing is increasing day by day as customers are more like to begin producing product at home [3]. The consumer's say will increase as he can give more input and specifications for a final product and have it according to his requirement. The use of 3D printing by Reebok for its sport shoes manufacturing is the latest and best example of the same. Establishing the 3D printing facilities near consumer will ensure responsive manufacturing process, better quality control as well as reduction of global transportation cost. The existing state of logistics in industries will change dramatically as a result of the adoption of 3D printing technology, as they will be able to provide comprehensive and end-to-end services more effectively.

Currently, 3D printing is used for manufacturing, (i) customised parts, fixtures and other tools in automobile industry, (ii) for production of artificial heart pump and 3D printed cornea in medical industry, (iii) for fine, nicely carved and customised jewellery, (iv) in aerospace industry with reducing the weight of different parts and PGA engine of rockets, (v) Amsterdam's steel bridge, (vi) for education industry by helping the students to create prototypes without expensive tolls and various products associated to aviation and food industry [4]. (vii) 3D printed jellyfish robots to monitor and protect fragile coral reefs [5]. There are enormous uses and application of 3D printing across various industries and sectors, which we will discuss further in the chapter.

In 2020, The global 3D printing market was valued USD 13.7 billion and it is expected to increase to USD 63.46 billion by 2026, with a CAGR of 29.48% during 2021–2026. 3D printing which is also known as additive manufacturing, is an object-creation technique that allows for the development, design, and performance of one-of-a-kind structural shapes, construction systems as well as use of materials. It is a more innovative, quick, and adaptable method of product creation and production.

The change shall always produce disequilibrium, obstacles and assuming great challenges before it offers its utility. Technological innovations have always resulted into reducing human efforts and enriching comfort zone. The adoption of AM on wider scale shall have great apprehension on adverse impact to use of labour and there by threat to existing manual employment. It may result in to making existing skill set redundant and thereby surfacing adverse consequences on significant component of economy in general and developing economies in particular [6]. Since 3D printing technology may be made applicable to any types of objects including those that may endanger human life, for example dangerous items like knives or guns, its use will always remain objectionable. Furthermore, the set of knowledge involve in 3D printing technique is very simple and easy to learn, just sketching and generating 3D objects in machine printing, that has great potential of being misused by anti-social elements, including criminals and terrorists. The state would be facing a challenge to restrict its access to those who have malafide intensions for its destructive use. The stage of development of this innovation is that of infant and therefore to have its whispered spread a long path of developing infrastructural facilities to be passed through which may require deploying exorbitant number of resources.

1.1 3D Printing Terminology

As the use of digital printing technology grows around the world, new terminologies for AM are emerging. With fast advancements in the world of 3D printing, standardisation of these terminologies is essential:

(i) **Mass Customizing:** This is a manufacturing and marketing method which combines customisation and personalisation for individual customers with the inherent benefit of cost savings associated with mass production.
(ii) **Rapid Prototyping:** It is amongst the initial terms used for describing the layer-upon-layer approach of creating 3D prototypes. The existing technologies were allowing to create one prototype, which can be used for mass production at a later stage. Even the changes in final prototype were difficult to make in those technologies. 3D printing allowed rapid prototyping at faster speed and cheaper cost.
(iii) **Customised Manufacturing:** often known as desktop manufacturing, this is the process of designing and manufacturing physical goods using a computer.
(iv) **Rapid Tooling:** Later it was manifested that the additive manufacturing technology could be used to make more than just prototypes, such as moulds, matrices, and tools, the term 'rapid tooling' was coined to distinguish it from rapid prototyping.

(v) **3D Printing:** The term refers to the creation of three-dimensional objects. This is the most widely used term in recent days. When we use printers that are affordable to residential or semi-professional users, the term "low-cost 3D printing" is frequently used.
(vi) **Flexible Fabrication:** It is a group of manufacturing techniques that allow parts to be made without the use of part-specific tooling. The component is being created as a digital model. It's computationally sliced, and the layer data is transferred to a fabricator, who then recreates the same in actual material.
(vii) **Additive Manufacturing:** It refers to the process of creating something with addition of material layer by layer, this is the latest terminology to be used and it refers to this technology as a whole. This technology was initiated from obtaining prototypes for mass manufacturing and later applied for mass customisation of various products.

Objectives of the Chapter

The objectives of the Chapter are as follows:

- To provide basic understanding of 3D printing including background, types of technologies, materials and process used in 3D printing,
- To elaborate Global and Indian Scenario of 3D Printing industry,
- To highlight Additive Manufacturing for Indian MSME sector,
- And, to illustrate applications of 3D printing technology during COVID-19 pandemic, and future challenges for Additive Manufacturing in India.

Organization of Chapter

The chapter is organized as: Sect. 2 enlightens background of Additive Manufacturing in terms of History, Types, Technologies, Material used and Product cycle. Section 3 highlights diverse applications of Additive manufacturing in various sectors. Section 4 enlists recent developments and key players. Section 5 gives overview of Additive Manufacturing as Game Changer for Indian MSME Sector. Section 6 gives a detailed view of contributions of Additive Manufacturing during COVID-19 pandemic. Section 7 enlists perspective contributions and operational constraints. Section 8 concludes the chapter with future scope.

2 Background of Additive Manufacturing

This section highlights history, types, technologies, material used and product cycle of Additive Manufacturing.

2.1 History of Printing in General and 3D Printing in Specific: (Timeline of Growth of Additive Manufacturing) [7, 8]

Various different additive manufacturing technologies are developed these days and they facilitate production of prototypes as well as final products. The timeline of printing technology is very interesting to study. These technologies were totally different from each other in terms of its manifestation, product they produce, principle used but they surprisingly share many functional commonalities. To develop the understanding of the subject from its origin, the chronological study of development of printing technology is carried out. This chronological study is based on the dates of publishing of works, the filing of patent applications, and the acceptance of these patents, keeping in mind that the dates of development are always anterior to those of a public character.

Printing technology's roots are grounded way back in (AD 200) with invention of wood block printing technology on China, used initially for cloth printing and later used for book and other texts. The next one is movable type (AD 1040) printing which uses movable components. The next in the sequence came after a huge gap, in around 1430, called intaglio representing group of printing and printing technologies followed by used of printing press in 1440, which was a mechanical device transferring ink to a print medium through applying pressure on it. Etching is a traditional technique developed in 1515 which is used still today. For creating designs incised in the metal, it uses acid and other chemicals to cut in to parts of metal surface. Relief printing which was used in 1690, where a block or matrix is bought in contact with surface of paper, which had ink applied on its recessed surface. Mezotint (1642), Aquatint (1772) also belong to intaglio family where image is incised into a surface and the incised line holds the ink, it is exactly opposite to relief printing. In 1796, German author Senefelder invented and used cheap method of printing—Lithography which is based on the immiscibility of oil and water, and printing is done on a smooth-surfaced stone and metal plate. Oleograph also known as Chromolithography was stemmed from lithography, it is most successful type of colour printing developed by 19th century. Rotary press, Hectograph, Offset printing, Hot metal type pressing, Mimeograph, Daisy wheel printing are the various printing techniques developed in 18th century. The first photocopy machine was developed by Rectigraph company in early 19th century, 'photostat' name was the trademark of the company which later genericized. The other techniques developed in 19th century for printing were Screen printing, Spirit duplicator—which was commonly used for rest of the 20th century used alcohol as a major component of solvent which were used as ink, Dot matrix printing is computerized printing which has print mechanism like line printer or typewriter using relatively low-resolution dot-matrix ink is applied to the printing surface. Xerography was invented in 1942 is a dry technique of photocopying which does not use chemicals. Today xerography technique is used in photocopy machines, digital presses and laser printers. Inkjet is most widely used printer today was invested in early 1950s. It's also a sort of computer printing that uses droplets of ink to generate a digital image on paper. Invented by Xerox in around 1970s Laser printing uses laser beam which repeatedly pass back and forth over a negatively charged cylinder called a drum and produces high quality graphics as well as text.

Finally, on March 9, 1983, Charles W Hull printed a teacup using the first additive manufacturing machine which he developed himself, the Stereolithography Apparatus (SLA-1). Earlier then, the technique of 3D printing was used mainly for developing functional prototypes and so the appropriate term was 'rapid prototyping'.

From then on, the range of material, degree of precision and repeatability has increased so now a days it is suitable for industrial production. In 1986, Carl Deckard, developed a method of 'Selective Laser Sintering' (SLS), which was an initial stage in development of additive manufacturing through it, as during that time all metal work was done by non-additive methods like casting, fabrication etc. In 1988, Michael Feygin and his team developed Laminated Manufacturing for producing integral parts through layers which then will form as prototype. Scott Crump developed the additive manufacturing technique through Fused Deposition Modelling (FDM) in 1988. Emanuel Sachs and his team created 3D printing techniques in 1989 by employing the injectors of a conventional ink-jet printer to inject binding agent and coloured ink onto a bed of powdered material. In mid 1990s, the new techniques including micro-casting and metal spraying, for metal disposition was developed. the research and development in the area of Additive manufacturing is very vibrant in past decade as AM has started responding to the needs of many industries across the globe. In 2014, Benjamin Cook and Manos demonstrated VIPRE, which allows 3D printers of functional electronics to work up to 40 GHz, similar to how Filabot developed a technique by closing a loop that allows the 3D printer to print with a wide spectrum of plastics in 2012. It is widely used technique of 3D printing these days is FDM as it is less expensive than SLA and SLS.

2.2 Types of 3D Printing

Various techniques of 3D printing are developed over a passage of time, each with its own set of capabilities. The American Society of Testing and materials (ASTM) Standard F42 [9] classifies main seven 3D printing processes which are (i) Vat Photopolymerization, (ii) Material Extrusion, (iii) Material Jetting, (iv) Binding Jetting, (v) Powder Bed Fusion, (vi) Sheet Lamination, and (vii) Directed Energy Deposition, [8, 10]. Though the term additive manufacturing is used to group all the technologies, each technology is different in terms of machines & methods as well as materials used for it.

There are no disagreements over whether equipment or technology performs better because each has its own set of uses. In today's era 3D printing technology has its application much beyond the prototyping. For example, Companies generally use the low-cost online model to market extrusion-based printers. Manufacturers of sinter or stereolithography printers, on the other hand, are more likely to use the technology export pattern. Therefore, the technology appears to have an impact on the selection of business models used for commercialization of technology [11].

Vat Photopolymerization—The most widely used 3D printing process is the curing of photo-reactive polymers with a laser, projector, LED or UV Light. One of the most populat technique of 3D printing—Stereo Lithography (SLA) belongs to this category. The other techniques are Digital Light Processing (DLP) and Continuous Digital Light Processing (CDLP). The material used in this method are plastics and UV curable Photopolymer Resins. Photopolymerization is ideal for creating a high-end product with fine details and a high-quality finish [14].

Material Extrusion—In material extrusion method, the material is deposited through a nozzle which can move vertically and horizontally. Fused Deposition Modelling (FDM) is extensively utilised in the industry and its expenditures are minimal. Furthermore, this method can provide completely functional product parts. The materials used in this method are plastics, polymers and composites.

Material Jetting—This 3D printing technique uses same technology as that of ink jet printer. The material is deposited on either drop by drop (DOD) or continuous bases and the nozzle can only move horizontally. The material is then solidify using ultraviolet (UV) light. The other methods of material jetting are Nano Particle Jetting and Drop by demand. It uses plastic, metal, polymer and wax materials [15].

Binder Jetting—This technique requires powder based material as well as a binder. Binder jetting method forms the layer by spraying a chemical binder onto the distributed powder. It uses materials like Gypsum, Sand, metals, polymers, ceramics. Binder jetting is also easy, quick, and inexpensive because powder particles are fused together. It has significant characteristic of printing products which are large.

Sheet Lamination—It includes Laminated Object Manufacturing (LOM) and Ultrasound Additive Manufacturing (UAM). It is a technique of 3D printing that contains bonding of sheets of materials together to create a component. UAM uses sheets of metals which are bounded together using ultra sonic welding and LOM uses paper and adhesive material instead of metal and welding. Sheet lamination has the advantages of being able to print in full colour, being reasonably economical, easy material handling, and the ability to recycle extra material.

Powder Bed Fusion—The 3D printing techniques used in power bed fusion are Selective Heat Sintering (SHS), Selective Laser Sintering (SLS), Direct Material Laser Sintering (DMLS), and Electron Beam Melting (EBM). In this technique, material powder is melted together using an electron beam or a laser. To melt thermoplastic material, SHS technology, on the other hand, uses a head thermal print and generate 3D printed objects. It uses materials like plastics and metals, mainly powder based. SLS is a 3D printing process that works rapidly, accurately, and has a variety of surface finishes.

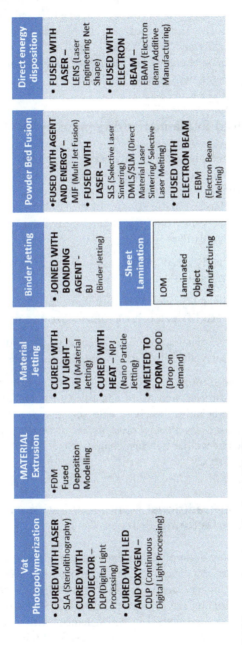

Fig. 1 Additive manufacturing technologies [12, 13]

Directed Energy Deposition (DED)—This technique is widely used for repairing or adding material to existing components. It has grain like structure control of high degree that create good object quality. In concept, the technique is similar to material extrusion, only the nozzle isn't restricted to one axis and can move in any direction. Although the process can be used with ceramics and polymers, it is most typically utilised with metals and metal-based hybrids in wire or powder form. The techniques used in this technology are Electronic Beam Laser Manufacturing (EBLM) and laser engineered net shaping (LENS).

2.3 Materials Used in Manufacturing Industry for 3D Printing

To manufacture consistent high-quality products, 3D printing, like any other manufacturing process, requires superior-quality materials that meets the specifications consistently. Procedures, requirements, and material control agreements are formed between traders, buyers, and end-users of materials to achieve the same. 3D printing technology can create fully functioning parts from various materials. This technology uses materials like:

Nylon—The most commonly used plastic material as it provides flexibility, durability and no fear of corrosion with it.

ABS (Acrylonitrile Butadiene Styrene)—This is a thermoplastic material which is easily available, low cost and lifespan is higher than other materials like nylon.

Resin—Though it is costly, many types of resin materials are widely used in 3D printing technologies. It has wide range of uses and chemical resistant.

Gold and Silver—The 3D printing has expanded its horizons from manufacturing industry to jewellery industry as well. Gold and silver materials are mainly used in this industry to prepare beautiful jewellery with very fine designs.

The other materials used are PLA (Polylactic Acid), various polymer, wide range of metal, ceramics as well as hybrids, composites, and Functionally Graded Materials (FGMs) [16] (Fig. 2).

2.4 Product Cycle: Traditional Manufacturing Process Versus Additive Manufacturing Process

This section compares the traditional product cycle with product cycle under additive manufacturing process. The traditional manufacturing techniques can be clubbed together into four main categories: Injection Moulding, CNC Machining, Plastic Forming, Plastic Joining. To begin with, CNC (Computer Numerical Control) machining is a production method in which industrial tools and machinery are controlled by pre-programmed computer software. Three-dimensional cutting tasks can be completed in a single set of instructions with CNC machining. Injection Moulding necessitates the use of a tool called a mould. It's a manufacturing

Fig. 2 Materials used in 3D printing technology [compiled by author] [16, 17]

method that involves pumping molten material into a mould to create pieces. Plastic Forming involves heating a sheet of plastic and draping it over a mould in some way. The sheet is then formed into a shape using air pressure and male plugs. Finally, Plastic Attaching is the process of joining semi-finished pieces together. Fastening, adhesive bonding, and welding are all examples of this (Fig. 3).

In additive manufacturing, the process starts with identifying the opportunity from the problem. To explain this process with the case study, we take the development of Dr Cardio by the sample company Engineering Technique in Vadodara. They started the process with the problem statement that, traditional ECG machines are bulky and mostly installed at hospitals, doctors cannot carry it as it is not portable, remote installation is not possible where hospital facilities are not available, portable ECG are not available in Indian Market as of now and with the implementation of portable ECG patient can be treated immediately. This led then to a manufacturing of Dr Cardio—World's smallest 12 channel ECG machine and now doctors can keep this device with them every time; like at home, in car, at

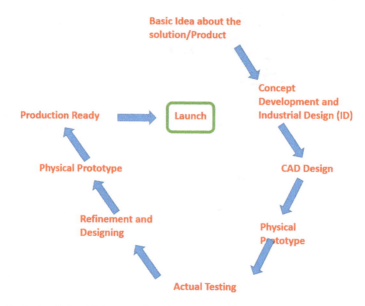

Fig. 3 Product cycle in additive manufacturing process [18]

their clinic/hospitals. After clearly defining the problem, the next step is to develop the concept and industrial design for the same followed by CAD design and physical prototype of the same. The important and inherent advantage of AM is the at designing stage only changes can be made after actual testing of the product which can be redesigned and redefined later before developing its final physical prototype. The product is then ready for final production (Fig. 4).

The advantages of AM product cycle are: (i) the product manufacturing cycle is faster as majority of the work is done at designing stage, (ii) waste can be avoided as AM manufactures layer by layer avoiding deposition of material at

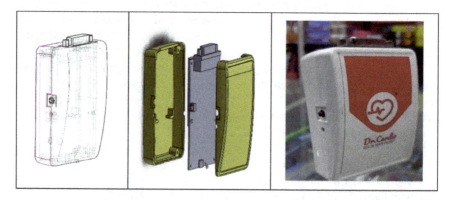

Fig. 4 Product—Dr Cardio, developed by Engineering Technique [18]

hollow places, (iii) it reduces or almost eliminates the risk of product failure as changes can be made at designing and prototyping stage easily and (iv) more innovations are possible in AM as customisation allows risk taking. In spite of all these significant advantages, there are still certain features which are provided by traditional manufacturing process and still are not available in AM, like (i) Mass production, (ii) the range of material used for AM is still limited compare to traditional manufacturing, (iii) manufacturing of large sized parts and (iv) when perfect smooth finish is required.

3 Application of Additive Manufacturing in Various Sectors [19, 20]

3.1 Education Sector

Academic institutions have started recognising its due significance and future need in terms of its prospects and therefore perhaps having and indispensable inclusion in modern curriculum. It is going to be instrumental for students while creating prototypes with inexpensive tooling. The ideas and imaginary pictures which were there on a screen or a page can be created in three-dimensional form in real world. Thus, it has great potential of being accepted not only at the higher educational institutions but likely to be used in school education to provoke thoughts and feed curiosity of aspiring students (Fig. 5).

3.2 Aeronautics and Aerospace Industries

The design optimisation for parts, tools, equipment and accessories has been always a great concern and challenge to the aeronautics and aerospace industries. The revolutionary advancement of 3D printing would essentially replace traditional, conventional means and processes with more accurate and efficient design for inputs. The same will have long lasting impact on cost and operational efficiency in the industry (Fig. 6).

3.3 Medical and Dentistry

The world has witnessed an astonishing impact of use of 3D printing during the prevailing pandemic situation in terms of developing variety of medical instruments and stuff, such as prototype of ventilator, face shields, PPE kits, masks, artificial human heart, bones to facilitate medical requirement and even with regard to bridging gap between demand and supply of the same. In the time to come it will have wider spread, by its inclusion, in teaching research pedagogy. Further versatile use of the technology will be a great support to the growth of artificial intelligence in probably all sectors with its special implications on healthcare

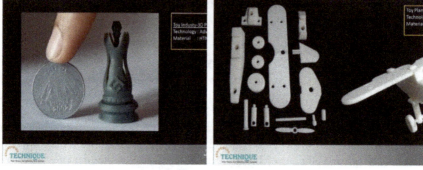

Fig. 5 Products developed by Engineering Technique, Vadodara [18]

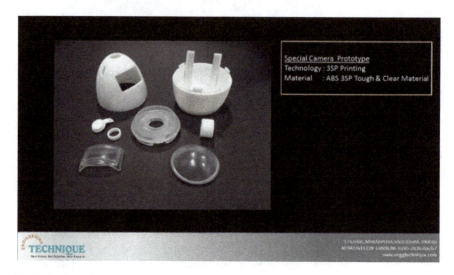

Fig. 6 Products developed by Engineering Technique, Vadodara [18]

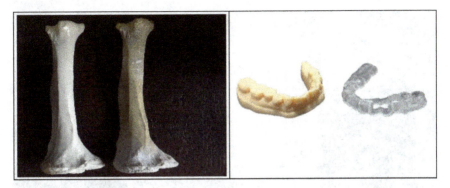

Fig. 7 Products developed by Engineering Technique, Vadodara [18]

sector [20]. Medical field is the earliest hotspot [21] in 3D printing and still it is extremely active due to application of 3D printing in the medical field during current pandemic [22] (Fig. 7).

3.4 Construction Sector

The development of AM has fetched drastic changes in the field of civil engineering and construction. It could replace conventional components, means and modality for performance with more appealing, cheaper and comfortable demonstration in the construction sector [23, 24]. Engineers, architects and site contractors have been enjoying pleasant use of the technology in developing designs and specifications, preparing models; precision in work with such other many aspects. In particular, use of 3D printing in construction comprises of extrusion, powder bonding as well as additive welding. More complexity and precision, cheaper labour costs, greater functional integration, and less waste are all evident benefits of these technologies [25] (Fig. 8).

3.5 Jewellery Sector

Wide spread of AM is proved to be disruptive in terms of enriched augmentation. The jewellery manufacturers will have better room to create variety of permutation and combinations while developing designs, ensuring uniformity and precision and thereby adding beauty to the final output. It has ample scope to incorporate hybrid mechanism that may provide bundle of traditional and modern components in jewellery making (Fig. 9).

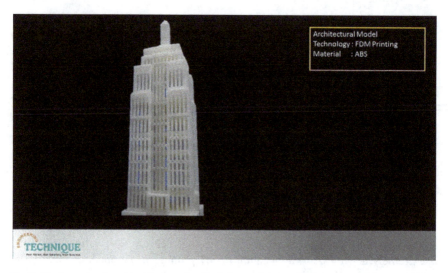

Fig. 8 Products developed by Engineering Technique, Vadodara [18]

Fig. 9 Products developed by Engineering Technique, Vadodara [18]

3.6 Cultural Heritage

In the different parts of world, especially in developed countries, 3D printing has been successfully used for preservation and restoration in the field of cultural heritage. For preserving ancient glory of culture and heritage, the technology can facilitate in recreating the missing parts of relics (Fig. 10).

Furthermore, Museums and various other avenues of display and demonstration have started selling 3D models of artifacts prepared through 3D scanners through their online platforms. These have been created in 3D printing supported file format, which can be 3D printed by visitors at their places.

Fig. 10 Products developed by Engineering Technique, Vadodara [18]

4 Recent Developments and Key Players

4.1 Global Scenario [26]

In 2020, the global market for 3D printing was worth 13.7 billion USD and predicted to grow to 63.46 billion USD by year 2026, the CAGR will be 29.48% in the forecasted period i.e., 2021–2026. 3D printing, also known as additive manufacturing, is an object-creation technique that provides a wide range of possibilities for the making, designing and preparing unique geometrical shapes, products as well materials. AM is product development and production technique that is more innovative, faster, and agile. AM has evolved as significant functioning aspect in fabrication from a simple prototyping tool. The reason behind this is significant developments in composition of material like increasing use of polymers and metals. Innovative materials, reduced lead time and new finishes are allowing it to be important part of production processes while conforming to regulations (FDA, ASTM, and ISO).

Additionally, the availability of this technology has expanded due to lower pricing of additive manufacturing-based machines, as well as increasing competence and awareness. In recent years, newer and more powerful fused deposition modelling technologies have enabled the use of a wider range of materials, resulting in widespread acceptance across a variety of industries. According to a March 2019 Essentium survey, various producers are employing AM as full-scale production in 2019. Due to the rising number of industrial solutions given to the market by producers, there has been a wider acceptance of 3D printing across sectors (Fig. 11).

Various technical breakthroughs, such as Artificial Intelligence (AI) and Machine Learning (ML), are also boosting the usage of 3D printing machines, since these technologies allow for computerized printing for more effective manufacturing. Governments all over the world have begun to invest for R&D in AM,

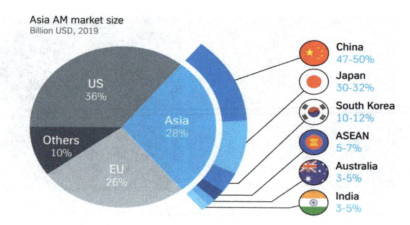

Fig. 11 Global market of AM [27]

which aided in the spread and use of the technology. Furthermore, as AM progresses, it can pose a threat to established systems of Intellectual Property (IP) protection, resulting in an increase in the unlawful use of printed weapons and pharmaceuticals, which is projected to stymie market growth. In addition, the market's standardised structure is hampered by absence of Body of International Standards that regulates producers using AM.

For meeting the enlarged demand for 3D printing applications in various fields like automotive, healthcare, aerospace as well as defence sectors, industry players keep focusing on improving the technology. By incorporating additive manufacturing into new product development processes, important players are recognising opportunities for business transformation. Market leaders like as Stratasys Ltd. are pushing beyond prototype to take advantage of the flexibility which AM provides for the whole Production Value Chain. The firm can swiftly and in a cost-efficient way build big finished components utilising AM with its revolutionary range of 3D printers that feature Fused Deposition Modelling (FDM) and Selective Absorption Fusion (SAF) technologies. The leading global players in the field of 3D printing are: Stratasys, Ltd. Materialise, Envision Tec, Inc., 3D Systems, Inc., GE Additive, Autodesk Inc., Made in Space etc.

4.2 Indian Scenario

Hewlett-Packard Inc (HP) launched 3D printers in India in January 2020, with prices starting at Rs. 2.5. The printers, according to the company, may manufacture things up to ten times faster than current printers while costing less than half as much. The output of inkjet printers is based on FDM printers, making them injection moulding's closest competitors. Customers don't seem to want to buy the printer because of its high price. Every college receives Rs 20 lakh for robotics, 3D printing, and other technology. ATLs (Atal Tinkering Laboratories) have helped to raise awareness about these structures.

Experts predict that the commercial 3D printing market will reach $7–10 billion by 2024, representing a compounded annual growth rate of more than 30% from today. According to current studies, the marketplace is rising economies alone might reach $4.5 billion by 2020 [28]. Government agencies, on the other hand, believe that there is a need to investigate the rise of 3D printing grey markets in which people make guns at home. (3D printing is building a world in 2019). Major players active in Indian 3D printing market (comprising of manufacturers as well as distributors) include Altem Technologies, Imaginarium, Brahma 3, KCbots, 3Ding (REDD Robotics) and JGroup Robotics.

5 Additive Manufacturing: Game Changer for Indian MSME Sector

5.1 Overall View of MSME Sector in India

As small drops of water can create a huge river likewise is the MSME works in an Indian economy. To attain the motive of self-reliance (AtmaNirbhar Bharat) set by our Honorable PM Shri Narendra Modi the MSME sector plays a significant role. This sector majorly contributes in socio-economic development of Indian economy that is development with increase in employment, enhancing rural areas, growth in small industries, reducing rural-urban migration. According to the MSME Annual Report 20–21, this sector is able to contribute in Indian GDP around 30% with 6.11% in manufacturing sector and 24.63% in service sector. Also, its contribution in overall exports is around 45% and provided employment to around 120 million people. The goal of government to boost Indian economy by USD 5 trillion can be achieved by enhancing the MSME sector [29].

Recently there was a change in definition of MSME w.e.f 1st July 2020 by the government with motive of removing fear in MSME entrepreneurs that if they expand their business, they may not enjoy the benefits of MSMEs because of low threshold limit. So, the new definition came with increased threshold limits and also with extra category of turnover along with investments and elimination of bifurcation between manufacturing and service sector (Fig. 12).

Classification	Micro	Small	Medium
Manufacturing and service sector	*Investment in P/M*-not more than Rs. 1 crore and *Annual turnover*-not more than Rs. 5 crores	*Investment in P/M*-not more than Rs. 10 crore and *Annual turnover*-not more than Rs. 50 crores	*Investment in P/M*-not more than Rs. 50 crore and *Annual turnover*-not more than Rs. 250 crore

According to the MSME industry report, there is 18.5% increase in number of units of MSMEs in India from 2019 to 2020 in terms of Compound Annual

Fig. 12 Udyam registrations up to December 2020 [30]

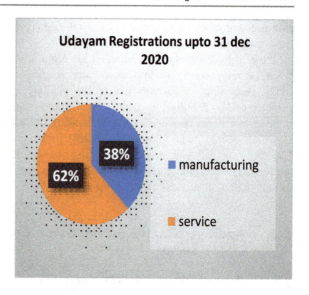

Growth Rate (CAGR). The Ministry of MSME has replaced the Udyog Aadhaar Registrations with Udayam Registrations from July 2020. As per the annual report of MSME 20–21 up to 31st December 2020 total 1,402,735 registrations were placed in Udayam Registrations with the bifurcations shown in Fig. 12

The Top 5 Industrial sectors registered were–Food Products, Textile, Apparel, Fabricated Metal products and Machinery & equipment's. According to ministry of MSME as on 28th august 2021 total registrations on Udayam portal are 4,548,619 units with top five states namely Maharashtra (21.33%), Tamil Naidu (10.86%), Gujarat (8.74%), Rajasthan (8.32%), Uttar Pradesh (7.62%) with majority units of micro enterprises. So, there is an increase of 224 times in no of units from dec 2020 to august 2021 in Udayam portal. Up to 30th June 2020 total no of units registered in Udyog Aadhaar portal were 10,228,619 with top five states Maharashtra, Tamil Naidu, U.P, M.P, Bihar. It interprets that from June 2020 till August 2021 Gujarat and Rajasthan have shown better results in MSME registrations.

Recent achievements in MSME sector includes advancement in technology through Jamshedpur (IDTR) which has innovated 8 Pitch Circle Diameter (PCD) for gauges checking with 5 μm accuracy in usage of inspection of items of export for MSME unit. Also, MSMEs are moving towards BSE exchange for raising funds. According to BSE SME 16 SMEs have entered the market through IPO route and raised amount of Rs. 100 crore in the year 2020 and further expecting more than 60 SMEs in the year 2021–22. The government initiative of Government e-marketplace (GeM) as an e-commerce site has achieved 1,16,291 crore orders. Under start-up India initiative around 50,000 start up with 5,50,000 jobs up to June 2021 were recognized. As per the annual report of Ministry of MSME 20–21, for ministry and its allied offices which includes KVIC, Coir Board, MGIRI, NIMSME, NSIC and DCMSME digital transactions has achieved growth up to

92.02% in amount and 90.19% in no of transactions. Also, this ministry has received the Open Data Champion Award (2020).

Recent achievements in policies and schemes of MSME Ministry up to May 2021 include:

i. Under PMEGP around Rs. 303.94 crore has been distributed as margin money with target of setting up 10,449 Micro enterprises to create jobs of around 83,592 persons.
ii. Under CGTMSE around 29,378 guarantees have been approved with amount of Rs. 1999.28 crores and during period of 1st April 2021–24th may 2021 50,868 guarantees were provided with amount of 3711.80 crores.
iii. Under various organizations like TCSP, NSIC, KVIC, MSME tool rooms around 13,857 people were imparted with skill trainings.
iv. To overcome the covid 19 pandemic situation by MSMEs and improve overall scenario in Union Budget 2021 the major packages were Government of India announced amount of Rs. 10,000 crores for Guarantee Emergency Credit Line scheme (GECL) to boost this sector. The allocation of Budget is more than doubled of previous year amounting to Rs. 15,700 crores. Also, government is planning to include the traders in the definition of MSME which will be a big relief to trading sector.
v. Role of MSMEs in Gujarat is also very significant. As per the RTI application applied under Udyog Aadhaar Portal up to 30th June 2020 the registered MSME units in Gujarat state are 8,67,508 stands around 9% of overall units in India and also according to data as on 28 august 2021 Gujarat state ranks on 3rd position after Maharashtra and Tamil Naidu in number of units registered. As per the Gujarat MSMEC report from Oct 2015 to Feb 2020 there has been an increasing trend. There has been increase in investment, employment and no of units approximately 9.64 times, 12.81 times and 15.01 times respectively.

Udyog Aadhaar Memorandum—UAM
See Table 1 and Fig. 13.

Table 1 Number of units registered [30]

	No. of units	Employment	Investment (Rs. Lakh)
FY 2015 (From Oct 2015–Mar 2016)	52,046	357,236	1,631,036
FY 2016	246,719	1,575,012	6,181,227
FY 2017	440,182	2,603,917	9,465,150
FY 2018	610,139	3,603,312	12,678,372
FY 2019 (April 2019–Feb 2020)	781,615	4,576,139	15,724,901

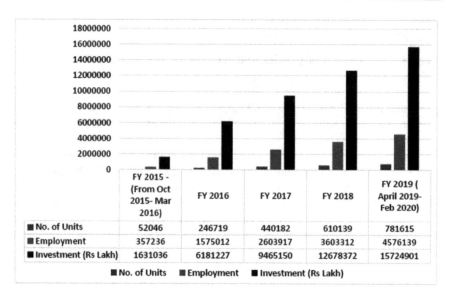

Fig. 13 Number of units registered [30]

5.2 Various Government Initiatives for Adoption of AM [31]

- Manpower development, providing knowledge facilities, availability of finance, technology, infrastructure facilities, easy access to market and ease of doing business are all areas where MSMEs can benefit from holistic growth and thus the government of India has launched various schemes to promote and boost the growth of MSME sector. These schemes include **Udyami Mitra Portal, MSME Sambandh, MSME Samadhaan, Digital MSME Scheme, Prime Minister Employment Generation Programme, Revamped Scheme of Fund for Regeneration of Traditional Industries (SFURTI),** *A Scheme for Promoting Innovation, Rural Industry & Entrepreneurship (ASPIRE),* **National Manufacturing Competitiveness Programme (NMCP), Micro & Small Enterprises Cluster Development Programme (MSE-CDP), Credit Linked Capital Subsidy Scheme (CLCSS)** etc.
- The Government has also realized the importance of 3D printing and its application in MSME sector in India. Thus, the Government of India has done several efforts for the same. For promoting 3D printing on commercial and industrial scale as well as to help local companies in overcoming technological and economical barriers, the Government of India's Ministry of Electronics and Information (MeitY) has formulated India's 3D printing policy.
- India's 3D printing strategy will aid in the establishment of a favorable environment for 3D printing and additive manufacturing design, development, and implementation. It includes not just the manufacturing side, but also the design and software aspects. The government will also try to persuade market leaders to set up worldwide hubs for 3D printing in India, while discouraging the import

of printed materials for national use. The auto and ancillary car as well as motor spare parts, such as engines, interior/exterior elements of luxury automobiles, or landing gear, complex brackets, and turbine blades, are also important areas of attention.
- The other initiatives include

    Atal Innovation Mission: Atal Tinkering Labs, 1200–1500 square feet dedicated innovation workspaces have been set up under the aegis of Atal Innovation Mission, where latest technology do-it-yourself (DIY) kits such as 3D Printers, Robotics, Internet of Things (IOT), Miniaturized electronics have been installed through financial support of government of Rs. 20 lakhs so that students from Grade VI to Grade X can learn about them.
- The Gujarat government has signed a Memorandum of Understanding (MOU) with the US Institute of 3D Technology (USI3DT) in California and OEM 3D Systems (a leading global 3D printing company) to establish seven 3D printing Centres of Excellence across the state's engineering colleges and technical institutes. The University of Wollongong and the Andhra Pradesh MedTech Zone cooperated to establish a 3D Bioprinting Lab.

Private Sector Initiatives [28, 30]

- INTECH Additive Solutions, based in Bangalore, is India's largest commercial setup of 3D printing using metal. INTECH also created OPTOMET, a unique software for developing new parameters and alloys. They've also created "AM Builder," a process programme for producing high-quality parts at a lesser cost and with ease of use. Under the MAKE IN INDIA project, they also designed and developed India's first metal printer. INTECH has also launched a digital academy, which will provide industry, academia, and R&D organisations with orientation and training.
- Wipro has introduced "Addwize—An additive technology Adoption & Acceleration programme" with the goal of enabling enterprises and institutions to systematically adopt and scale metal Additive Manufacturing (AM) for measurable business benefits.
- Accreate Labs & Innovation, a Bengaluru-based start-up, has announced that it will develop user interface panels for the ISRO-operated GSLV.

Collaborative Efforts [28, 30]

- IISc Bengaluru's Department of Heavy Industries' COE (Additive Manufacturing for High Performance Metallic Alloys) joined with Wipro to create India's first industrial 3D printer. HP Inc. has signed a Memorandum of Understanding with the Government of Andhra Pradesh to establish a 3D printing Centre of Excellence.
- Stratasys and NTTF (Nettur Technical Training Foundation) have announced a partnership to establish India's first additive manufacturing certification course.

The training programme seeks to assist students in learning new 3D printing technologies and filling skill gaps in the industry.

5.3 Adoption Model of AM by MSMEs and Role Enrichment of Facilitating Agencies with Revitalisation Government Concern

Opening doors of economy of our country, through liberalisation, privatisation and globalisation, since 1991, has resulted into not only in an intensified competition but imposition of international competitive standards too. The final output needs to be cost competitive and should have international standards of quality for its being acceptable. Fast technological changes resulted in to Modernization and automation of production processes that has increased productivity and reduce unit cost. The value addition, in all possible context, has become an entry pass and an inevitable condition for success. The major trouble, in this respect, lies with regard to small and medium-scale enterprises (SME). The large-scale manufacturers of tools and moulds who are engaged for producing parts, components, systems, and final products, often uses them through outsourcing. When it is required to be carried out by SMEs, with due constrains of financial resources and skill set or competence, it obviously has lot of inadvertence and challenges. Therefore, if wider spread of 3D printing technology to be ensured, necessary support system had to be built up.

In order to meet with the bottom-line requirement, initiative was taken by Ministry of MSME, Government of India in association with German Technical Cooperation—GIZ and State Government. The said initiative could achieve significant outcome in terms of information exposure and spread of awareness among many deserving SMEs. It could also provide relevant consultancy services to impart needed skill and knowledge to have their independent functioning. Furthermore, through making provision for 'Indo German Tool Room' at various places across the country, the said effort could result into imparting needed practical training to make people understand detail operational aspect of newly invented technology. So far it has setup almost 10 such Tool Rooms to provide assistance to SSI units for their technical upgradation and to provide better quality tools to meet the growing demand. These tool rooms are located at Indore, Ahmedabad, Ludhiana, Hyderabad, Bhubaneshwar, Jamshedpur, Calcutta, Jallandhar & Nagpur. State Govt. run tool rooms are at Lucknow, Delhi, Bangalore, Mysore and Goa. These Tool Rooms and Technology Development Centres are now known as Technology Centres (TC). The Ministry of MSME has developed 18 such Technology Centres and planning to develop more 15 centres in future.

Under consultancy Services, an attempt is made to provide information exposure for new developments in the area of Tools and Die manufacturing technology. With due concern for small and medium units an attempt is made to provide guidance and support in the area of Product& Process Development, Training and Skill upgradation, spread of entrepreneurship, shaping Turnkey projects, Counselling for Lean Manufacturing together with Total Tooling Solutions.

The provision for Tool Room is indeed a great support to the learners who have constraints of resources and infrastructural facilities that may be needed to learn and get demonstration of the technology. The Tool Room could provide all possible facilities and solution to major common trouble that operators may be facing in die making. To name few, in terms of its contribution, the tool rooms have added precision in use of technology, imparted training to infuse culture of using new technology, enriched learning through reverse engineering and rapid prototyping etc.

5.4 Development of Additive Manufacturing Centre

In line with the modality adopted by the initiative from the government, in consultation with German Technical Cooperation, for development of Indo German Tools Room, similar modality can be adopted for the wider spread of 3D printing technology with establishment of centres taking care of Additive Manufacturing (AM). In order to make it acceptable and convenient in use among small and medium enterprises (SMEs), it is essential to spread its exposure and to provide needed support to make users comfortable with the same. The government is expected to establish such centres at selected locations who may promote its usage by cultivating due understanding on various dimensions of 3D printing technology together with imparting training to make the same convenient to the new users. During the phase of its introduction, the centres should provide support in terms of practical training, counselling on making provision for needed infrastructural facilities to successfully adopting the same in units and if possible, by making instrumental to make provision for financial support for its wider adoption in organisations.

The Additive Manufacturing Centres can provide services at various levels: (i) seminars and workshops with MSMEs for spreading awareness regarding use and future of 3D technology, (ii) Participating in Science Fairs and Tech-Fests of schools and colleges to educate students regarding this technology, (iii) providing consultancy services for MSMEs on preparing prototypes and uses of 3D printing in their existing manufacturing technology, (iv) preparing AM Tools Room where costly 3D Printers are installed, which can be used by MSMEs at low cost, project to project basis, so that they don't have to incur huge capital investment in purchasing 3D Printers at initial stage.

The researches and figures show that, AM is the future of manufacturing industry, no company and country can avoid this fastest growing technology. The development of AM Centre by centre and state governments will lead to spreading awareness about 3D printing faster and making it available for MSMEs to uses without investing in them.

5.5 Engineering Technique—A Case Study

The case study of Engineering Technique is analysed in the present section. Engineering Technique is amongst the major leading players in Gujarat providing 3D CAD and 3D Printing solutions to various companies. The company has business alliance with Solid works, Envision Tec, Mark forged and Shining 3D. The company is Authorized Value-Added Reseller of Dassault Systems SolidWorks Corporation for their entire range of CAD/CAM/CAE/PLM Solutions. Engineering Technique being a leading Engineering Technology solution provider has been actively engaged with Education institutes across Gujarat to supply various technologies, solutions, and services. Engineering Technique is Authorized Distributors of Leading OEMs like: i. Envision TEC for technologies like, DLP, 3SP, CDLM, SLCOM, Bio-fabrication, RAM, etc. ii. Mark forged for 3D Printing composite materials (i.e., Carbon fibre, Fibre Glass, Kelvar, etc.) and Metals. iii. Shining3D for technologies like, SLA, SLS, SLM, 3D Scanning, etc. Engineering Technique has also setup 3D Printing Service Bureau at Baroda facility to provide 3D Printing Services to Industries. The company also offers IoT solutions & Robotics solutions for Education and Industries.

It has been observed in the previous study [32], that due to lack of satisfaction on reliability and repeatability front, the current entry level 3D printers used by small business houses forced to put it in a second raw behind production grade AM system even for prototyping but the industry wise profile of the company makes it very clear that additive manufacturing technology is not limited to prototypes, large scale companies, selective raw materials or particular product category. It is widely applicable from large to medium to small enterprises. This emphasises the need of spreading awareness about 3D printing in the industry, which the company Engineering Technique is already doing and that to from school and college level to various industries. It is rightly said that 3D printing is future of manufacturing technology and thus it is essential for all the industries to get acquainted to it and for government to make it affordable for MSMEs (Fig. 14).

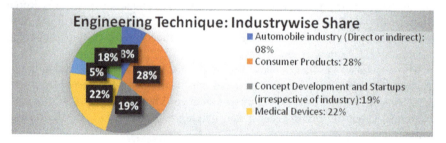

Fig. 14 Industry wise share of Engineering Technique, Vadodara [18]

6 Contribution of AM During COVID-19 Pandemic

The year 2020 put the entire globe to a halt, all the industries and sectors were standstill for almost a year due to this pandemic of COVID-19. Same was the case with manufacturing industry in India, lockdown across the nation and migration of laborer led to closing the factories and less or no production. Due to which, supply chain was hampered and there was a gap between demand and supply of the products.

Amid this difficult situation, various reports shown that due to this pandemic the AM has received 'mainstream exposure' because of the role it played in helping and supporting Corona Warriors by speedily producing Personal Protective Equipment (PPE), prototypes and various components for ventilators etc. The pandemic has disconnected the countries and practically it was impossible for our country to import various parts. During this critical situation, India was looking for local and faster solutions, 3D printing helped the country by, manufacturing the prototypes and end-use components locally and rapidly, which also boosted our prime Minister's dream of 'AtmaNirbhar Bharat'.

The case study of Engineering Technique—A leading Technology Solution Provider, having headquarter in Vadodara District, Gujarat State have been analyzed to evaluate the production of prototypes as well as end-use parts during COVID-19 times.

The company is manufacturing "Medical Face Shield" for Doctors & Nurses, Medical Staff of Hospitals, Police & Security Personnel, Government Officers, etc., which help them to protect eyes, nose and mouth (Fig. 15).

(i) Company has designed & 3D Printed Door Opener to avoid Touching palm (Fig. 16).

Fig. 15 Products developed by Engineering Technique, Vadodara [18]

Fig. 16 Products developed by Engineering Technique, Vadodara [18]

Fig. 17 Products developed by Engineering Technique, Vadodara [18]

(ii) Mass Production of 3D printed NP swabs by the company for COVID-19 testing (Fig. 17).
(iii) The company also helped its clients like Jyoti CNC, Rajkot in manufacturing affordable ventilator called 'Dhaman', Sahjanand Laser—PranSaiyam, Gandhinagar to prepare indigenous ICU ventilators and UD Laboratory, Rajkot to develop disinfectant chamber.

7 Perspective Contribution and Operational Constrains

7.1 Perspective Contribution of 3D Printing

1. **Adding edge in corporate operations:** The 3D printing has a capacity to move to the closest to an idea in terms of thought. It shall have amazing implications for the corporate firms in developing ideas along with its deserving precision. All such aspiring change would take shape faster unlike introducing change in conventional/traditional context.

2. **Long term positive cascading impact on operational efficiency:** Even though initial investment seems to be exorbitant but as the technology has its wider spread with multiple application in an organisation, will certainly generate positive cascading impact on operational efficiency. It may be apparently visible in variety of components that may decide efficiency such as reduction in wastage, down word movement of marginal cost or precise quality control and many others.
3. **Dilution of an element of risk through pre testing:** Unlike the past practices of spending time and resources to test expensive modelling, the technology tends to be far cheaper to carry out pre-testing of prototypes and suggesting corresponding appropriate alteration in designs.
4. **Better and less costly 'Test Marketing':** As against conventional practice of Test Marketing, before mass commercialisation of product/design, the technology shall offer an effective and closer approach to solicit preference for proposed product or design from the perspective target audience. The marketer can have a close eye on perspective acceptance or otherwise of the idea without deploying a great number of resources. The technology may prove to be a new starting phase of 'Close Customisation', moving closer to not only individual latent demand but hidden aspirations.
5. **Feel without actual purchase experience:** When perspective customer can have vivid picture or virtual prototype on electronic medium, even without actually purchasing or having physical ownership, of product, he can have feel of the product in terms of its look, comfort, augmentation etc. This unique aspect may ignite more genuine concern and involvement in purchase behaviour. This may lead to generating brand loyalty among the target audience in long run. Furthermore, personalised solution to any need has always proved itself as a unique selling proposition and there by offers core competitive advantage in the market place. The technology has its stretch in this regard.
6. **Realisation moving closer to remote imagination:** The technology related to 3D printing certainly providing kick start where whatever creative people may have incubated can reached to the stage of realisation. Earlier, probably, imagination was limitless but with the 3D printing advancement, realisation would also be limitless. Any thought, idea, desire or even aspiration may afford to have temptation to reach to its reality. It is also observed that designer and product developers are focusing on development of useful as well as pleasing as well as well performing 3D printed products [33].
7. **Deep implications in product innovation with less expensive route:** The technology would certainly provide a new product developer an understanding as to whether the product idea sounds with adequate potential or otherwise, at an early stage which is less expensive as it may results into early withdrawal from the process.

7.2 Operational Constraints and Challenges [34]

1. **Protecting Intellectual Property Right:** The emergence and development of an innovation, related with 3D printing technology, with the passage of time may have great potential for its imitation and thereby unauthorised use too. This may lead to an obvious challenge to protect the same with appropriate copyright protection. The issue may crop-up with variety of its dimensions as the exposure pertaining to the technology increases. If not curb, with appropriate precautionary and curative measures would lead to unauthorised use and monetary gain out of it [35].
2. **Constraints for Large Size Objects:** Until further innovation in the same direction take shape, the present 3D printing technology would face constraints in term of not giving desirable outputs for large sized objects. This may limit its application and simultaneously igniting room for further innovations.
3. **Constraints on Inputs:** Even though the 3D printing technology is incorporating around hundreds of various raw materials in order to create desirable outputs, but still it is lacking so many in numbers and variety, as compared to traditional mechanism, which is constraining in nature. Furthermore, users shall have to wait for future development to make the technology apt with regard to the leftovers.
4. **Implications on cost as a constrain:** The technology is very appealing in terms of serving the corporate entities in a versatile manner but simultaneously likely to impose an exorbitant cost for its adoption. In particular, small and medium size units may not find it lucrative to adopt as a permanent feature due to the said limitation. Future development or mass adoption may likely to fetch the technology at a level that may be affordable to the said kind of units.
5. **Resistance on account of obsolesce of skill:** The technology has an inherent feature to make skill of operators redundant and with the passage of time may be resulted into a great apprehension for loss of employment irrespective of size of units or scale of operations. Therefore, there may be resistance on the part of those who had skill and proficiency to do the same nature of work with traditional tools and techniques.
6. **Apprehension against Civil Protection:** If 3D printing technology is found in unwanted hands that may use the same against the genuine interest of the society i.e., for criminal purposes, may be counter-productive. The legislative measures are required to be propounded to provide means and ways that may capable of restricting use of the technology for antisocial/dangerous purposes.
7. **Time and Efforts to Unlearn and Relearn:** As the technology is new and revolutionary, it requires the existing operators to unlearn existing modality and way of performance and are required to relearn new skills and competence to make use of the same. It may, even for a temporary time period creates disequilibrium even for those who are inclined towards its adoption.
8. **Requirement of Advanced Quality Control Method:** Due to advancement on AM technologies, 3D printed parts are directly used by some industries as finished products. However, to make them up to mark advanced quality control

methods are required. For real time monitoring and quality analysis, the majority of AM processes employ in-situ sensors and cameras to collect processing data. However, some approaches, such as BJP, are not qualified to use this technique due to the closed geometry of the 3D printing system. To achieve a higher quality final product and a faster process, more complex process monitoring and rapid feedback control are required [36, 37].

8   Conclusion and Future Scope

The five stages of technology adoption are: awareness, assessment, acceptance, learning, and usage. The perceived usefulness of technology and its simplicity of use are the two most essential characteristics for technology adoption.

The era of fourth Industrial revolution, popularly known as Industry 4.0 is bringing in new aspects in various industrial practices as well as management of value chain [38, 39]. Additive Manufacturing (AM), popularly known as 3D printing is playing transformative role in Industrial Revolution 4.0 and bringing revolutionary changes in manufacturing landscape of Indian industries. 3D painting uses Computer Aided Designs (CAD) to create three dimensional objects through layering materials. Robots and Machines can communicate with one another, make independent judgments, and self-update & adapt to changing circumstances [40–42].

Indian MSME sector is comprising of 51 million units and form backbone of industrial landscape. As the world is heading towards Industry 4.0, the MSME sector is required to pay utmost attention to manufacture superior quality and technological products. The Zero effect and Zero-defect goal can be achieved by use of 3D printing which can minimise the cost as well time for MSMEs.

Circular economy practices are being combined with Industry 4.0 technologies, in order to develop business model for use of 3D printing in industry. To develop a circular business model for recycling waste and supplying new products, resulting in considerable reductions in consumption of resources and better utilization of natural resources. With the suggested combination of web technologies, reverse logistics, and AM as a technical platform, the circular business model can be used to recycle e-scrap in the first stage [43]. These have societal repercussions in terms of the environment, sociotechnical issues, and economics [44].

According to the 3D Hub's survey [45], 3D printing will "continue on its path of growth" in 2021, with 73% of engineering firms expecting to produce or source more 3D printed parts this year than in 2020. Cost, part quality, limited experience, and material choice constraints, according to respondents to 3D Hubs' survey, are all problems holding back the use of additive manufacturing. 3D printing's higher design flexibility can lead to superior strength-to-weight ratios and more efficient parts, which can help overcome these perceived hurdles. In addition, the research views the development of new metal 3D printing materials and procedures as a way to reduce the technology's cost. There will be a continual trend toward more end-use parts, and that the cost of the technology will gradually decrease, allowing it to be used to more applications. Hybrid manufacturing, as well as the advancement of "material freedom," workflow software, and surface treatments and alloys in 3D printing, are expected to "make significant strides" this year. AM manufacturing is a reality in some industries, not limited to selected customised goods. The performance demonstrates that AM adequate for small, medium to large objects and not only suitable for prototypes but for end-use product also. A substantial GDP increase has been attained within Indian local economic circumstances. Furthermore, AM systems can be used to replace traditional and common production methods.

Future Scope

Consumers nowadays are far more concerned with personalisation, customization, and delivery speed, and traditional manufacturing models, no matter how modern, cannot compete with 3D printers in both categories [46].

The adoption of 3D printing on domestic as well as industrial scale has changed the approach of manufacturing products. Until recently, costly, sophisticated and advanced equipment and tooling were necessary for producing high-quality actual physical products; but, with innovations of low-cost 3D modelling tools, and industry assistance, this is no longer the case. The ideal canvas for making Do It Yourself (DIY) production a revolution in and of itself is 3D modelling. While it is uncertain how 3D printing will impact traditional production, emerging trends suggest that a paradigm shift is taking place. Though all the business models used by entrepreneurs for AM are different from each other to various extent and there is no generic business model [47], something is certain—as a greater number of businesses become producers, the line between consumer and producer may become increasingly blurred.

The introduction of 3D printing could have a variety of outcomes. Products will need to be promoted more swiftly; outsourcing may become obsolete; designs will be open and community-oriented; customisation will become the norm; and the actual delivery of finished goods will raise the printing industry's carbon impact. It can be cut down by selling models all over the world rather than shipping them [48, 49].

The relevant materials can be published in the area's room stations. Traditional subtractive methods such as CNC milling, in which material is removed to build the component; formative operations (fabrication or forging), in which a mould is typically required; and joining processes like as welding and bonding are all in opposition to this technology. Although there are many distinct forms of 3D printing, they all work in layers. It enables use Rs to make sophisticated structures that would be impossible to shape or mould otherwise. As a result, the correct amount of material is used during the procedure, and no waste is generated [50].

3D printing, according to economists, will have a huge impact on productivity since it will redefine financial services by making it cheaper to make as many products as possible. The amount of metal required is dictated by the degree of coating, and just the bare minimum is required. As a result, there is no waste generated during the process. Typically, economists believe that 3D printing will have a large impact on productivity because it would surely rewrite financial systems in order to cut the cost of as many things as feasible.

As a futuristic concept, 3D printing is being investigated as a way for building extra-terrestrial houses, such as habitats on the Moon or Mars. Using building-construction 3D printer technology, it has been proposed to create lunar building structures with enclosed inflatable shelters for hosting human residents inside the hard-shell lunar structures. For these ecosystems, only 10% of the structure would need to be relocated.

Food additive manufacturing is being created by compressing food into three-dimensional structures layer by layer. A wide range of foods, including chocolate and confectionery, as well as flat foods like crackers, pasta, and pizza, are suitable options. Because of the concept's adaptability, NASA has given the Systems and Materials Research Consultancy a contract to investigate the feasibility of printing food in space. NASA is also investigating the technology in order to generate 3D printed meals in order to reduce food waste and create food tailored to an astronaut's dietary requirements. Novameat, a food-tech business based in Barcelona, 3D-printed a steak using peas, rice, seaweed, and other components put together criss-cross to mimic intracellular proteins. The nature of a dish's texture is one of the issues with food printing. Foods that aren't sturdy enough to be filed, for example, aren't suitable for 3D printing [51].

References

1. D.J. Horst, C.A. Duvoisin, R.A. Viera, 'Additive manufacturing at Industry 4.0: a review' International Journal of Engineering and Technical Research, 8(8), pp. 1–8, 2018.
2. A. Rindfleisch, A. Malter, G. Fisher, 'Self-Manufacturing via 3D Printing: Implications for Retailing thought and Practice', Review of Marketing Research, 16, pp. 167–188, 2019, https://doi.org/10.1108/S1548-643520190000016011.
3. H. J. Steenhuis, L. Pretorius, 'Consumer additive manufacturing or 3D printing adoption: an exploratory study', Journal of Manufacturing Technology Management, 27(7), pp. 990–1012, 2016, https://doi.org/10.1108/JMTM-01-2016-0002.
4. N, Shahrubudin, Lee, R. Ramlan, 'An overview on 3D printing technology: technological, materials, and applications', Procedia Manufacturing, 35, pp. 1286–1296, 2019.
5. Thomas, '3D printed jellyfish robots created to monitor fragile coral reefs', 3D Printer and 3D Printing News, Available: http://www.3ders.org/articles/20181003-3d-printed-jellyfish-robots-created-to-monitor-fragile-coral-reefs.html. Last Seen: 26/09/2021, 2019.
6. A. Pirjan, & D.M. Petrosanu, 'The impact of 3D printing technology on the society and economy', Journal of Information Systems & Operations Management, pp. 1–11, 2013.
7. https://en.wikipedia.org/wiki/History_of_printing.
8. M. Jiménez, et.al., 'Additive Manufacturing Technologies: An Overview about 3D Printing Methods and Future Prospects', Complexity, 30 Pages, 2019, https://doi.org/10.1155/2019/9656938.
9. ASTM F2792-12a, Standard terminology for additive manufacturing technologies. ASTM International. West Conshohocken, PA, 2012. https://www.astm.org/.

10. V. Gokhre, D. Raut, D. Shinde, 'A Review paper on 3D-Printing Aspects and Various Processes Used in the 3D-Printing', International Journal of Engineering Research & Technology, 6(06), pp. 953–958, 2017.
11. P. Holzmann, R. J. Breitenecker, E. J. Schwarz, 'Business model patterns for 3D printer manufacturers', Journal of Manufacturing Technology Management, 31(6), pp. 1281–1300,2019, https://doi.org/10.1108/JMTM-09-2018-0313.
12. https://www.sculpteo.com/en/.
13. https://www.think3d.in/selective-laser-sintering-sls-3d-printing-service-india/.
14. W. Yuanbin, Blache, X. Xun, 'Selection of additive manufacturing processes', Rapid Prototyping Journal, 23(2), pp. 434–447, 2017.
15. I. Gibson, D. Rosen, B. Strucker, 'Additive Manufacturing Technologies', Springer, https://doi.org/10.1007/978-1-4419-1120.
16. L. Jian-Yuan, A. Jia, K. C. Chee, 'Fundamentals and applications of 3D printing for novel materials', Applied Materials Today, 7, pp. 120–133, 2017.
17. A. Goulas, R. J. Friel, '3D printing with moondust' Rapid Prototyping Journal, 22(6), pp. 864–870, 2016.
18. https://www.enggtechnique.com/.
19. R. Bogue, 3D printing: the dawn of a new era in manufacturing?' Assembly Automation, 33(4), pp. 307-311, 2013. https://doi.org/10.1108/AA-06-2013-055.
20. M. H. Ali, S. Batai, D. Sarbassov, '3D printing: a critical review of current development and future prospects', Rapid Prototyping Journal, 25(6), pp. 1108–1126, 2019. https://doi.org/10.1108/RPJ-11-2018-0293.
21. Y. Jin, et.al., 'Visualizing the hotspots and emerging trends of 3D printing through scientometrics', Rapid Prototyping Journal, 24(5), pp. 801–812, 2017, https://doi.org/10.1108/RPJ-05-2017-0100.
22. J. W. Stansbury, M. J. Idacavage, '3D Printing with polymers: Challenges among expanding options and opportunities', Dental Materials, 32, pp. 54–64, 2016.
23. M. Sakin, Y. C. Kiroglu, '3D Printing of Buildings: Construction of the Sustainable Houses of the Future by BIM' Energy Procedia, 134, pp. 702–711, 2017.
24. I. Kothman, N.Faber, 'How 3D printing technology changes the rules of the game: Insights from the construction sector', Journal of Manufacturing Technology Management, 27(7), pp. 932–943, 2016, https://doi.org/10.1108/JMTM-01-2016-0010.
25. S.C. Paul, et. al., 'A review of 3D concrete printing systems and materials properties: current status and future research prospects', Rapid Prototyping Journal, 24(4), pp. 784–798, 2018, https://doi.org/10.1108/RPJ-09-2016-0154.
26. X. et. al., 'The pattern of technological accumulation: The comparative advantage and relative impact of 3D printing technology. Journal of Manufacturing Technology Management, 28(1), pp. 39–55, 2017, https://doi.org/10.1108/JMTM-10-2016-0136.
27. https://www.idc.com/getdoc.jsp?containerId=US46296520.
28. https://www.meity.gov.in/writereaddata/files/National%20Strategy%20for%20Additive%20Manufacturing.pdf.
29. https://msme.gov.in/.
30. https://udyamregistration.gov.in/UA/UA_VerifyUAM.aspx.
31. https://www.meity.gov.in/content/national-strategy-additive-manufacturing.
32. B. P. Conner, G. P. Manogharan, K. L. Meyers, 'An assessment of implementation of entry-level 3D printers from the perspective of small businesses', Rapid Prototyping Journal, 21(5), pp. 582–597, 2015, https://doi.org/10.1108/RPJ-09-2014-0132.
33. A. Perry, 'Factors comprehensively influencing acceptance of 3D-printed apparel', Journal of Fashion Marketing and Management: An International Journal, 21(2), 2017, https://doi.org/10.1108/JFMM-03-2016-0028.
34. A. M. Syed, et. al., 'Additive manufacturing: scientific and technological challenges, market uptake and opportunities', Materials Today, 1, pp. 1–16, 2017.

35. G. I. Ogoh, N. B. Fairweather, 'The state of the responsible research and innovation programme: A case for its application in additive manufacturing', Journal of Information, Communication and Ethics in Society, 17(2), pp. 145–166, 2019, https://doi.org/10.1108/JICES-12-2018-0093.
36. H. C. Wu, T.C.T. Chen, 'Quality control issues in 3D-printing manufacturing: a review', Rapid Prototyping Journal, 24(3), pp. 607–614, 2018, https://doi.org/10.1108/RPJ-02--0031.
37. H. Kim, L. Yirong, B.T. Tzu-Liang B. T. 'A review on quality control in additive manufacturing', Rapid Prototyping. Rapid Prototyping Journal, 24(3), pp. 645–669, 2018, https://doi.org/10.1108/RPJ-03-2017-0048.
38. A. Nayyar, A. Kumar, (Eds.), 'A roadmap to industry 4.0: Smart production, sharp business and sustainable development', Springer, 2020.
39. A. Kumar, A. Nayyar, 'si 3-Industry: A Sustainable, Intelligent, Innovative, Internet-of-Things Industry. In A Roadmap to Industry 4.0: Smart Production, Sharp Business and Sustainable Development' Springer, Cham, pp. 1–21, 2020.
40. F. Baldassarre, F. Ricciardi, 'The Additive Manufacturing in the Industry 4.0 Era: The Case of an Italian FabLab', Journal of Emerging Trends in Marketing and Management, 1(1), pp. 1–11, 2017.
41. A. Nayyar, Rameshwar, R. U. D. R. A., A. Solanki, 'Internet of Things (IoT) and the Digital Business Environment: A Standpoint Inclusive Cyber Space, Cyber Crimes, and Cybersecurity. The Evolution of Business in the Cyber Age', 10, 9780429276484-6, 2020.
42. M.D. Ugur, et al., 'The role of additive manufacturing in the era of Industry 4.0. Procedia Manufacturing, 11, pp. 545–554, 2017.
43. K.K. Singh, A. Nayyar, S. Tanwar, M. Abouhawwash, 'Emergence of Cyber Physical System and IoT in Smart Automation and Robotics: Computer Engineering in Automation' Springer Nature, 2021.
44. D.L.M. Nascimento, et. al., 'Exploring Industry 4.0 technologies to enable circular economy practices in a manufacturing context: A business model proposal', Journal of Manufacturing Technology Management, 30(3), pp. 607–627, 2019, https://doi.org/10.1108/JMTM-03-2018-0071.
45. https://www.hubs.com/get/trends/.
46. M. Attaran, 'The rise of 3-D printing: The advantages of additive manufacturing over traditional manufacturing', Business Horizon, 1, pp. 1–12, 2017.
47. P. Holzmann, et al., 'User entrepreneur business models in 3D printing', Journal of Manufacturing Technology Management, 28(1), pp. 75–94, 2017, https://doi.org/10.1108/JMTM-12-2015-0115.
48. H. Rogers, N. Baricz, K.S. Pawar, '3D printing services: classification, supply chain implications and research agenda', International Journal of Physical Distribution & Logistics Management, 46(10), pp. 886–907, 2016, https://doi.org/10.1108/IJPDLM-07-2016-0210.
49. A. Sasson, J.C. Johnson, 'The 3D printing order: variability, super centers and supply chain reconfigurations', International Journal of Physical Distribution & Logistics Management, 46(1), pp. 82–94, 2016, https://doi.org/10.1108/IJPDLM-10-2015-0257.
50. J. Yuan, et al., 'Review on processes and colour quality evaluation of colour 3D printing. Rapid Prototyping Journal, 24(2), pp. 409–415, 2018, https://doi.org/110.8/RPJ-11-2016-0182.
51. Z. Liu, et al., '3D printing: Printing precision and application in food sector. Trends in Food Science & Technology, 2(1), pp. 1–36, 2017.